IB DIPLOMA PROGRAMME

Chemistry

Course Companion

Geoffrey Neuss

OXFORD

OXFORD
UNIVERSITY PRESS

Great Clarendon Street, Oxford OX2 6DP

Oxford University Press is a department of the University of Oxford.
It furthers the University's objective of excellence in research,
scholarship, and education by publishing worldwide in

Oxford   New York

Auckland   Cape Town   Dar es Salaam   Hong Kong   Karachi
Kuala Lumpur   Madrid   Melbourne   Mexico City   Nairobi
New Delhi   Shanghai   Taipei   Toronto

With offices in

Argentina   Austria   Brazil   Chile   Czech Republic   France   Greece
Guatemala   Hungary   Italy   Japan   Poland   Portugal   Singapore
South Korea   Switzerland   Thailand   Turkey   Ukraine   Vietnam

Oxford is a registered trade mark of Oxford University Press
in the UK and in certain other countries

© Geoffrey Neuss 2007

The moral rights of the author have been asserted

Database right Oxford University Press (maker)

First published 2007

British Library Cataloguing in Publication Data

Data available

10 9 8 7 6

ISBN 978-0-19-915146-2

Printed by Bell and Bain Ltd., Glasgow

Acknowledgements

We are grateful for permission to reprint the following copyright
material:

G Hill and J Holman: extract from *Chemistry in Context*
(5e, Nelson Thornes, 2000), reprinted by permission of the
publishers; Geoffrey Lean: 'Acid rain polluting one-third of
China', *The Independent on Sunday*, 27.8.06, copyright © The
Independent 2006, reprinted by permission of Independent
News & Media Ltd; C P Snow: extract from the Rede lecture 1959,
copyright © C P Snow 1959, reprinted by permission of Curtis
Brown Group Ltd, London on behalf of the Estate of C P Snow.

We are grateful for permission to reprint the following
photographs:
Cover: Nick Rowe/Photodisc; p9: NASA; p10: JPL/NASA; p23:
Science Museum/Science & Society Picture Library; p25 (top):
Maximilien Brice/CERN; p25 (middle right): Phototake Inc/
Photolibrary; p25 (bottom): D.M. Eigler, E.K. Schweizer/IBM;
p27: Gianni Tortoli/Photo Researchers/Photolibrary; p28:
Saskatchewan Archives Board for "Cobalt unit and patient in
position for treatment. University Hospital, Univ. of Sask., Nov.
1951"; p61: Andrew Lambert Photography/SPL/Photolibrary; p84:
National Library of Medicine/SPL/Photolibrary; p87: Matthieu
Paley/Corbis; p108 (top): Ieva Geneviciene; p108 (bottom): Roger
Hiorns/Corvi-Mora, London; p111: Ford Motors; p115: professor
Rolf K. Eckhoff, University of Bergen, Norway; p135 (top):
SPL/Photolibrary; p135 (bottom): Corbis; p136: David Bishop/
Foodpix/Photolibrary; p142: SPL/Photolibrary; p148: Charles D
Winters/Photo Researchers, Inc/Photolibrary; p149: Ted Mead/
Photolibrary.Com; p162: By permission of The Royal Society of
Chemistry and Warwick Bailey; p163: Charles & Josette Lenars/
Corbis; p176: First published in Financial Times How To Spend
It magazine, Superior Interiors October 2003; p195: NASA; p207:
Manceau/BSIP/Photolibrary; p241 (top) M.C. Escher's "Waterfall"
© 2007 The M.C. Escher Company-Holland. All rights reserved.
www.mcescher.com; p241 (bottom left & right): Bettmann/Corbis;
p275: Thomas Nixon, Karolinska Institutet University Library;
p292: MA. Ansary Dr/SPL/Photolibrary; p310: Cordelia Molloy/
Science Photo Library/Photolibrary; p312 (left): SPL/Photolibrary;
p312 (right): Bettmann/Corbis; p339: McPHOTO/Still Pictures;
p341: Argus/Still Pictures; p343 (top): Digital Vision/Photolibrary;
p343 (bottom): Bettmann/Corbis; p344: Werner H. Muller/Still
Pictures; p346: Sean Sprague/Still Pictures; p357: Clouds Hill
Imaging Ltd./Corbis.

We have tried to trace and contact all copyright holders before
publication. If notified the publishers will be pleased to rectify
any errors or omissions at the earliest opportunity.

Mixed Sources
Product group from well-managed
forests and other controlled sources
www.fsc.org   Cert no. TT-COC-002769
© 1996 Forest Stewardship Council
FSC

Meet the author

Geoffrey Neuss has been teaching IB Chemistry for the past twenty-five years at The
United World College of the Atlantic in Wales, UK. He also teaches TOK. A graduate of
Warwick University, he gained his Ph.D. in Organometallic Chemistry at Manchester
University. He has been Deputy Chief Examiner and was heavily involved in the last
three curriculum reviews of the programme. He has led over fifty workshops worldwide
for IB teachers. Currently he is the Principal Examiner for all extended essays in
Chemistry.

He is married and all of his three children gained the IB Diploma before continuing their
education at university. Apart from Chemistry his main passions in life are photography
and mountaineering. He has climbed in the UK, Norway, the Alps, the Andes, and the Himalaya and is a member of the
Alpine Club.

Course Companion definition

The IB Diploma Programme Course Companions are resource materials designed to provide students with extra support through their two-year course of study. These books will help students gain an understanding of what is expected from the study of an IB Diploma Programme subject.

The Course Companions reflect the philosophy and approach of the IB Diploma Programme and present content in a way that illustrates the purpose and aims of the IB. They encourage a deep understanding of each subject by making connections to wider issues and providing opportunities for critical thinking.

These Course Companions, therefore, may or may not contain all of the curriculum content required in each IB Diploma Programme subject, and so are not designed to be complete and prescriptive textbooks. Each book will try to ensure that areas of curriculum that are unique to the IB or to a new course revision are thoroughly covered. These books mirror the IB philosophy of viewing the curriculum in terms of a whole-course approach; the use of a wide range of resources; international-mindedness; the IB learner profile and the IB Diploma Programme core requirements; theory of knowledge; the extended essay; and creativity, action, service (CAS).

In addition, the Course Companions provide advice and guidance on the specific course assessment requirements and also on academic honesty protocol.

IB mission statement

The International Baccalaureate aims to develop inquiring, knowledgeable, and caring young people who help to create a better and more peaceful world through intercultural understanding and respect.

To this end the IB works with schools, governments, and international organizations to develop challenging programmes of international education and rigorous assessment.

These programmes encourage students across the world to become active, compassionate, and lifelong learners who understand that other people, with their differences, can also be right.

The IB learner profile

The International Baccalaureate aims to develop internationally minded people who, recognizing their common humanity and shared guardianship of the planet, help to create a better and more peaceful world. IB learners strive to be:

Inquirers They develop their natural curiosity. They acquire the skills necessary to conduct inquiry and research and show independence in learning. They actively enjoy learning and this love of learning will be sustained throughout their lives.

Knowledgeable They explore concepts, ideas, and issues that have local and global significance. In so doing, they acquire in-depth knowledge and develop understanding across a broad and balanced range of disciplines.

Thinkers They exercise initiative in applying thinking skills critically and creatively to recognize and approach complex problems, and make reasoned, ethical decisions.

Communicators They understand and express ideas and information confidently and creatively in more than one language and in a variety of modes of communication. They work effectively and willingly in collaboration with others.

Principled They act with integrity and honesty, with a strong sense of fairness, justice, and respect for the dignity of the individual, groups, and communities. They take responsibility for their own actions and the consequences that accompany them.

Open-minded They understand and appreciate their own cultures and personal histories, and are open to the perspectives, values, and traditions of other individuals and communities. They are accustomed to seeking and evaluating a range of points of view, and are willing to grow from the experience.

Caring They show empathy, compassion, and respect towards the needs and feelings of others. They have a personal commitment to service, and act to make a positive difference to the lives of others and to the environment.

Risk-takers They approach unfamiliar situations and uncertainty with courage and forethought, and have the independence of spirit to explore new roles, ideas, and strategies. They are brave and articulate in defending their beliefs.

Balanced They understand the importance of intellectual, physical, and emotional balance to achieve personal well-being for themselves and others.

Reflective They give thoughtful consideration to their own learning and experience. They are able to assess and understand their strengths and limitations in order to support their learning and personal development.

A note on academic honesty

It is of vital importance to acknowledge and appropriately credit the owners of information when that information is used in your work. After all, owners of ideas (intellectual property) have property rights. To have an authentic piece of work, it must be based on your individual and original ideas with the work of others fully acknowledged. Therefore, all assignments, written or oral, completed for assessment must use your own language and expression. Where sources are used or referred to, whether in the form of direct quotation or paraphrase, such sources must be appropriately acknowledged through the use of footnotes and bibliographies.

Footnotes (placed at the bottom of a page) or endnotes (placed at the end of a document) are to be provided when you quote or paraphrase from another document, or closely summarize the information provided in another document. You do not need to provide a footnote for information that is part of a "body of knowledge". That is, definitions do not need to be footnoted as they are part of the assumed knowledge.

Bibliographies should include a formal list of the resources that you used in your work. "Formal" means that you should use one of the several accepted forms of presentation. This usually involves separating the resources that you use into different categories (e.g. books, magazines, newspaper articles, Internet-based resources, CDs, and works of art) and providing full information as to how a reader or viewer of your work can find the same information. A bibliography is compulsory in the extended essay.

What constitutes malpractice?
Malpractice is behaviour that results in, or may result in, you or any student gaining an unfair advantage in one or more assessment component. Malpractice includes plagiarism and collusion.

Plagiarism is defined as the representation of the ideas or work of another person as your own. The following are some of the ways to avoid plagiarism:

- words and ideas of another person to support one's arguments must be acknowledged
- passages that are quoted verbatim must be enclosed within quotation marks and acknowledged
- CD-Roms, email messages, web sites on the Internet, and any other electronic media must be treated in the same way as books and journals
- the sources of all photographs, maps, illustrations, computer programs, data, graphs, audio-visual, and similar material must be acknowledged if they are not your own work
- copying works of art, whether music, film, dance, theatre arts, or visual arts, and where the creative use of a part of a work takes place, must be acknowledged.

Collusion is defined as supporting malpractice by another student. This includes:

- allowing your work to be copied or submitted for assessment by another student
- duplicating work for different assessment components and/or diploma requirements.

Contents

Introduction

One of the strengths of the IB Diploma Programme is that subjects are not studied in modules or in a vacuum. This book encourages you, the student to make relationships between the different topics within Chemistry and links between Chemistry and the other five different subject areas of the IB "hexagon". It places the emphasis throughout on examining the different ways of knowing in Chemistry and encourages you to question the basis of factual knowledge in Chemistry and the assumptions that you make. Examples of genuine laboratory accounts written by IB students are included and analyzed with advice on how to fulfil the internal assessment criteria. A separate chapter is devoted to the extended essay with guidance on how to research and write such a demanding 4000-word piece of individual work.

Chemistry is more than just the academic. It deals with the very stuff of matter and affects all our lives. It brings great benefits to individuals and societies and contributes much to the quality of life. However, it can also cause considerable harm to individuals and damage to the natural environment. Chemists need to be alert to the possible outcomes of their investigations and be aware of their responsibilities to society. They need to have inquiring minds, be able to face challenges, acknowledge and build upon the achievements of others and be innovative and take calculated risks in their problem solving. These are exactly the attributes encouraged in the IB learner profile. This book provides many examples and many thought-provoking questions to emphasize and re-enforce these attributes. Inherent too in the book is the cultural heritage of Chemistry. Credit is given to the individuals behind many of the great discoveries in Chemistry and short biographies are given of some of these chemists, particularly those whose discoveries or lives involved them in difficult ethical decisions. Throughout the book,

the international nature of Chemistry is stressed—whether this is the collaboration between scientists working in different countries, the international nature of the language of chemistry, or the way in which the products of the chemical industry know no boundaries—global warming affects us all.

The distinction between the material required for Standard and Higher level is clearly made. Because SL students must study all of the Core and HL students all of the Core and Additional Higher Level material (AHL) most of the emphasis on critical thinking has been placed in these sections. Some extension material has also been included for those who wish to extend their knowledge and understanding a little further. As each student needs to study only two of the seven options these have been covered in less detail than the Core and AHL although they do address all the chemical content in the programme.

This is not a book that teaches just to the IB examination. Many of the questions are open-ended and have no "right" answer. They are intended to get you to discuss and think critically about issues and to develop the qualities of open-mindedness and caring–two other attributes of the IB learner profile.

I hope this book will enthuse you with a genuine interest in Chemistry and encourage you to enquire further. I should like to thank the many IB students and teachers who have stimulated my interest and understanding of Chemistry. I should also like to thank Christopher Talbot, an IB Chemistry and TOK teacher at the Anglo-Chinese School in the Republic of Singapore, who has not only helped to proof read the early drafts but also contributed many suggestions and ideas. My colleague John and my wife Chris have also been extremely supportive throughout. This book is dedicated to the memory of my son Jack.

Quantitative chemistry

Introduction

A student in the southern hemisphere looking up at the night sky will see a very different set of constellations from those seen by a student in the northern hemisphere. But, as far as we know, in whichever direction we look, matter in the whole of the universe is made up from the same basic building blocks.

These basic building blocks, which are called **elements**, are the simplest form of matter, as they cannot be broken down by chemical means into anything simpler. Here on Earth, whether we consider the rocks beneath our feet, the air we breathe, the water we drink, the clothes we wear, even our own physical being—everything is made up of some combination of these chemical elements, which number just over 100. Some of the less reactive elements, such as copper, gold, tin and lead, have been known for centuries; others, such as fluorine, needed the discovery of electricity before they could be isolated from their compounds.

One element was actually discovered first on a nearby star, and is named after it. When heated strongly, each element emits light to form a spectrum with lines at particular and characteristic frequencies. In 1868 the British astronomer Joseph Lockyer (1836–1920) analysed the light coming from the Sun. The frequencies of some of the lines fitted no known element, and he named the element helium, from the Greek word for "sun", *helios*. Later, in 1895, William Ramsay (1852–1916) isolated and identified the element here on Earth.

The elements can be arranged in order in a form known as the Periodic Table, which is explained in Chapter 3. The names of the first 20 elements are listed in Table 1.

Chemistry is concerned with the properties and reactions of these elements, and the compounds they form when they combine with each other. Over the past 200 years the study of chemistry has made an enormous contribution to the quality of our lives. Much of this is positive. For example, life expectancy in the developed world has increased significantly.

Table 1 Names of the first 20 elements are listed below.

Atomic Number	Name	Symbol	Relative atomic mass
1	hydrogen	H	1.01
2	helium	He	4.00
3	lithium	Li	6.94
4	beryllium	Be	9.01
5	boron	B	10.81
6	carbon	C	12.01
7	nitrogen	N	14.01
8	oxygen	O	16.00
9	fluorine	F	19.00
10	neon	Ne	20.18
11	sodium	Na	22.99
12	magnesium	Mg	24.31
13	aluminium	Al	26.98
14	silicon	Si	28.09
15	phosphorus	P	30.97
16	sulfur	S	32.06
17	chlorine	Cl	35.45
18	argon	Ar	39.95
19	potassium	K	39.10
20	calcium	Ca	40.08

life expectancy
- > 73
- 70 to 73
- 64 to 70
- 51 to 64
- < 51

Figure 1 Current worldwide life expectancy

Reasons for this increase in life expectancy include the disinfection of water to produce water safe to drink, the synthesis of antibiotics and other drugs to combat previously lethal diseases, and the use of fertilizers to grow more food. At the same time we face the challenges caused by the negative influences of chemistry. Some of these are: pollution of water, air, and the ground; the increase in carbon dioxide emissions, leading to global warming; and the indiscriminate use of explosives such as landmines in many areas of the world.

Chemistry is truly global and international. The exhaust gases coming from a factory chimney are no respecter of national borders. The relocation of heavy industry, such as plants producing sulfuric acid, to less developed countries where the environmental controls are less strict is an increasing international problem.

Ethical issues abound. For example, should pharmaceutical companies and governments provide low-cost medicines to combat HIV/AIDS in poorer countries where the population is being ravaged by the disease? By studying chemistry, we can not only fulfill the natural human curiosity to try to understand the world around us, but also maximise the positive benefits that chemistry can offer, and find solutions to minimise and overcome the problems chemistry can also cause.

Although each chapter of this Course Companion covers a particular aspect, chemistry does not fit neatly into separate compartments, because all the areas of knowledge of chemistry interact. In order to understand quantitative chemistry fully, some very basic knowledge of atomic theory is helpful. Atomic theory and the evidence for atoms and sub-atomic particles are covered in proper detail in Chapter 2, so that you can always refer to this if you need further clarification about a particular term or statement.

1 List five major achievements of chemistry that have contributed to improving the quality of life.
2 List five problems caused by chemistry which are important to the quality of life in your own locality.

Elements and compounds

The smallest unit of an element is an **atom**. Some substances—the noble gases, for example—are made up of single atoms of the same elements. Most substances, however, contain atoms joined together to form the smallest particles, known as **molecules**, of that substance. For example, oxygen gas molecules have the formula O_2. Each molecule of oxygen contains two atoms of oxygen, and the molecule is said to be **diatomic**. Other examples of diatomic molecules are hydrogen gas (H_2), nitrogen gas (N_2), and chlorine gas (Cl_2). A molecule of phosphorus contains four atoms of phosphorus (P_4), and a molecule of sulfur contains eight atoms of sulfur (S_8). Figure 2 shows some simple molecules.

Compounds are made up of more than one element, chemically bonded together. Carbon dioxide molecules, for example, each contain one carbon atom and two oxygen atoms and have the formula CO_2. Carbon dioxide is a compound, not an element, because it can be broken down chemically into its constituent elements carbon and oxygen.

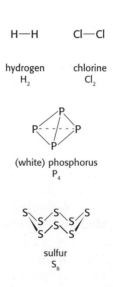

Figure 2 Simple molecules.

The mole concept and Avogadro's constant

Although there are just over 100 different elements, all the atoms of all the elements are made up of the same basic sub-atomic particles. Atoms contain different numbers of **protons** and **neutrons**, and so the atoms of the various elements have different masses. For example, a carbon-12 atom has a mass 12 times greater than the mass of one hydrogen-1 atom, and an oxygen-16 atom has a mass 16 times greater than the mass of one hydrogen-1 atom. It is impossible to measure the mass of individual atoms using a balance, because their mass is so small. Chemists have developed their own unit, called a **mole** (symbol mol), which enables them to deal with manageable amounts of different substances, all of which contain the same number of particles of the substance in question.

One gram of hydrogen-1 atoms contains 6.02×10^{23} hydrogen-1 atoms. This number is known as **Avogadro's constant** (symbol L). If we wanted to weigh out exactly the same number of carbon-12 atoms then, as each carbon atom has a mass 12 times greater than that of a hydrogen-1 atom, we would need to weigh out 12 g of carbon-12 atoms. Similarly, an oxygen-16 atom has a mass 16 times greater than that of a hydrogen-1 atom, so 6.02×10^{23} oxygen atoms have a mass of 16 g.

One mole of any substance contains 6.02×10^{23} particles of that substance. Thus a mole of oxygen atoms has a mass of 16 g and contains 6.02×10^{23} atoms of oxygen-16. One mole of carbon dioxide molecules contains one mole of carbon atoms and two moles of oxygen atoms. The mass of one mole of carbon dioxide will therefore be equal to $12 + (2 \times 16) = 44$ g.

1 How many atoms are there in:
   (a) 2 g of hydrogen?
   (b) 8 g of oxygen?
   (c) 36 g of carbon?
2 What is the mass of:
   (a) 4 mol of hydrogen atoms?
   (b) 4 mol of hydrogen gas molecules?
   (c) 1 mol of methane, CH_4, molecules?
3 How many carbon atoms are there in:
   (a) one molecule of butane, C_4H_{10}?
   (b) 1 mol of butane?

Critical considerations

Just how big is Avogadro's constant?

Some students find the mole concept difficult. In fact, in many areas of life we use different units to measure the same quantity. For example, how old are you? If you answered with just a number, e.g. 17, then you have not given a very good answer. It should of course be 17 years. For other questions involving time we will often give a different unit. What is the world athletic record for running 100 m? Currently the answer to this is 9.77 seconds. However, the gestation period for a human foetus is 9 months or 38 weeks. In each case we choose the unit that is most appropriate.

Other examples of different units are the use of a pair when buying shoes, and—in some countries—using one dozen instead of twelve when buying eggs. In India one lakh is used to represent 100 000 when dealing with large numbers of people or large amounts of money.

We can understand the numbers represented by one pair, one dozen or one lakh, but can we really understand how large one mole is? You might like to work out how old you are in seconds. If we assume you are exactly 17 years old, and we ignore leap years, then you are $17 \times 365 \times 24 \times 60 \times 60 = 5.36112 \times 10^8$ seconds, which is considerably less than one mole of seconds. In fact, if we work out the age of the Earth in seconds we still end up with considerably less than one mole of seconds (Figure 3).

Figure 3 The Earth is estimated to be 4.54 billion years old: that is, approximately 1.43×10^{17} seconds.

How can we envisage how large a mole is? Let's try to "guesstimate" the number of grains of sand that there are in the Sahara Desert. Surely this will be more than one mole?

A quick look at a map of Africa will give us the information that the Sahara Desert is about 1500 km from north to south and about 4000 km from east to west (Figure 4). Let's assume that the average depth of the sand throughout the desert is 10 m. Then the volume of the sand must be $1.5 \times 10^6 \times 4.0 \times 10^6 \times 10 = 6.0 \times 10^{13}$ m^3. If you pick up 1 cm^3 of sand you can guess that it contains about 1000 grains of sand. This means that 1 m^3 of sand contains approximately 1×10^9 grains of sand, and that the Sahara Desert therefore contains 6×10^{22} grains of sand: that is, one tenth of a mole! It is difficult even for experienced chemists to comprehend that there are more atoms of carbon in just 12 grams of carbon than there are grains of sand in the whole of the Sahara Desert.

Figure 4 How many grains of sand are there in the Sahara Desert?

Relative atomic mass A_r, and relative molecular mass M_r

Although each carbon-12 atom has a mass exactly 12 times greater than that of a hydrogen-1 atom, carbon atoms are not exactly 12 times more massive than hydrogen atoms in general. This is because the atoms of an element are made up of a mixture of **isotopes**. Isotopes are atoms of an element with the same number of protons but a different number of neutrons, so that the atoms will have a slightly different mass.

The **relative atomic mass** of an element, A_r, is defined as the weighted mean mass of all the naturally occurring isotopes of an element relative to one twelfth of the mass of a carbon-12 atom. This explains why the relative atomic masses given in data books are not exactly whole numbers. For example, chlorine has a relative atomic mass of 35.45, because normal chlorine gas contains approximately 75% chlorine atoms that have 18 neutrons in their nucleus, and approximately 25% that have 20 neutrons in their nucleus. The units of atomic mass are g mol^{-1}, but relative atomic mass has no units. Similarly for molecules the measure **relative molecular mass**, M_r, is used; this too has no units.

For any substance or species the general measure **molar mass** is used, and this has the units of g mol^{-1}. If the term "relative" is added, then we omit the units.

Formulas of compounds

Empirical formula

Traditionally, the chemical composition of a compound was determined empirically by burning it in oxygen to form the oxides of all the elements in the compound. The amount of each oxide was then determined, and hence the original amount of each element in the compound. This formula became known as the **empirical formula**, but this is now defined as the simplest whole number ratio of atoms of each element in a particle of the substance.

❓
(To answer these questions use the data from Table 5 – The Periodic Table – in the IB Chemistry Data Booklet.)

1 How many atoms are there in 30.97 g of phosphorus?
2 What is the mass, to two decimal places, of one mole of magnesium hydroxide, $Mg(OH)_2$?
3 How many moles of ammonia are there in 28.08 g of ammonia, NH_3?
4 What is the mass of 0.100 mol of caffeine, $C_8H_9O_2N_4$?

An experiment to determine the empirical formula of magnesium oxide

An empty crucible and lid were weighed accurately. A coil of magnesium ribbon about 10 cm long was placed in the crucible and its mass determined. The crucible was then placed on a clay triangle and heated strongly (Figure 5). When the magnesium started to burn, tongs were used to lift the lid slightly to allow more air to enter, and the heating was continued until all the magnesium had burned. After cooling, the crucible and its contents were weighed again.

The following results were obtained.

	Mass / g (Uncertainty ± 0.001 g)
Mass of crucible + lid	29.809
Mass of crucible + lid + magnesium	29.935
Mass of crucible + lid + magnesium oxide	30.004

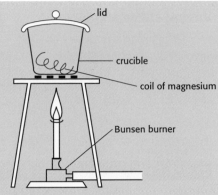

Figure 5

Observations.

After the magnesium ribbon had been heated strongly for a few minutes it began to melt, and then glow with a bright light. Each time the lid was lifted the magnesium glowed more brightly, and some white smoke was seen to escape. The heating was stopped when the magnesium no longer glowed brightly when the lid was lifted. A greyish-white powder remained in the crucible.

Mass of magnesium = 29.935 − 29.809 = 0.126 g

Mass of magnesium oxide = 30.004 − 29.809 = 0.195 g

Mass of oxygen combining with magnesium = 0.195 − 0.126 = 0.069 g

Amount of magnesium = $\dfrac{0.126}{24.31}$ = 5.2 × 10^{-3} mol

Amount of oxygen = $\dfrac{0.069}{16.00}$ = 4.3 × 10^{-3} mol

Ratio of Mg to O = 5.2 × 10^{-3} to 4.3 × 10^{-3}

To simplify this ratio divide each amount by the smallest amount. This gives a ratio of 1.2:1. To convert this to the simplest whole number ratio it is necessary to multiply by 5 to give a ratio of 6:5.

Empirical formula of magnesium oxide = Mg_6O_5.

These are the genuine results from a student's actual experiment, and the student was disappointed that the result was not closer to the generally accepted answer, which is MgO.

However, it is worth evaluating this experiment and looking at the assumptions that have been made in calculating the result. See the table alongside.

Assumption	Possible effect on the result
None of the product escaped when the lid was lifted.	If some *did* escape, the product would weigh less, so it would appear that less oxygen had combined with the magnesium. This would give a greater ratio of magnesium to oxygen in the answer.
All the magnesium burned.	If some of the magnesium did not react then the product would weigh less, so again the ratio of magnesium to oxygen would be higher.
The magnesium was pure.	If impure magnesium was used then other products apart from magnesium oxide would be formed.
The only substance reacting with the magnesium was oxygen.	The magnesium was burned in air, not pure oxygen. In fact magnesium also reacts with the nitrogen in the air under these conditions to give some magnesium nitride, Mg_3N_2.
The crucible did not react.	There is some evidence that, during the high temperatures reached, the crucible *does* react. This will affect the product(s) formed.

? Can you now suggest how the experiment could be improved to best take account of all these assumptions and obtain a more accurate result?

Percentage composition

Often the composition of a compound is given as the percentage by mass of each element in the compound. It is wise to check that all the percentages add up to 100%, because by tradition the percentage of oxygen is often omitted as it does not burn to form an oxide. To find the empirical formula from percentage composition, convert the percentages directly into a mass by assuming that there is 100 g of the compound.

Worked example

An organic compound contains 60.00% carbon and 4.48% hydrogen by mass. Calculate the empirical formula of the compound.

1 As the percentages do not add up to 100% it can be assumed that the compound also contains 35.52% oxygen by mass.

2 Assume that there are 100 g of compound, and calculate the amount of each element by dividing the mass of the element by its atomic mass:

Element	Mass / g	Amount / mol
Carbon	60.00	60.00/12.01 = 5.00
Hydrogen	4.48	4.48/1.01 = 4.44
Oxygen	35.52	35.52/16.00 = 2.22

3 Find the simplest ratio by dividing through by the smallest amount and then converting any decimals to give the simplest whole-number ratio.

Element	Amount / mol	Simplified ratio	Simplest whole number ratio
Carbon	5.00	2.25	9
Hydrogen	4.44	2.00	8
Oxygen	2.22	1.00	4

4 Write the empirical formula, using subscripts for the numbers in the ratio.
The empirical formula of the compound is $C_9H_8O_4$.

❷ Calculate the empirical formula of the following compounds.
1 The compound formed when 1.00 g of phosphorus combines with 5.72 g of chlorine.
2 A compound that contains 40.4% carbon, 7.93% hydrogen, 15.7% nitrogen and 35.9% oxygen by mass.

Molecular formula

The empirical formula gives only the simplest ratio of the atoms of the different elements in a compound. The **molecular formula** gives the actual number of atoms of each element in a molecule of the substance. This is illustrated well by the compounds methanal, CH_2O ($M_r = 30$), ethanoic acid, $C_2H_4O_2$ ($M_r = 60$), and glucose, $C_6H_{12}O_6$ ($M_r = 180$). All three have the same empirical formula, CH_2O, but each has a different molecular formula. To obtain the molecular formula from the empirical formula it is necessary to also know the relative molecular mass of the compound. This can be determined in a variety of ways. These include the direct weighing of gases, titration of acids and bases, and mass spectroscopy.

❷ A compound contains 12.79% carbon, 2.15% hydrogen and 85.06% bromine by mass. Its relative molecular mass is 187.9. Determine (a) the empirical formula and (b) the molecular formula of the compound.

Structural formula

Although the molecular formula gives much information about a compound, it is still possible that two or more different compounds can have the same molecular formula. The **structural formula** shows the arrangement of atoms and bonds within a molecule. Two compounds with the same molecular formula but different structural formulas are known as **isomers**. For example, both 1,1-dichloroethane and 1,2-dichloroethane have the same empirical formula, CH_2Cl, and the same molecular formula, $C_2H_4Cl_2$, but their structural formulas are different. In 1,1-dichloroethane both the chlorine atoms are bonded to the same carbon atom, whereas in 1,2-dichloroethane each of the two carbon atoms has one chlorine atom bonded to it (Figure 6).

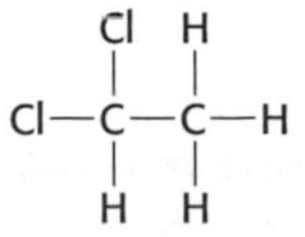

The use of structural formulas is particularly important in organic chemistry. Compounds with the same molecular formula but with different structural formulas are called **structural isomers**. Structural isomers often have very different physical and chemical properties. For example, methoxymethane, CH_3OCH_3, and ethanol, C_2H_5OH, both have the same molecular formula, C_2H_6O (Figure 7), but, unlike methoxymethane, ethanol is completely miscible with water and is generally much more reactive chemically.

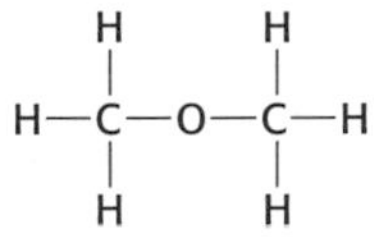

Chemical equations

Once the correct formulas of all the reactants and products are known, it is possible to write a chemical equation to describe a reaction taking place. All chemical reactions have the following properties.

- New substances are formed.
- Bonds in the reactants are broken and bonds in the products are formed, resulting in an energy change between the reacting system and the surroundings.
- There is a fixed relationship between the number of particles of reactants and products, resulting in no overall change in mass.

In a chemical equation the reactants are written on the left-hand side, and the products are written on the right-hand side. As there is no overall change in mass, the total amount of each element must be the same on the two sides of the equation.

For example, consider the reaction between zinc metal and hydrochloric acid to produce zinc chloride and hydrogen gas. The correct formulas for all the reacting species and products are first written down.

$$Zn + 2HCl \rightarrow ZnCl_2 + H_2$$
reactants   products

The equation is then balanced by writing the correct coefficients in front of each species. When the correct coefficients are in place, the reaction is said to be **stoichiometrically balanced**. The stoichiometry tells us that in this reaction two moles of hydrochloric acid react with one mole of zinc to form one mole of zinc chloride and one mole of hydrogen gas. Note the difference between the coefficient of 2 in front of the hydrochloric acid, HCl, and the

1,1-dichloroethane

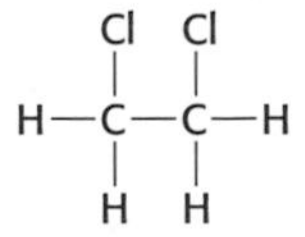

1,2-dichloroethane

Figure 6 Isomers.

methoxymethane

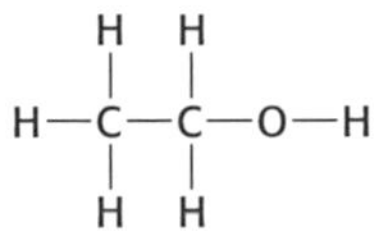

ethanol

Figure 7 Structural isomers.

subscript 2 in zinc chloride, $ZnCl_2$. The coefficient tells us how many moles of substance are involved in the reaction; the subscript tells us how many chloride ions there are in one unit of zinc chloride.

If the compound contains a complex ion or polyanion derived from an acid, then the subscript is placed outside the brackets. For example, $Ba(OH)_2$ or $(NH_4)_2SO_4$ show that each barium hydroxide formula unit contains two hydroxide ions, and each ammonium sulfate formula unit contains two ammonium ions.

It is good practice to also include **state symbols** in a chemical equation. The common state symbols are (s) for *solid*, (l) for *liquid*, (g) for *gas*, and (aq) for *in aqueous solution*. So the above equation is better written:

$$Zn(s) + 2HCl(aq) \rightarrow ZnCl_2(aq) + H_2(g)$$

In this example a single arrow has been used, because this reaction goes virtually to completion. Many reactions describe systems in equilibrium, where both reactants and products are present in an equilibrium mixture. In this case reversible arrows are used. For example:

$$2SO_2(g) + O_2(g) \rightleftharpoons 2SO_3(g)$$

Sometimes only certain parts of the reactants actually react, and other parts remain unreacted. This is particularly true with ionic compounds (see Chapter 4), where often some of the ions simply remain in the solution as **spectator ions**. For example, consider the reaction between an aqueous solution of silver nitrate and an aqueous solution of potassium chloride. The reaction could be written

$$AgNO_3(aq) + KCl(aq) \rightarrow AgCl(s) + KNO_3(aq)$$

But in fact silver nitrate, potassium chloride and potassium nitrate are all fully ionized in solution, so the equation could also be written

$$Ag^+(aq) + NO_3{}^-(aq) + K^+(aq) + Cl^-(aq)$$
$$\rightarrow AgCl(s) + K^+(aq) + NO_3{}^-(aq)$$

As the potassium and nitrate ions are spectator ions, and do not change, the ionic equation becomes

$$Ag^+(aq) + Cl^-(aq) \rightarrow AgCl(s)$$

This equation now tells us that one mole of silver ions from any soluble silver salt will react with one mole of chloride ions from any soluble chloride salt to form one mole of silver chloride.

Measurement of molar quantities

Once the equation is stoichiometrically balanced, it is possible to use the chemical equation to work out how much product will be formed from known amounts of reactants. The equation represents the amounts in moles; however, in the laboratory, amounts are normally measured using either mass or volume. It is therefore important to be able to convert both masses and volumes into molar quantities or amounts.

Solids

A solid is normally weighed to obtain its mass. The SI (Système International) unit of mass is the **kilogram**, although chemists often use grams.

$$1000 \text{ g} = 1.000 \text{ kg}$$

When using a balance it is usual to indicate the precision of the balance by including the correct number of decimal places. For example, exactly 24.5 g of a substance would be recorded as 24.50 g on a balance weighing to ± 0.01 g ($\pm$ means *plus or minus*), and as 24.500 g on a balance weighing to ± 0.001 g.

Liquids

Liquids may be weighed, or their volume may be measured. If the density of the liquid is known, then the mass can be obtained from the volume using the expression

$$\text{density} = \frac{\text{mass}}{\text{volume}}$$

Density is usually measured in g cm^{-3} or kg dm^{-3}.

In the laboratory, volume is measured using different apparatus depending on how precisely the volume is required (Figure 8). For very approximate volumes a beaker or a conical flask can be used. Measuring cylinders are more precise, but still have a large amount of uncertainty. For fixed volumes, volumetric flasks or pipettes are used for precise measurements, and for variable volumes a burette or graduated pipette is used. The uncertainty associated with the burette, pipette and volumetric flask can vary. In schools, grade B equipment is usually used, but more expensive grade A (Analytical) equipment can be accurate to ± 0.01 cm^3.

Although chemists make much use of volumes, it is worth remembering that probably the most accurate measuring instrument in a school laboratory is the analytical balance (Figure 9). However, it measures mass, not volume. The mass of a liquid can be converted into its volume if the density of the liquid is known.

Aqueous solutions

A **solution** consists of a **solute** dissolved in a **solvent**. The **concentration** of a solution can be expressed in g dm^{-3}, but it is more usefully expressed in mol dm^{-3}. In other words a one molar solution (1.00 mol dm^{-3}) will contain one mole of solute dissolved in one litre (1.00 dm^3) of solution. Normally aqueous solutions are measured using volume.

$$1 \text{ litre} = 1000 \text{ cm}^3 = 1.000 \text{ dm}^3$$

Note that square brackets around a solute are sometimes used to represent its concentration. For example a 1.00 mol dm^{-3} solution of hydrochloric acid can be represented as [HCl(aq)] = 1.00 mol dm^{-3}.

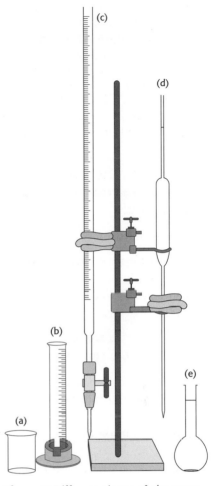

Figure 8 Different pieces of glassware for measuring 50.0 cm^3 of a liquid. (a) A beaker; very inaccurate ($\pm$ 5.0 cm^3). (b) A measuring cylinder; quite inaccurate ($\pm$ 1.0 cm^3). (c) A burette; accurate to $\pm$ 0.05 cm^3. (d) A pipette; accurate to $\pm$ 0.05 cm^3. (e) A volumetric flask; accurate to $\pm$ 0.05 cm^3 Note that because the amount delivered from the burette involves two readings the actual uncertainty is 2 $\times$ 0.05 cm^3 = 0.1 cm^3

Worked example

Find the concentration of a solution containing 2.00 g of sodium hydroxide in 125 cm³ of solution. (Note that to make up this solution you would dissolve 2.00 g of sodium hydroxide in a small volume of distilled water, and then make up the total volume to 125 cm³ with distilled water. If 125 cm³ of water is simply added to 2.00 g of sodium hydroxide then the final volume of the solution will not be 125 cm³.)

Step 1: Find the M_r for sodium hydroxide.

M_r(NaOH) = 22.99 + 16.00 + 1.01 = 40.00

Step 2: Find the amount of sodium hydroxide present in 125 cm³ of solution.

$$\text{Amount} = \frac{2.00}{40.00} = 0.0500 \text{ mol}$$

Step 3: Find the amount of sodium hydroxide present in 1.00 dm³ (1000 cm³). (One way of doing this is to find the amount in 1.00 cm³ by dividing by 125, then multiply by 1000 to find the amount in 1000 cm³.)

Amount in 1.00 dm³ = 0.0500 × 1000/125 = 0.400 mol
Concentration = **0.400 mol dm⁻³**
or [NaOH(aq)] = **0.400 mol dm⁻³**

If the concentration of the solution is known, then it is not necessary to involve the mass of the solute in order to find out the amount of solute in moles.

Worked example

What amount of sulfuric acid is present in 25.0 cm³ of 2.00 mol dm⁻³ sulfuric acid, H_2SO_4(aq) ?

By definition, 1000 cm³ of 2.00 mol dm⁻³ H_2SO_4 contains 2.00 mol of H_2SO_4
1 cm³ of 2.00 mol dm⁻³ H_2SO_4 contains 2.00/1000 mol of H_2SO_4
25.0 cm³ of 2.00 mol dm⁻³ H_2SO_4 contains 2.00 × 25.0/1000 mol of H_2SO_4
Amount of H_2SO_4 in 25.0 cm³ of 2.00 mol dm⁻³ H_2SO_4 = **0.0500 mol**

? 1 Calculate the amount of hydrochloric acid present in:
  (a) 2.00 dm³ of 1.50 mol dm⁻³ HCl(aq)
  (b) 150 cm³ of 1.35 mol dm⁻³ HCl(aq)
  (c) 23.65 cm³ of 0.100 mol dm⁻³ HCl(aq)

2 What mass of potassium manganate(VII) (also known as potassium permanganate), $KMnO_4$(s), is present in:
  (a) 500 cm³ of 0.0200 mol dm⁻³ $KMnO_4$(aq) ?
  (b) 25.55 cm³ of 0.35 mol dm⁻³ $KMnO_4$ (aq) ?

3 What will be the concentration of the solution formed by mixing 200 cm³ of 3.00 mol dm⁻³ HCl(aq) with 300 cm³ of 1.50 mol dm⁻³ HCl(aq)?
  (Hint: find the total amount present in moles and the total volume)

Gases

Mass or volume may be used for gases. Normally it is easier to measure the volume of a gas. However, as well as the amount of gas present, the volume of a gas also depends on the pressure and the temperature. The physical behavior of all gases is governed by the ideal gas equation:

$$pV = nRT$$

where:
p represents the pressure. Pressure is equal to force divided by area and so is measured in newtons per square metre (N m⁻²); one N m⁻² is also known as one pascal (symbol Pa). 1 atmosphere = 101 000 Pa.
V represents the volume measured in m³. (Note: 1.000 m³ = 1000 dm³ = 1.000 × 10⁶ cm³)
n represents the amount of gas in moles = mass of gas/M_r
T represents the absolute temperature measured in kelvin (symbol K). (Note: 0 °C = 273 K and 100 °C = 373 K)
R represents the gas constant.

Figure 9 An analytical balance measuring to ± 0.001 g. (Photo by G. Neuss)

The units of the gas constant R can be derived from the equation:

$$p \times V = \mathrm{N\ m^{-2} \times m^3 = N\ m = J}$$
$$n \times T = \mathrm{mol\ K}$$

so the SI units of R must be $\mathrm{J\ K^{-1}\ mol^{-1}}$.

R has the value $8.314\ \mathrm{J\ K^{-1}\ mol^{-1}}$, and is one of the best known constants in science.

Thinking about science

All the gas laws have been found by experiment. That is, they are empirical and cannot be proved. If certain assumptions are made they can be derived mathematically, but this is not proof. To prove any gas law an infinite number of experiments would have to be done on every single gas in the universe under every possible condition. An Austrian philosopher called Karl Popper (1902–1994) realized this in the 1950s, and suggested that, instead of trying to devise experiments to prove such a scientific law, it would be more sensible to try to devise experiments to falsify or disprove the law. If only one such experiment could be found in which the gas laws did not work then the laws would not be true. The more failures there are at attempts to falsify a law, then the more credible the law becomes, but it can never be actually proven.

Popper illustrated this with his famous example of swans. Consider the proposed law "All swans are white". This certainly appears to be true for all swans that occur naturally in the northern hemisphere. However, one only needs to go to the southern hemisphere to find a black swan and so disprove the law (Figure 10).

Figure 10 Black swan in Ibirapuera Park, São Paulo, Brazil. (Photo by G. Neuss)

The equation $pV = nRT$ is only true for an ideal gas. Unlike ideal gases, real gases such as oxygen or hydrogen do not obey the ideal gas law equation exactly. This is because there are still some weak attractive forces between the molecules in the gas, and the molecules themselves occupy some space even though most of the volume of a gas is empty space. However, for practical purposes we can use this ideal gas equation to describe the behavior of real gases.

1 Under what conditions of temperature and pressure would you expect a real gas to behave most like an ideal gas?

2 If we measured pressure in atmospheres, volume in $\mathrm{dm^3}$ and temperature in kelvin, what would be the units for the gas constant R?

An experiment to determine the relative molecular mass and hence the identity of the gas in the laboratory

A large flask full of air was fitted with a bung and weighed. In the fume cupboard gas from the gas tap was then blown into the flask so that the air was displaced. The flask was then re-weighed. This process was repeated until there was no further change in mass: that is, until all the air had been displaced. The flask was then filled with water, and the water poured into a measuring cylinder to find the volume of the flask. During the experiment the pressure and temperature were measured.

Data collection:

	Mass / g (uncertainty ± 0.01 g)
Mass of flask + air	739.50
Mass of flask + gas (first attempt)	740.89
Mass of flask + gas (second attempt)	741.78
Mass of flask + gas (third attempt)	742.10
Mass of flask + gas (fourth attempt)	742.10

Temperature = 19.0 °C ± 0.5 °C = 292.0 K ± 0.5 K
Pressure = 1 atmosphere = 1.01×10^5 Pa (assumed, not actually measured)
Volume of flask = 4043 cm³ ± 20 cm³ = 4.043×10^{-3} m³ ± 0.020×10^{-3} m³

Data processing
The amount of gas in the flask can be calculated from the ideal gas equation:

$$n = \frac{pV}{RT} = \frac{101\ 000 \times 4.043 \times 10^{-3}}{8.314 \times 292.0} = 0.168 \text{ mol}$$

Air is approximately 80% N_2 and 20% O_2 by volume: therefore

$$M_r \text{ of air} = \frac{(80 \times 28) + (20 \times 32)}{100} = 28.8$$

Hence mass of air in flask = $28.8 \times 0.168 = 4.84$ g

Mass of flask if it contained no air = $739.50 - 4.84 = 734.66$ g
Mass of gas in flask = $742.10 - 734.66 = 7.44$ g
M_r of gas = 7.44/0.168 = **44.29**

Uncertainties:

$$\text{Temperature} = \frac{0.5 \times 100}{292} = 0.17\%$$

$$\text{Volume} = \frac{20 \times 100}{4043} = 0.49\%$$

$$\text{Mass} = \frac{0.01 \times 100}{7.46} = 0.13\%$$

Total experimental uncertainty = 0.17% + 0.49% + 0.13% = 0.79%

Hence M_r of gas = 44.29 ± 0.79% = 44.29 ± 0.35

Conclusion and percentage error
The actual gas was propane, $M_r = 44.11$.
The percentage error for this experiment is thus

$$\frac{44.29 - 44.11}{44.11} = 100 = 0.40\%$$

This error lies within the uncertainty of the experiment, so even though very simple equipment was used, a reliable and accurate result was obtained.

Avogadro's law and molar volume of a gas

Consider two gases, A and B:

$$\frac{P_A V_A}{RT_A} = n_A \text{ and } \frac{P_B V_B}{RT_B} = n_B$$

If the gases occupy equal volumes, and the temperature and pressure are the same, then

$$\frac{P_A V_A}{RT_A} = \frac{P_B V_B}{RT_B}$$

and therefore

$$n_A = n_B$$

This is a derivation of **Avogadro's law**, which states that equal volumes of different gases at the same temperature and pressure contain the same number of particles. From this it must follow that, at a certain temperature and pressure, for any gas there will be a certain volume that contains one mole of particles of that gas. This volume is known as the **molar volume of a gas** and is equal to $22.4 \text{ dm}^3 \text{ mol}^{-1}$ ($22\,400 \text{ cm}^3 \text{ mol}^{-1}$) at 273 K and 1 atmosphere pressure (1.01×10^5 Pa).

Using chemical equations: limiting and excess reagents

The chemical equation for a reaction provides the stoichiometry of that reaction. If we know the amounts of reactants we start with, we can work out the maximum amount of product(s) that can be formed.

Consider the following reaction:

$$CaCO_3(s) + 2HCl(aq) \rightarrow CaCl_2(aq) + CO_2(g) + H_2O(l)$$

The equation tells us that two moles of hydrochloric acid are required to react completely with one mole of calcium carbonate. If we do have one mole of calcium carbonate then, if we have more than two moles of hydrochloric acid, it is said to be in **excess**, because some of it will be unreacted at the end of the reaction. Similarly, if we have less than two moles of hydrochloric acid then the hydrochloric acid is said to be the **limiting reagent**, and now the calcium carbonate will be in excess because not all of it can react. The **yield** is the amount of product that is actually obtained experimentally. From this we can work out the **percentage yield**, which will be the actual yield in moles divided by the theoretically possible yield, and expressed as a percentage.

$$\% \text{ yield} = \frac{\text{actual yield}}{\text{theoretical yield}} \times 100$$

Worked example

What is the maximum volume of ammonia gas (measured at 273 K and 1.01×10^5 Pa) that could be obtained from the reaction between 10.0 g of hydrogen and 30.0 g of nitrogen? If the actual amount of ammonia obtained is 7.05 dm³ calculate the percentage yield for the reaction.

Step 1: Write the equation for the reaction. (Assume the reaction goes to completion to give the maximum amount of products.)

$$3H_2(g) + N_2(g) \rightarrow 2NH_3(g)$$

Step 2: Work out the limiting reagent.
Amount of hydrogen = 10.0/2.02 = 4.95 mol
Amount of nitrogen = 30.0/28.00 = 1.07 mol
For all the nitrogen to react it requires $3 \times 1.07 = 3.21$ mol of H_2
Thus the nitrogen is the limiting reagent, and the hydrogen is present in excess.

Step 3: Work out the amount and hence the volume of product formed.
As 1 mol of nitrogen reacts to give 2 mol of ammonia:
Maximum amount of ammonia that can be formed = $2 \times 1.07 = 2.14$ mol
As one mole of any gas occupies 22.4 dm^3 at 273 K and 1.01×10^5 Pa:
Maximum volume of ammonia that can be formed = $2.14 \times 22.4 = 47.9$ dm^3

Step 4: Calculate the percentage yield.
Percentage yield = 7.05/ 47.9 $\times$ 100 = 14.7%

 1 What volume of 0.100 mol dm^{-3} HCl(aq) is required to react exactly with 10.0 g of calcium carbonate?
2 What is the maximum volume of carbon dioxide (measured at 273 K, 1.01×10^5 Pa) that can be formed when 25.0 cm^3 of 0.800 mol dm^{-3} nitric acid, $HNO_3(aq)$, is added to 1.22 g of sodium carbonate, $Na_2CO_3(s)$?

An example of an acid–base titration

In an acid–base titration a known volume of a solution of known concentration (known as a **standard solution**) is reacted with a solution of unknown concentration(see Chapter 8). This can be exemplified by using a real experiment.

An experiment to determine the number of moles of water of crystallization in crystals of hydrated ethane-1,2-dioic acid (oxalic acid), $(COOH)_2.xH_2O$

Approximately 1.5 g of oxalic acid crystals was weighed accurately and dissolved in distilled water. The solution was then transferred to a volumetric flask and the total volume was made up to 250 cm^3. A burette was then filled with this acid solution. 25.0 cm^3 of 0.100 mol dm^{-3} sodium hydroxide solution was pipetted into a conical flask and two drops of phenolphthalein were added. The sodium hydroxide solution was titrated with the acid solution until the faint pink colour due to the indicator just disappeared. The titration was repeated to give two accurate values.

Data collection:

Mass of oxalic acid crystals = 1.529 g $\pm$ 0.001 g
Burette: oxalic acid solution made using a 250 cm^3 volumetric flask with an uncertainty of $\pm$ 0.5 cm^3
Pipette: 25.0 cm^3 of 0.100 mol dm^{-3} NaOH(aq) $\pm$ 0.04 cm^3
Indicator: 2 drops of phenolphthalein
End point: Taken to one drop when the faint pink colour of the indicator just disappeared.

Titration results: Uncertainty = $\pm$ 0.05 cm^3

	Rough	First accurate	Second accurate
Initial burette/ cm^3	0.00	0.00	0.00
Final burette/ cm^3	25.55	25.25	25.25
Volume used/ cm^3	25.55	25.25	25.25

Average volume of oxalic acid solution used $= 25.25$ cm^3 $\pm$ 0.10 cm^3

Summary:

25.0 cm^3 of 0.100 mol dm^{-3} sodium hydroxide, NaOH(aq), reacted exactly with 25.25 cm^3 of a solution of oxalic acid containing 1.529 g of oxalic acid crystals in 250 cm^3 of solution.

Data processing:

Equation for the reaction:

$(COOH)_2(aq) + 2NaOH(aq) \rightarrow (COONa)_2(aq) + 2H_2O(l)$

Amount of sodium hydroxide used $= 25.0/1000 \times 0.100 = 2.50 \times 10^{-3}$ mol
As 1 mol of oxalic acid reacts with 2 mol of sodium hydroxide:
Amount of oxalic acid in 25.25 cm^3 $= 1.25 \times 10^{-3}$ mol
Amount of oxalic acid in 250 cm^3 $= 250/25.25 \times 1.25 \times 10^{-3} = 1.238 \times 10^{-2}$ mol
M_r for hydrated oxalic acid $= 1.529/1.238 \times 10^{-2} = 124$
But M_r for hydrated oxalic acid, $(COOH)_2.xH_2O = 90.04 + 18x$
Hence $x = (124 - 90.04)/18 = 1.89 = 2$ to the nearest whole number.
Hence the formula for hydrated oxalic acid is $(COOH)_2.2H_2O$

Uncertainties:

Balance $= \dfrac{0.001}{1.529} \times 100 = 0.06\%$

Burette $= \dfrac{0.10}{25.25} \times 100 = 0.40\%$

Pipette $= \dfrac{0.04}{25.0} \times 100 = 0.16\%$

Volumetric flask $= \dfrac{0.50}{250} \times 100 = 0.20\%$

Total uncertainty $= 0.06\% + 0.40\% + 0.16\% + 0.20\% = 0.82\%$

Hence M_r for oxalic acid $= 124 \pm 0.82\% = 124 \pm 1$

Percentage error $= \dfrac{126-124}{126} \times 100 = 1.6\%$

The percentage error lies just outside the uncertainties inherent in the experiment. If you were writing up this experiment and being assessed for conclusion and evaluation, you would now need to evaluate the method employed, and suggest how the experiment could be improved.

1 What volume of 0.250 mol dm^{-3} sulfuric acid, H_2SO_4(aq), will be required to react exactly with 23.0 cm^3 of 0.390 mol dm^{-3} potassium hydroxide solution, KOH(aq)?

2 Aspirin ($M_r = 180$) has the formula $C_9H_8O_4$. One mole of aspirin reacts in the cold with one mole of sodium hydroxide. A tablet of a commercial painkiller has a mass of 0.483 g and is known to contain aspirin. It was found that the tablet reacted with exactly 16.65 cm^3 of 0.100 mol dm^{-3} sodium hydroxide solution, NaOH(aq). Calculate the percentage by mass of aspirin in the tablet.

3 A student wished to determine the percentage by mass of calcium carbonate in an eggshell. She reacted 3.953 g of the eggshell with 50 cm^3 of 1.00 mol dm^{-3} hydrochloric acid. After all the shell had reacted, the resulting solution was placed in a volumetric flask and the volume made up to 100 cm^3. 10.0 cm^3 of this solution reacted exactly with 12.65 cm^3 of 0.100 mol dm^{-3} sodium hydroxide solution, NaOH(aq). Calculate the percentage by mass of calcium carbonate in the eggshell.

4 What assumptions have you made in arriving at your answers to the two previous questions?

The relative abundance of all the isotopes can be calculated in a similar way (Table 4).

The relative atomic mass of germanium is given by

$$A_r = \frac{(70 \times 20.5) + (72 \times 27.4) + (73 \times 7.8) + (74 \times 36.5) + (76 \times 7.8)}{100}$$

$$= 72.7$$

Table 4

Isotope	Relative abundance / %
70	20.5
72	27.4
73	7.8
74	36.5
76	7.8

Electron arrangement

When a sodium salt such as sodium chloride, NaCl, is placed in a hot Bunsen burner flame the flame becomes a characteristic yellow colour. Glass also contains sodium ions, so that when a test-tube is heated strongly in a Bunsen burner the flame turns the same colour yellow. Other metals give different characteristic colours: for example, lithium gives red, potassium gives lilac, and barium gives green. These colours are obtained when the gaseous metal ions are given energy. This may be in the form of heat, as in a Bunsen burner, or electrical energy, as in a discharge tube. Examples of discharge tubes include: fluorescent lights, which contain a mixture of gases to give a light that approximates as closely as possible to daylight; fog lamps, which contain sodium ions; and neon lights used for advertising. All these different colours are electromagnetic radiation. Each colour has a particular wavelength and frequency, and all of them lie in the visible region of the spectrum. Each colour is also associated with a particular amount of energy.

The electromagnetic spectrum

Electromagnetic waves can travel through space or matter. The velocity of travel, c, is related to its wavelength λ and its frequency f. Velocity is measured in *metres per second* (m s^{-1}), wavelength in *metres* (m), and frequency in *per second* (s^{-1}), so it is easy to remember the relationship between them:

$$c = \lambda \times f$$
$$\text{(m s}^{-1}) \text{ (m)} \text{ (s}^{-1})$$

Electromagnetic radiation is a form of energy. The energy of electromagnetic radiation is related both to its frequency and to a constant known as **Planck's constant**:

$$E = hf$$

where f represents the frequency and h is Planck's constant.

As energy is measured in joules, and frequency is measured in the reciprocal of seconds (s^{-1}), Planck's constant has the units of joule-seconds, J s. It has the value 6.63×10^{-34} J s.

The smaller the wavelength, and thus the higher the frequency, the more energy the wave possesses. Electromagnetic waves have a wide range of wavelengths ranging from low-energy radio waves to

?

1 Calculate the relative atomic mass of iron from its mass spectrum shown in Figure 13.

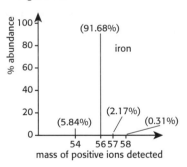

Figure 13

2 An approximation is often made that naturally occurring chlorine contains 75% chlorine-35 and 25% chlorine-37. This would give a relative atomic mass of 35.5. However, according to the IB Data Booklet the relative atomic mass of chlorine is 35.45. Calculate to two decimal places the actual percentage of chlorine-35 in naturally occurring chlorine.

very high-energy gamma (γ) radiation. Visible light occupies a very narrow part of the spectrum (Figure 14).

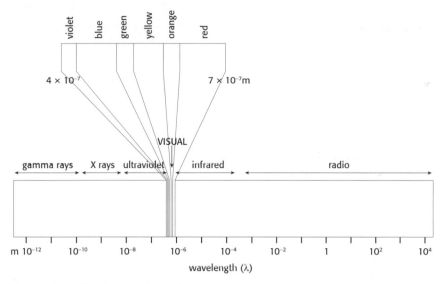

Figure 14 Visible light, and the electromagnetic spectrum.

Emission spectra

When evaluating an experiment it is common to look at the uncertainties associated with the apparatus being used. However, the uncertainties in one piece of apparatus that is used all the time are often ignored. The flame tests described above are observed using the human eye. The eye is an amazing optical instrument, but it does have its physical limitations. Apart from the fact that the eye can see only in the visible region of the electromagnetic spectrum, it does not scan each wavelength in turn but selects only the most obvious features. If yellow sodium light is passed through the prism or diffraction grating in a spectrometer then a line corresponding to the wavelength of green light can also be seen. In fact each of the more than 100 elements gives its own characteristic set of colours or lines, each at a fixed wavelength, when observed through a spectrometer. This is known as the **emission spectrum** of the element. Emission spectra are not continuous but consist of a series of separate (discrete) lines, which become closer together (converge) towards the higher-energy end of the spectrum (Figure 15).

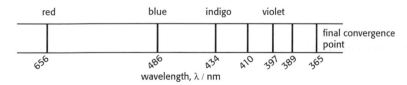

Figure 15 The visible hydrogen spectrum.

Figure 15 shows the series of lines that occurs in the visible region of the hydrogen spectrum. There is another series of lines in the ultraviolet region of the electromagnetic spectrum and several series in the infrared region. Each series consists of separate lines converging towards the higher-energy end of the spectrum.

Explanation of emission spectra

Niels Bohr (1885–1962), who studied with Rutherford, published his explanation for emission spectra in his model of atomic structure in 1913.

Figure 16 Niels Bohr depicted on a Danish banknote.

He proposed that electrons travel in orbits around the nucleus of the atom, and that each orbit is in a fixed energy level or shell. If an electron is given energy it is promoted to a higher energy level. As it drops back to a lower energy it emits a packet of light called a **quantum** with a particular amount of energy. This energy corresponds to light of a particular wavelength, and shows up in the spectrum as one of the discrete lines. A continuous spectrum is not formed, because electrons can exist only in certain fixed energy levels, and not in the spaces between them. Electrons dropping back to the lowest energy level ($n = 1$) emit the most energy, so this produces the series in the ultraviolet region of the hydrogen spectrum. The visible spectrum occurs when the electrons fall to the second level ($n = 2$), and the first series in the infrared occurs when electrons fall to the third level ($n = 3$). The lines in the spectra converge as the energy levels themselves converge. The value of the letter n used to describe the levels is known as the **principal quantum number** (Figure 17).

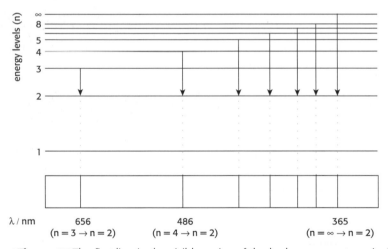

λ / nm	656	486	365
	($n = 3 \rightarrow n = 2$)	($n = 4 \rightarrow n = 2$)	($n = \infty \rightarrow n = 2$)

Figure 17 The first line in the visible series of the hydrogen spectrum is due to the electron transition from $n = 3$ to $n = 2$. The second line is due to the transition from $n = 4$ to $n = 2$, and the lines converge at $n = \infty$ to $n = 2$. The first line in the ultraviolet region will be due to electrons falling from $n = 2$ to $n = 1$ etc.

Electron arrangement

Each of the energy levels described by the principal quantum number can contain only a certain number of electrons. Generally, when the level is full, extra electrons begin to fill up the next available energy level. In the case of the third level, after eight electrons have been added the next two electrons occupy the fourth level before the next ten electrons then complete the third level.

Principal quantum number	Maximum number of electrons
1	2
2	8
3	8 or 18

As the atomic number determines how many electrons there are in a neutral atom, we can use these numbers to describe the electron arrangement of atoms. The **electron arrangements** of the first 20 atoms are listed in Table 5.

Table 5

Element	Electron arrangement	Element	Electron arrangement
H	1	Na	2.8.1
He	2 (first level full)	Mg	2.8.2
Li	2.1	Al	2.8.3
Be	2.2	Si	2.8.4
B	2.3	P	2.8.5
C	2.4	S	2.8.6
N	2.5	Cl	2.8.7
O	2.6	Ar	2.8.8 (third level "full")
F	2.7	K	2.8.8
Ne	2.8 (second level full)	Ca	2.8.8.2

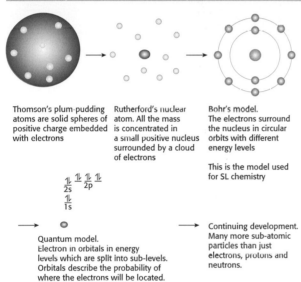

Thomson's plum-pudding atoms are solid spheres of positive charge embedded with electrons

Rutherford's nuclear atom. All the mass is concentrated in a small positive nucleus surrounded by a cloud of electrons

Bohr's model. The electrons surround the nucleus in circular orbits with different energy levels

This is the model used for SL chemistry

Quantum model. Electron in orbitals in energy levels which are split into sub-levels. Orbitals describe the probability of where the electrons will be located. Nucleus contains protons and neutrons. This is the model used for HL Chemistry - see below

Continuing development. Many more sub-atomic particles than just electrons, protons and neutrons.

Figure 18 The evolving models of an atom; each model builds on the previous model.

HL : Further physical evidence to support Bohr's description of electrons in energy levels can be seen from **ionization energies**. The first ionization energy of an element is defined as the energy required to remove one electron from an atom in its gaseous state. It is measured in kilojoules per mole.

$$X(g) \rightarrow X^+(g) + e^-$$

A graph of first ionization energies plotted against atomic number has the pattern shown in Figure 19. The highest value is for helium, an atom that contains two protons and two electrons. The two electrons are in the lowest level and are held very strongly by the two protons. For lithium it is relatively easy to remove an electron, which suggests that the third electron in lithium is in a higher energy level than the first two. The value then generally increases until element 10, neon, is reached, and then it drops sharply for sodium. The ionization energy then generally rises again until eight electrons in the third level have been added and argon is reached. The next element, potassium, has a very low first ionization energy as the fourth level begins to be occupied.

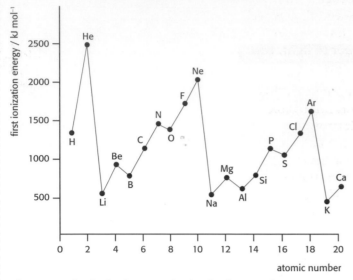

Figure 19 First ionization energies for the first 20 elements.

Successive ionization energies for the same element also provide evidence for the electron arrangement in each energy level.

$$X^+(g) \rightarrow X^{2+}(g) + e^- \text{ second ionization energy}$$
$$X^{2+}(g) \rightarrow X^{3+}(g) + e^- \text{ third ionization energy etc.}$$

As more electrons are removed, the electrostatic "pull" of the protons holds the remaining electrons more tightly, so considerably more energy is required to remove them: hence a logarithmic scale is usually used (Figure 20).

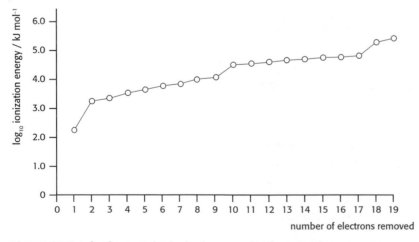

Figure 20 Graph of successive ionization energies for potassium.

The large "jumps" occur each time a level has been filled. By counting back to the first large jump you can determine the number of electrons in the outer energy level (and hence the group in the Periodic Table to which the element belongs). In the next chapter we shall see that the chemical properties of elements are very much determined by the number of electrons in their outer shell.

1 How are astronomers able to know the composition of stars that are light years away from the Earth?
2 The first eight successive ionization energies (in kJ mol⁻¹) for an element are: 1060, 1900, 2920, 4960, 6280, 21 200, 25 900 and 30 500. How many electrons does this element have in its outer energy level (shell)?
3 Use the "image" tool of a search engine on the Internet to find the emission spectra of lithium, sodium and potassium, and identify the line in each of their spectra that is responsible for their flame colours.

The nature of the electron

The Bohr model describes electrons as very small particles occupying fixed energy levels or shells. In the 1920s Louis de Broglie (1892–1987) proposed that electrons possess a dual nature, and that as well as behaving like particles they also have the properties of waves. Evidence for this is that electrons, in a similar manner to light, can be diffracted by passing them through very narrow slits.

Shortly afterwards, Werner Heisenberg (1901–1976) published his **uncertainty principle**. This principle states that it is impossible to know the exact position and momentum of an electron. As momentum is related to time, what this implies is that it is impossible to know the exact location of an electron at an exact moment in time. The more precisely the time is known, the less precisely the location is known, and vice versa. Heisenberg's uncertainty principle actually applies to any particle with mass, but the more massive the particle the less the uncertainty.

This concept has been extended by philosophers to the whole of the physical universe, which can be seen to exist as a collection of probabilities, because no individual particle or event can be located precisely in time. This led Albert Einstein to retort, "I cannot believe that God would choose to play dice with the universe." The new science of **quantum mechanics** was further developed in the 1920s by the Austrian physicist Erwin Schrödinger (1887–1961). The mathematical solutions to the Schrödinger wave equation describe the three-dimensional shapes of the atomic orbitals where there is a high probability that electrons are located.

Sub-levels and types of orbital

If you look closely at the emission spectrum of sodium the yellow colour is not actually one single discrete line, but two lines very close together. One has a wavelength of 5.889×10^{-7} m and the other 5.895×10^{-7} m. This is evidence that the energy levels can be split into sub-levels. The difference in energy between the two lines will be equivalent to the energy difference between the split sub-levels. This also explains why the graph of the first ionization energy against atomic number does not show a regular increase as the electrons are added.

From quantum theory it can be shown that electrons in an atom exist in **orbitals**, and that each orbital can hold a maximum of two electrons. Each electron in an atom is uniquely described by four different quantum numbers. The first, or principal, quantum number describes the main energy level. The second quantum number describes the type of orbital in that level. The first level contains just one type of orbital, called an s orbital. The second level contains s and p orbitals, the third level contains s, p and d orbitals, and the fourth level contains s, p, d and f orbitals.

The third quantum number determines the number of orbitals of each type in each level. Each level contains just one s orbital. Each level apart from the first contains three p orbitals, and each level apart from the first two contains five d orbitals. There are seven f orbitals in the fourth energy level in addition to the one s orbital, three p orbitals and five d orbitals.

Finally, the fourth quantum number describes the spin of the electron. Electrons in an orbital can either be spinning in one direction or in the opposite direction. The **Pauli exclusion principle** states that no two electrons in the same atom can have exactly the same four quantum numbers. Hence each orbital can contain a maximum of only two electrons, because if they held more, at least two of the electrons would have the same four quantum numbers. See Table 6.

Table 6

Principal level (shell)	Number of each type of orbital				Maximum number of electrons in level
	s	p	d	f	
1	1	–	–	–	2
2	1	3	–	–	8
3	1	3	5	–	18
4	1	3	5	7	32

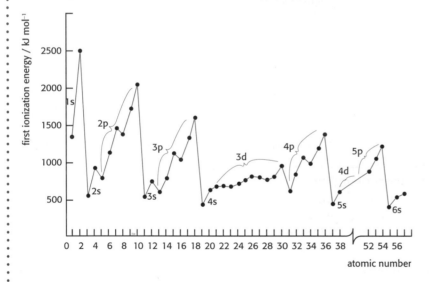

Figure 21 Graph of first ionization energies against atomic number, showing the sub-level from which the electron has been removed.

By looking at the graph of first ionization energies the relative energies of the orbitals and sub-levels within an atom can be deduced. The 1s level is the lowest in energy and is closest to the nucleus. Once this level is full it is very hard to remove an electron, so helium has the highest first ionization energy of all the elements. There is then a large energy difference to the second level, with the 2s orbital slightly lower than the three 2p orbitals. This explains why there is a slight decrease in ionization energies between beryllium ($Z = 4$) and boron ($Z = 5$), because the electron being removed from boron is a 2p electron, which is at a slightly higher energy than the 2s electron being removed from beryllium. The third level is much closer to the second level than the second is to the first, and is split into the one 3s, three 3p and five 3d sub-levels. The gap between the third and fourth levels is even smaller, and from the first ionization energy graph shown in Figure 22 it can be seen that the 4s orbital is actually lower than the five 3d orbitals.

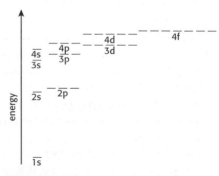

Figure 22 Relative energies of sub-levels within an atom (only the first four levels are shown).

Shapes of s and p orbitals

A straight line is the solution to an equation in the form $y = mx + c$, and a parabola is the solution to an equation in the form $y = mx^2 + c$. In a similar way, the solutions to the Schrödinger wave equation give the three-dimensional shapes of the atomic orbitals. These shapes represent the probability that the one or two electrons in the orbital will be located somewhere within the three-dimensional shape with about 90% certainty. We cannot state the radius of the electron cloud with absolute certainty, because of the uncertainty principle. It should be stressed that the shapes of atomic orbitals are mathematical models, and a useful tool to describe the location of electrons, but they cannot be observed directly.

s orbitals are spherical, and are symmetrical about the nucleus, but p orbitals are shaped rather like a figure of eight (Figure 23). The three p orbitals are all of equal energy: that is, they are said to be degenerate. Their two lobes are arranged orthogonally: that is, they are at right angles to each other. One lies along the x axis and is known as p_x; one lies along the y axis, and is known as p_y; and the third, p_z, lies along the z axis. Each orbital has a region in the middle centred on the nucleus where there is virtually zero probability of locating the electron.

Electron configuration

The electron arrangement describes the number of electrons in each main energy level. The **electron configuration** is more detailed; it also describes how the electrons are distributed in sub-levels. When building up the electronic structure, the lowest available energy levels (those that are closest to the nucleus) are always filled with electrons first. This is known as the **Aufbau principle**. There are two ways in which electron configurations are shown. The usual way is to write the number of the principal quantum number followed by the letter of the sub-level, with a superscript to indicate the number of electrons in the sub-level: for example, $1s^2 2s^2 2p^3$ (Table 7). Sometimes it is helpful to also show the spin of each electron; if so, boxes are used to represent orbitals.

For carbon ($Z = 6$) there are actually two possible ways in which the electrons could occupy the 2p orbitals. These can be seen clearly if we use boxes to represent the orbitals (Figure 24).

Electrons occupy orbitals of equal energy singly before they are paired up: so the p_x and p_y orbitals both contain one electron each rather than the p_x containing both electrons. This is known as **Hund's rule**. It occurs because electrons with opposite spins in the same orbital tend to repel each other to some extent, so it would be less energetically favourable to pair them up. For the next element nitrogen ($Z = 7$) each of the three 2p orbitals will contain one electron each, so the electronic configuration can be written $1s^2 2s^2 2p^3$ or, more precisely, $1s^2 2s^2 2p_x^1 2p_y^1 2p_z^1$. This explains the shape of the ionization energy graph for the second period, as there is a regular increase in the first ionization energies going from boron to nitrogen. The graph then dips slightly for oxygen as one of the p orbitals now contains a pair of

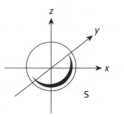
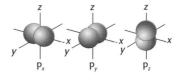

Figure 23 Shapes of s and p orbitals.

Table 7

Atomic number, Z	Element	Electron configuration
1	He	$1s^1$
2	He	$1s^2$
3	Li	$1s^2 2s^1$
4	Be	$1s^2 2s^2$
5	B	$1s^2 2s^2 2p^1$
6	C	$1s^2 2s^2 2p^2$

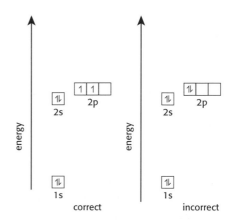

Figure 24 Two possible electron configurations for carbon: only one is correct.

electrons, and it is easier to remove one of these spin-paired electrons in oxygen than it is to remove one of the unpaired electrons in nitrogen.

This same pattern is repeated in the third level. To avoid writing out all the different levels there is a convention that builds on the electron configuration of the last noble gas (Table 8).

The ionization energy graph then clearly provides evidence that the 4s orbital is filled before the 3d. The 3d orbitals are then filled up as expected, but there are two notable exceptions. Chromium and copper atoms both contain only one 4s electron, because it is more energetically favourable to half-fill and completely fill the 3d sub-level respectively (Table 9).

It is not necessary to remember the orders of the sub-levels, because, as we shall see in the next chapter, the electronic configuration of any element can be deduced from its position in the Periodic Table. For example, take the element tin, Sn (Figure 25). It is element number 50 ($Z = 50$) and occurs in the p block region of the Periodic Table. It is in group 4, so its outer configuration will be ns^2np^2. If hydrogen and helium are assumed to form the first period, then tin is in the fifth period, so $n = 5$. The configuration is therefore $1s^22s^22p^63s^23p^64s^23d^{10}4p^65s^24d^{10}5p^2$. As the last noble gas was krypton ($Z = 36$), tin can also be written $[Kr]5s^24d^{10}5p^2$.

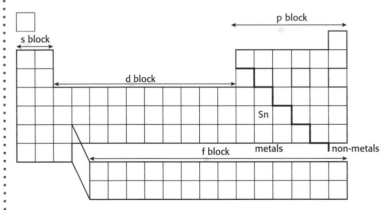

Figure 25 The position of tin in the Periodic Table.

When writing electronic configurations, the sum of the superscripts must always total the number of electrons in the atom or ion. Thus chloride ions, Cl^-, and potassium ions, K^+, both contain 18 electrons and have the configuration $1s^22s^22p^63s^23p^6$. When the d block elements form positive ions the energy levels are attracted more strongly to the nucleus, and now the 3d sub-level drops below the 4s sub-level in energy, so the 4s electrons are removed first. Thus the electronic configuration of the iron(II) ion, Fe^{2+}, is $1s^22s^22p^63s^23p^63d^6$ or $[Ar]3d^6$, and the electronic configuration of the copper(II) ion, Cu^{2+}, is $1s^22s^22p^63s^23p^63d^9$ or $[Ar]3d^9$.

Table 8

Atomic number, Z	Element	Electron configuration
11	Na	$1s^22s^22p^63s^1$ or $[Ne]3s^1$
12	Mg	$1s^22s^22p^63s^2$ or $[Ne]3s^2$
13	Al	$1s^22s^22p^63s^23p^1$ or $[Ne]3s^23p^1$
14	Si	$[Ne]3s^23p^2$
15	P	$[Ne]3s^23p^3$
16	S	$[Ne]3s^23p^4$
17	Cl	$[Ne]3s^23p^5$
18	Ar	$[Ne]3s^23p^6$

Table 9

Atomic number, Z	Element	Electron configuration
19	K	$[Ne]3s^23p^64s^1$ or $[Ar]4s^1$
20	Ca	$[Ne]3s^23p^64s^2$ or $[Ar]4s^2$
21	Sc	$[Ne]3s^23p^64s^23d^1$ or $[Ar]4s^23d^1$
22	Ti	$[Ar]4s^23d^2$
23	V	$[Ar]4s^23d^3$
24	Cr	$[Ar]4s^13d^5$
25	Mn	$[Ar]4s^23d^5$
26	Fe	$[Ar]4s^23d^6$
27	Co	$[Ar]4s^23d^7$
28	Ni	$[Ar]4s^23d^8$
29	Cu	$[Ar]4s^13d^{10}$
30	Zn	$[Ar]4s^23d^{10}$

❓

1 Rhodium ($Z = 45$) is a second-row transition element. It can form the Rh^+ ion and the Rh^{3+} ion.
  (a) Give the electronic configuration of the rhodium atom.
  (b) Give the electronic configurations for the Rh^{3+} ion.
2 (a) Explain why the first ionization energy of rubidium is lower than that of strontium.
  (b) Explain why the second ionization energy of strontium is lower than the second ionization energy of rubidium.

Extension material

Mendeleyev and the Periodic Table

One of the five Areas of Interaction for the IB Middle Years Programme is *homo faber*—man the toolmaker. The Periodic Table is probably the best example of *homo faber* in chemistry. If you understand how the Periodic Table is constructed and works, it enables you to predict much about the chemical behaviour and physical properties of all the elements.

The Periodic Table was first proposed in 1869 by Dmitri Mendeleyev (1834–1907), a Russian chemist. The discovery of electricity at the end of the eighteenth century and its use in electrolysis led to the rapid discovery and characterization of many more new elements (Figure 1).

By the middle of the nineteenth century it was clear that some elements displayed similar properties. Thus lithium, sodium and potassium showed similarities, as did chlorine, bromine and iodine. These groups of three, known as **triads**, were first recorded in 1829 by a German chemist, Johann Döbereiner (1780–1849) (Figure 2).

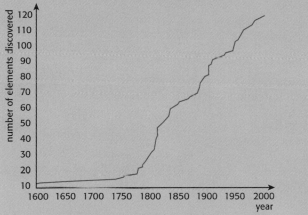

Figure 1 A graph showing the number of known elements from the year 1600.

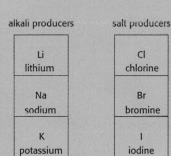

Figure 2 Two of Döbereiner's triads.

In 1864 a British chemist, John Newlands (1837–1898), arranged the elements in order of atomic weight and found that there was a repeating pattern, beginning with each eighth element. Newlands was also a musician, and he called his pattern the "law of octaves", because the notes in musical octaves also show a repeating pattern at each eighth note. See Table 1.

Table 1 Part of Newland's octaves

H	Li	Be	B	C	N	O
F	Na	Mg	Al	Si	P	S
Cl	K	Ca	Cr	Ti	Mn	Fe

However, his paper reporting this was rejected by the *Journal of the Chemical Society*. This was because the pattern was not complete. For example, the metal iron, Fe, does not have similar properties to the non-metals oxygen and sulfur. Similarly, manganese, Mn, does not fit with nitrogen and phosphorus. Newlands was ridiculed by his peers, one of whom went so far as to ask why he had not simply arranged them alphabetically.

It took the genius of Mendeleyev to devise the first form of the Periodic Table that we would still recognize today. Unlike Newlands, Mendeleyev was ready to take a risk and assume both that some of the information, such as atomic weight, might not be correct for all the elements, and also that some elements still remained to be discovered. At the March meeting of the Russian Chemical Society in 1869 he stated: "The elements, if arranged according to their atomic weights, exhibit an apparent periodicity of properties." In his table he placed the known elements in order of atomic weight, but he sometimes changed the order to help the chemistry fit, and —crucially—he left spaces for undiscovered elements (Figure 3).

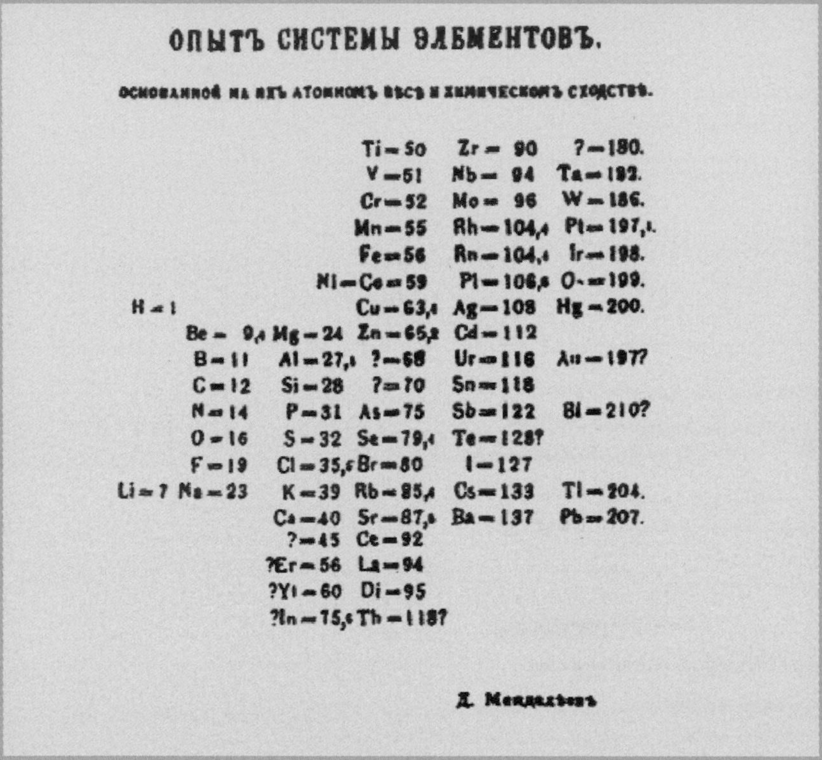
ОПЫТЪ СИСТЕМЫ ЭЛЕМЕНТОВЪ,

основанной на ихъ атомномъ вѣсѣ и химическомъ сходствѣ.

			Ti—50	Zr—90	?—180.
			V—51	Nb—94	Ta—182.
			Cr—52	Mo—96	W—186.
			Mn—55	Rh—104,4	Pt—197,4.
			Fe—56	Ru—104,4	Ir—198.
			Ni—Co—59	Pl—106,6	Os—199.
H—1			Cu—63,4	Ag—108	Hg—200.
	Be—9,4	Mg—24	Zn—65,2	Cd—112	
	B—11	Al—27,4	?—68	Ur—116	Au—197?
	C—12	Si—28	?—70	Sn—118	
	N—14	P—31	As—75	Sb—122	Bi—210?
	O—16	S—32	Se—79,4	Te—128?	
	F—19	Cl—35,5	Br—80	I—127	
Li—7	Na—23	K—39	Rb—85,4	Cs—133	Tl—204.
		Ca—40	Sr—87,6	Ba—137	Pb—207.
		?—45	Ce—92		
		?Er—56	La—94		
		?Yt—60	Di—95		
		?In—75,6	Th—118?		

Д. Менделѣевъ

Figure 3 Mendeleyev's original Periodic Table, published in 1869.

Where there were spaces, Mendeleyev was able to predict the properties of the as yet undiscovered elements. Once chemists knew what to look for, the elements were easier to isolate, and the agreement between the predicted properties and the actual properties was remarkably accurate. An example is germanium, the element that lies between silicon and tin. Mendeleyev called this element *eka*-silicon, and predicted its chemical and physical properties in 1871. It was finally discovered in 1886. See Table 2.

By 1905 Mendeleyev had revised his Periodic Table to include new elements, including a whole new group, the noble gases, which were isolated only in the late nineteenth century. However, the only truly significant change between Mendeleyev's early table and the modern form of the Periodic Table is that, in the modern

Table 2 Predicted and actual properties of germanium

Property	*eka*-silicon (predicted)	germanium (actual)
Atomic mass	72	72.59
Density / g cm^{-3}	5.5	5.35
Melting point / °C	high	947
Appearance	grey metallic	grey metallic
Reaction with acid	very slow	very slow
Formula of oxide	EkO$_2$	GeO$_2$
Density of oxide / g cm^{-3}	4.7	4.7
Nature of oxide	weakly basic	weakly basic
Formula of chloride	EkCl$_4$	GeCl$_4$
B.p. of chloride / °C	under 100	86 (GeCl$_4$)
Density of chloride / g cm^{-3}	1.9	1.9

form, the elements are arranged in order of **atomic number**, not atomic mass (known as atomic weight to Mendeleyev). Mendeleyev's gamble of re-ordering elements where necessary was inspirational, because the atomic weights had been calculated correctly. Because of the presence of isotopes some elements, such as argon and potassium, have their order in terms of atomic mass reversed. Mendeleyev's Periodic Table is one of the best examples of how inspiration, risk-taking and creativity as well as logical thought and determination are the essence of a great scientist.

The internationalism of the language of chemistry

The language of chemistry transcends cultural and national boundaries. If you have students in your class who speak a language with a different script ask them whether they have a Periodic Table in their own language. The description may well be in their own language, but the symbols for the elements are universal. The example in Figure 4 is from a Chinese Periodic Table written in Mandarin.

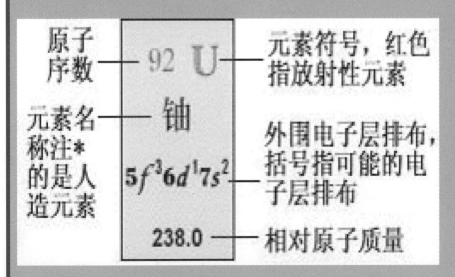

Figure 4 Chinese Periodic Table, written in Mandarin.

Chemistry and literature

Primo Levi (1919–1987), an Italian Jew, attributed his survival of the Auschwitz concentration camp in the Second World War to his knowledge of chemistry. Instead of hard physical labour he was assigned to one of the synthetic rubber factories attached to the camp. In his book *The Periodic Table*, published in 1975, he transforms chemical concepts into metaphysical comments on his life. Each of the 21 chapters is devoted to a particular element and relates to a different part of his life, including his response to the Holocaust. In the chapter on hydrogen Primo Levi writes:

"For me chemistry represented an indefinite cloud of future potentialities which enveloped my life in black volutes torn by fiery flashes, like those which had hidden Mount Sinai. Like Moses, from that cloud I expected my law, the principle of order around me, and in the world … I would watch the buds swell in spring, the mica glint in the granite, my own hands, and I would say to myself: 'I will understand this, too, I will understand everything.'"

The modern Periodic Table

There are actually several versions of the modern Periodic Table. Some are three-dimensional, others are circular; some have eight groups, others have 18 groups. But all of them have the elements arranged in order of increasing atomic number, and arranged in groups and periods.

In the IB version (known as the long form), which can be found in the IB Data Booklet, elements are placed vertically in the same **group**, and there are eight different groups numbered 1 to 7 with the eighth group numbered zero. All the elements in a particular group show similarities in their chemical properties. The number of the group corresponds to the number of electrons in the outer shell (valence electrons) of atoms of the elements in that group.

The elements are arranged horizontally in **periods**. Elements in each period have electrons in the same outer shell. Thus hydrogen and helium are in the first period, as the first level or shell is filled; the elements from lithium ($Z = 3$) to neon ($Z = 10$) are in the second period as the second energy level is being filled. Across a period there is a noticeable change in the chemical and physical properties of the elements. There is a gap between Group 2 and Group 3. From the fourth period onwards this gap is filled by 10 transition elements in each period.

HL For higher level students it can be seen that this block arrangement follows the order of the sub-levels. Groups 1 and 2 are s block elements, Groups 3, 4, 5, 6, 7 and 0 are p block elements, and the transition elements make up the d block elements. Although not on the IB programme there is a further block of elements known as the f block or lanthanides (first row, Ce to Lu) and actinides (second row, Th to Lr). See Figure 5.

1 Look at a Periodic Table and find other examples, apart from argon and potassium, where the elements do not follow the order of atomic mass.

2 Find out why some elements, e.g. 104 to 107, have two different names:

104 Rutherfordium, Rf, *or* unnilquadium, Unq
105 Dubnium, Db, *or* unnilpentium, Unp
106 Seaborgium, Sg, *or* unnilhexium, Unh
107 Bohrium, Bh, *or* unnilseptium, Uns

(a) Why do some scientists still disagree about these names?
(b) Should elements be named after particular scientists or places, or should they follow a logical, systematic naming system?
(c) How does national pride come into this?

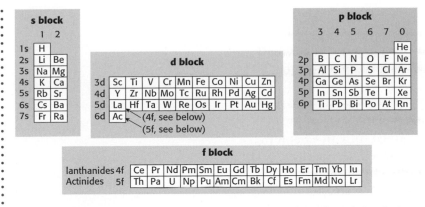

Figure 5 The Periodic Table rearranged into blocks to show the sub-levels (HL only).

The Periodic Table and physical properties

Ionization energies

Periodicity refers to the repeating pattern of physical and chemical properties that is seen at regular intervals in the Periodic Table. A graph of the first ionization energies of the elements against atomic number illustrates periodicity very clearly (Figure 6).

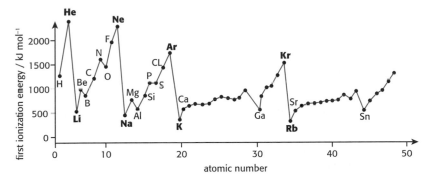

Figure 6 First ionization energies plotted against atomic number.

As we saw in Chapter 2, the first ionization energy of an element refers to the energy required to remove one electron from an atom of the element in the gaseous state. It is measured in kilojoules per mole.

$$M(g) \rightarrow M^+(g) + e^-$$

The elements in Group 1 (the alkali metals) have the lowest values in each period. As we descend Group 1 from lithium to caesium the values decrease, because the outer electron is further away from the nucleus and is therefore already in a higher energy level, so less energy is required to remove it. As each energy level is successively filled with electrons an equal number of protons are also being added to the nucleus. As each electron is added the level is attracted closer to the nucleus and therefore it becomes lower in energy, so that ionization energies generally increase across a period.

The exceptions to the general increase (e.g. boron and oxygen) are due to the presence of sub-levels within the main energy levels. The elements with the highest first ionization energy in each period are the Group 0 elements, the noble gases (He, Ne, Ar, Xe and Rn).

Electronegativities

A covalent bond is formed when one or more pairs of electrons are shared between two atoms. This is explained fully in Chapter 4 on bonding. When an atom is covalently bonded, its relative ability to attract a bonding pair of electrons to itself is known as **electronegativity**. Electronegativity is a relative value, not an absolute value, and so there are different scales of electronegativity in use. The values used by the IB are attributed to the North American chemist Linus Pauling (1901–1994) (Figure 7).

H	He																
2.1																	
Li	Be											B	C	N	O	F	Ne
1.0	1.5											2.0	2.5	3.0	3.5	4.0	
Na	Mg											Al	Si	P	S	Cl	Ar
0.9	1.2											1.5	1.8	2.1	2.5	3.0	
K	Ca	Sc	Ti	V	Cr	Mn	Fe	Co	Ni	Cu	Zn	Ga	Ge	As	Se	Br	Kr
0.8	1.0	1.3	1.5	1.6	1.6	1.5	1.8	1.8	1.8	1.9	1.6	1.6	1.8	2.0	2.4	2.8	
Rb																I	Xe
0.8																2.5	
Cs																At	Rn
0.7																2.2	

Figure 7 Pauling's scale of electronegativities

Electronegativity values also exhibit periodicity. Apart from helium and the other Group 0 elements, which have no values because the noble gases either form no compounds or form them only with difficulty, the electronegativity values decrease down each group. The values increase across each period. The most electronegative element is fluorine and the least electronegative element is caesium. Metals in Groups 1 and 2 are sometimes described as **electropositive** elements, as they have relatively low electronegativity values, whereas non-metals on the right of the Periodic Table in Groups 5, 6 and 7 are often described as **electronegative** elements. As you will see in Chapter 4, it is generally the difference in electronegativity between two bonding atoms that is important, rather than the precise value that each atom has. The value of the electronegativity is related to the size of the atoms. As the atoms become smaller, the nucleus will tend to attract a pair of electrons more strongly. However, it cannot be as simple as this: apart from helium, which has no value, hydrogen is the smallest atom, and on this basis it might be expected to have the highest value.

Some books explain this by referring to the **effective core charge**. This (also known as the effective nuclear charge) is the negative charge of the inner electron shells plus the positive charge due to the protons in the nucleus. Thus the core charge of fluorine is +7, because there are nine protons in the nucleus, two of which are "cancelled out" by the inner shell of two electrons. Chlorine, too, will have a core charge of +7, but because fluorine is smaller it will have a higher electronegativity. This is not really much better as an explanation: for example, it does not explain why oxygen, with a core charge of +6, has a higher value than chlorine, with a core charge of +7, unless size is by far the predominating factor.

Atomic radii

Logically, the atomic radius should be defined as the distance from the centre of the nucleus to the outermost electron. However, this is impossible to measure directly, partly because the precise position of an electron can never be known at a fixed point in time (**Heisenberg's uncertainty principle:** see Chapter 2). What can be measured, by a technique known as **X-ray diffraction**, is half the distance between two nuclei of bonded atoms (see Chapter 4). If they are bonded covalently it is known as the **covalent radius** of an atom, whereas if metallic bonding is involved it is the **metallic radius**. See Figure 8.

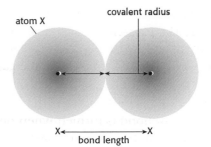

Figure 8 Covalent radius.

The covalent radius is defined as half the distance between the nuclei of two identical atoms that are covalently bonded together. If the radius is doubled it is equal to the bond length between the two atoms.

As a group is descended in the Periodic Table, the atomic radius increases. For example, the atomic radius of the atoms in Group 1 increases from 0.152 nm for lithium to 0.262 nm for caesium. This increase in atomic radius upon descending a group occurs because the outer electron is in an energy level that is progressively further away from the centre of the nucleus: for example, Li 2.1, Na 2.8.1 and K 2.8.8.1. See Figure 9.

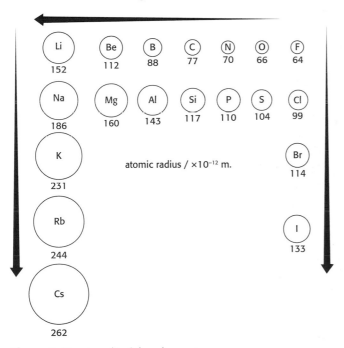

Figure 9 Atomic radii of the elements.

The atomic radius decreases across a period. As the number of outer electrons increases upon moving across a period, so the number of protons in the nucleus also increases. This increase in nuclear charge increases the attraction to the outer shell, so that the outer energy level progressively becomes closer to the nucleus. This decrease in size is quite considerable: a chlorine atom, for example, has a radius that is only about half that of a sodium atom.

Ionic radii

Positive ions (cations)

When an atom of a Group 1 element such as sodium loses an electron, the ion that is formed has a much smaller radius—almost half the value, in fact. There are two reasons for this. First, there is now one fewer electrons than there are protons, so the nucleus attracts the remaining electrons much more strongly. Second, there is one fewer energy level, because the outer shell has effectively been removed, and the remaining electrons have the noble gas electron arrangement of the preceding element. The size of the ions increases as Group 1 is descended, as the outer energy level becomes progressively further from the nucleus.

It is difficult to compare positive ions going across a period directly, because, apart from Group 1, unipositive ions tend not to be formed in compounds, as more than one electron is lost when ions in Groups 2 and 3 are formed. However, it is easy to see a trend if we consider isoelectronic ions—that is, those that contain the same number of electrons. Sodium ions, Na^+, magnesium ions, Mg^{2+}, and aluminium ions, Al^{3+}, all contain 10 electrons and have the electron configuration of neon (2.8), with the second shell completely full. However, sodium has 11 protons in the nucleus, magnesium has 12, and aluminium has 13. The 13 protons in the aluminium nucleus will attract the eight electrons in the outer shell much more strongly than the 11 protons in the nucleus of the sodium ion. As we move across the period, the isoelectronic ions will become much smaller: see Figure 10.

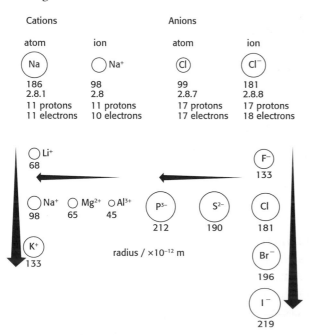

Cations		Anions	
atom	ion	atom	ion
Na	Na$^+$	Cl	Cl$^-$
186	98	99	181
2.8.1	2.8	2.8.7	2.8.8
11 protons	11 protons	17 protons	17 protons
11 electrons	10 electrons	17 electrons	18 electrons

Li$^+$ 68
F$^-$ 133

Na$^+$ 98   Mg^{2+} 65   Al^{3+} 45   P^{3-} 212   S^{2-} 190   Cl 181

K$^+$ 133   radius / ×10^{-12} m   Br$^-$ 196

I$^-$ 219

Figure 10 Radius of ions.

Negative ions (anions)

When the atoms of elements in Group 7 (the halogens) gain one electron to form a negative ion, there will be one more electron in the outer shell and hence more electron–electron repulsion. As the number of protons in the nucleus is unchanged, each of the electrons will be attracted less strongly, and the radius of the ion increases to almost twice the radius of the atom. The size of the negative ions increases down the group as the outer shell is further away from the nucleus. Across a period we again need to compare isoelectronic ions. Phosphide ions, P^{3-}, sulfide ions, S^{2-}, and chloride ions, Cl^-, all have the electron configuration of argon (2.8.8). However, the 18 electrons will be less attracted by the 15 protons in the nucleus of the phosphide ion than they will be by the 16 protons in the sulfide nucleus. Similarly, the sulfide ion will be larger than the chloride ion, where the 18 electrons are more strongly attracted by the 17 protons in the chloride ion nucleus.

Melting points

There are essentially two factors that determine the melting point of a crystalline substance. When a substance melts, the attractive forces holding the particles together in the crystal structure of the solid are overcome, and the particles are free to move around in the liquid state. The temperature at which this happens will depend both on the strength of the attractive forces and on the way in which the particles are packed in the solid state.

Within each group of the Periodic Table the forces of attraction tend to be similar. This is certainly true for Group 1, in which the elements all have a metallic structure, and for Group 7, in which there are only weak forces of attraction between the separate halogen molecules. It is less clear cut in other groups. For example, in Group 4, carbon and silicon have strong covalent bonds between the atoms to form giant covalent molecules, whereas tin and lead—at the bottom of the group—have metallic structures. In Group 1 the melting points decrease down the group (Table 3). Lithium melts at 181 °C, whereas the melting point of caesium is only just above room temperature. The melting points decrease because, as the atoms get larger, the forces of attraction between them, which are proportional to the inverse of the distance squared, decrease.

Table 3 Melting points of Group 1 elements (alkali metals)

Element	Li	Na	K	Rb	Cs
Melting point / °C	181	98	64	39	29

Table 4 Melting points of Group 7 elements (halogens)

Element	F_2	Cl_2	Br_2	I_2
Melting point / °C	−220	−101	−7.2	114

In Group 7 the melting points show the opposite trend, and increase down the group (Table 4). This is because the solid crystals of the halogens contain non-polar diatomic molecules, which are only weakly attracted to each other. As you will read in Chapter 4, these weak forces, which are known as van der Waals' forces, increase as the mass of the molecules increases.

Across the period there is a large change in the pattern of melting points as the bonding type changes from metallic (Na, Mg and Al), to giant covalent (Si), to weak van der Waals' attraction between simple molecules (P_4, S_8, Cl_2) and monatomic molecules (Ar). What is striking is the way the trend shows periodicity, as the pattern repeats itself with the next period. See Figure 11.

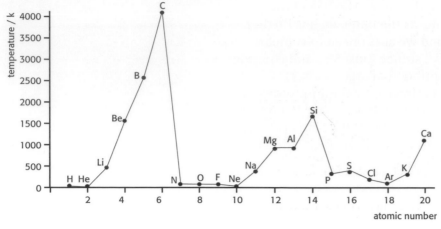

Figure 11 Melting points of the first 20 elements.

Thinking about science

How perfect is the Periodic Table?

Mendeleyev's Periodic Table and its subsequent revisions to include new elements rank as one of the greatest achievements in science. The Periodic Table is very much the chemist's tool, and a chemist is able to use it to predict the properties of elements (and their compounds) that he or she is unfamiliar with or which have not yet been synthesized. It fulfils Popper's criterion for a scientific theory, that it is capable of being tested by falsification (see Chapter 1). The graph of first ionization energies against atomic number is a perfect example of periodicity, and can be explained by the way in which the elements are set out in the Table. However, sometimes the trends down a group or across a period are not as perfect as we might expect. Let us look at the formulas for the highest fluorides formed by the elements in Period 3. Argon is left out, because it does not form a stable compound with fluorine.

NaF MgF_2 AlF_3 SiF_4 PF_5 SF_6

There is obviously a very clear trend here, and we would expect that the highest fluoride of chlorine would have the formula ClF_7. In fact the trend breaks down, and the highest fluoride of chlorine has the formula ClF_5. Sometimes, when the Periodic Table appears to give us the "wrong" answer, it can lead to an even greater understanding. The clue to the explanation as to why chlorine does not form ClF_7 lies in the fact that bromine also forms BrF_5 as its highest fluoride, but iodine does form a fluoride with the expected formula of IF_7. If we look at the covalent radii of the chlorine, bromine and iodine atoms we can see that the chlorine and bromine atoms are too small to accommodate seven fluorine atoms around them, whereas they are able to fit around the larger iodine atom.

However, not all the anomalies are so easy to explain. We have seen that the values for the electronegativities of the elements increase across a period and up a group, so that fluorine is the most electronegative element. Electronegativity is a measure of the relative ability of an atom of the element to attract a bonding pair of electrons. When a single electron is added to a gaseous atom of the element we can physically measure the energy change. This is known as the **first electron affinity**, and it is measured in kilojoules per mole. The value is always exothermic when just one electron is added to an atom of any gaseous element,

1 Use the graph of first ionization energies against atomic number (up to $Z = 50$) to predict what the value for the first ionization energy of caesium ($Z = 55$) will be. Check your prediction with the value given in the IB Data Booklet.

2 The trans-uranium elements with an atomic number greater than 92 are all radioactive and hence unstable. Many have been made artificially, often only in very small amounts. Element 119, eka-francium or ununennium, Uue, has not yet been made. If it follows the expected periodic trend it should have one electron in its outer shell, which will place it below francium in Group 1. Use your knowledge of the Periodic Table to predict the following about ununennium, Uue:
   (a) If it could be obtained in sufficient quantities, what would it be expected to look like at room temperature?
   (b) What ion would it be expected to form when it reacts?

3 Can you suggest a possible reason why the electron affinity of fluorine is lower than that of chlorine? Hint: look at the values for the other elements in the second period compared with the corresponding elements in the third period of the Periodic Table.

because the attraction between the nucleus and the added electron is greater than the electron–electron repulsion.

The values for chlorine, bromine and iodine are

Element	Cl	Br	I
Electron affinity/ kJ mol^{-1}	−364	−342	−314

There is a clear trend, and if we try to explain it we would probably refer to the size of the atoms and the ions formed, and perhaps to the number of protons in the nucleus. Almost certainly our explanation would lead us to predict that the value for fluorine would be even greater, perhaps in the region of −380 kJ mol^{-1}. In fact the value for fluorine is −348 kJ mol^{-1}. The trend has broken down, and there does not at present seem to be an obvious explanation for this, although several theories have been proposed.

It is when a hypothesis does not explain all the observed information that one needs to question and perhaps modify the hypothesis. Scientific theories continue to evolve, and perhaps the Periodic Table will one day evolve to become even better and closer to the truth.

Chemical properties of elements in the same group

Group 1: the alkali metals

The Group 1 elements are all shiny, silvery metals, which react vigorously with both water and air. You will most probably be shown lithium and sodium, and possibly potassium, all of which are stored under liquid paraffin to stop them reacting with the air. The metals are all soft, and can be cut easily with a knife. Once they are cut they rapidly tarnish as they react with the oxygen in the air to form the metal oxide. All of the Group 1 elements have just one electron in their outer shell, and when they react they lose this outer electron to form the unipositive ion. The reactivity increases down the group as the outer electron becomes successively easier to remove. Rubidium and caesium are so reactive that they are usually kept sealed in a glass tube under an inert atmosphere of argon. If you search the Internet you can find several websites that have videos that demonstrate the properties and reactions of all the alkali metals, including rubidium and caesium.

Reaction with water

All the alkali metals react exothermically with water to form hydrogen and ions of the metal hydroxide in water:

$$2M(s) + 2H_2O(l) \rightarrow 2M^+(aq) + 2OH^-(aq) + H_2(g)$$

where M represents Li, Na, K, Rb or Cs.

Lithium reacts the most slowly, and retains its shape as it reacts. The heat given out when sodium reacts is enough to melt the sodium into a ball so that it darts about on the surface of the water, and occasionally the hydrogen released may ignite and spark with the characteristic yellow flame due to the presence of sodium ions. Potassium is much more reactive, and the hydrogen evolved usually burns steadily with a lilac flame because of the presence of potassium ions. Rubidium is even more reactive, and the reaction with caesium may be so violent that the glass reaction vessel explodes.

Reaction with halogens

The alkali metals also react readily with the halogens, with the reactivity of the metals increasing down the group. If a piece of heated sodium is lowered into a gas jar of chlorine it will burst into flames, and white fumes will be seen, caused by the formation of the ionic salt sodium chloride.

$$2Na(s) + Cl_2(g) \rightarrow 2NaCl(s)$$

Similarly, if bromine vapour or iodine vapour is passed over heated sodium, a vigorous reaction occurs, and the two elements combine to produce sodium bromide and sodium iodide respectively.

$$2Na(s) + Br_2(g) \rightarrow 2NaBr(s)$$
$$2Na(s) + I_2(g) \rightarrow 2NaI(s)$$

Redox reactions are covered fully in Chapter 9. All the alkali metals are good reducing agents, because they can readily donate one electron. Similarly the halogens are good oxidizing agents, because they can readily accept one electron.

Group 7: the halogens

The four halogens fluorine (F_2), chlorine (Cl_2), bromine (Br_2) and iodine (I_2) all exist as diatomic molecules. When they react, the single bond between the atoms in the halogen molecule is broken, and the two atoms then each gain one electron to form halide ions.

Fluorine is the strongest oxidizing agent known, and is so reactive that its use is banned in school laboratories. The reactivity decreases down the group, but chlorine, bromine and iodine are all very reactive and need to be handled carefully, because they are all also poisonous.

The poisonous nature of chlorine was utilized in the First World War, when it became the first of the "war gases". The English poet Wilfred Owen (1893–1918) describes the use of chlorine in his poem "Dulce et Decorum Est", written a few months before he himself was machine-gunned to death one week before the war officially ended.

Dulce et Decorum Est

Bent double, like old beggars under sacks,
Knock-kneed, coughing like hags, we cursed through sludge,
Till on the haunting flares we turned our backs
And towards our distant rest began to trudge.
Men marched asleep. Many had lost their boots
But limped on, blood-shod. All went lame; all blind;
Drunk with fatigue; deaf even to the hoots
Of gas shells dropping softly behind.

Gas! GAS! Quick, boys!—An ecstasy of fumbling,
Fitting the clumsy helmets just in time;
But someone still was yelling out and stumbling,
And flound'ring like a man in fire or lime…
Dim, through the misty panes and thick green light,
As under a green sea, I saw him drowning.

In all my dreams, before my helpless sight,
He plunges at me, guttering, choking, drowning.

If in some smothering dreams you too could pace
Behind the wagon that we flung him in,
And watch the white eyes writhing in his face,
His hanging face, like a devil's sick of sin;
If you could hear, at every jolt, the blood
Come gargling from the froth-corrupted lungs,
Obscene as cancer, bitter as the cud
Of vile, incurable sores on innocent tongues, –
My friend, you would not tell with such high zest
To children ardent for some desperate glory,
The old Lie: Dulce et decorum est
Pro patria mori.

(Note: *Dulce et decorum est pro patria mori*, when translated from Latin, means "It is sweet and honourable to die for your country.")

Oxidizing ability of chlorine, bromine and iodine

The oxidizing ability—that is, the ability to accept electrons—decreases down the group. This can be shown by reacting chlorine dissolved in water with an aqueous solution of sodium bromide. It will be observed that the solution turns yellow. The chlorine accepts an electron from the bromide ion and a dilute solution of bromine is formed, which accounts for the yellow colour.

$$Cl_2(aq) + 2Br^-(aq) \rightarrow 2Cl^-(aq) + Br_2(aq)$$

Similarly, if chlorine water is added to a solution of potassium iodide the iodide ions are oxidized by the chlorine to form a red colour due to the presence of iodine, and the chlorine atoms each gain one electron to form chloride ions.

$$Cl_2(aq) + 2I^-(aq) \rightarrow 2Cl^-(aq) + I_2(aq)$$

It is easy to confirm that the oxidizing ability of bromine lies between that of chlorine and iodine by reacting bromine water with a solution of potassium iodide. The bromine will oxidize the iodide ions to form bromide ions and iodine.

$$Br_2(aq) + 2I^-(aq) \rightarrow 2Br^-(aq) + I_2(aq)$$

Thus the oxidizing ability of the halogens follows the following order:

$$F_2 > Cl_2 > Br_2 > I_2$$

Chemical properties of elements in the same period

There is a very noticeable change in the properties of the elements across the period. Sodium, magnesium and aluminium are metals. They are shiny, and are excellent conductors of heat and electricity. Silicon is a semiconductor of electricity, and is called a **metalloid**, as it possesses some of the properties of a metal and some of a non-metal. Phosphorus, sulfur, chlorine and argon are all non-metals, and do not conduct electricity.

Metals can also be distinguished from non-metals by their chemical properties. The oxides of metals tend to consist of the metal ions and oxide ions, and so are said to be ionic. Thus the formulas of sodium oxide, magnesium oxide and aluminium oxide are $(Na^+)_2O$, $Mg^{2+}O^{2-}$ and $(Al^{3+})_2(O^{2-})_3$ respectively, and they are white solids at room temperature, with relatively high melting and boiling points. When they are in the liquid state they will conduct electricity, and are decomposed to their elements in the process. Silicon dioxide, SiO_2, has a giant covalent structure and so has a very high melting and boiling point, but it does not conduct electricity when molten. The oxides of the non-metals sulfur, phosphorus and chlorine are all covalent, and have relatively low melting and boiling points. Argon does not form an oxide.

Metal oxides tend to be basic, whereas non-metal oxides tend to be acidic. Sodium oxide and magnesium oxide are both basic, and react with water to form aqueous solutions of the metal hydroxides:

$$Na_2O(s) + H_2O(l) \rightarrow 2NaOH(aq)$$
$$MgO(s) + H_2O(l) \rightarrow Mg(OH)_2(s)$$

Aluminium is a metal, but its oxide is **amphoteric**: that is, it can be either basic or acidic, depending on whether it is reacting with an acid or a base.

The remaining elements in Period 3 have acidic oxides. For example, phosphorus pentoxide, P_4O_{10}, and sulfur trioxide, SO_3, react with water to form phosphoric acid and sulfuric acid respectively:

$$P_4O_{10}(s) + 6H_2O(l) \rightarrow 4H_3PO_4(aq)$$
$$SO_3(g) + H_2O(l) \rightarrow H_2SO_4(aq)$$

The oxides of Period 3 are summarized in Table 5.

Table 5 Summary of the oxides of Period 3

	Element							
	Na	Mg	Al	Si	P	S	Cl	Ar
Formula of highest oxide	Na_2O	MgO	Al_2O_3	SiO_2	P_4O_{10}	SO_3	Cl_2O_7	none
Nature of oxide	basic	basic	amphoteric	acidic	acidic	acidic	acidic	none

HL More detailed trends across Period 3

Oxides

Sodium oxide, Na_2O, magnesium oxide, MgO, and aluminium oxide, Al_2O_3, are all ionic. This accounts for their high melting points and electrical conductivity when molten. Silicon dioxide has a diamond-like macromolecular structure with a high melting and boiling point. The oxides of phosphorus, sulfur and chlorine are covalent because of the relatively small difference in electronegativity values between the elements and oxygen. These oxides have simple covalent molecular structures with low melting and boiling points. Because they are covalent they do not conduct electricity in the liquid state. All three of the elements phosphorus, sulfur and chlorine form more than one oxide with oxygen. Both the oxides of phosphorus exist as dimers. Phosphorus forms phosphorus(III) oxide, P_4O_6, and phosphorus(V) oxide, P_4O_{10}. Sulfur forms sulfur dioxide, SO_2, and sulfur trioxide, SO_3, and chlorine forms several oxides including chlorine(I) oxide, Cl_2O, and chlorine(VII) oxide, Cl_2O_7.

?

1 What would you expect to see if a brown solution of iodine dissolved in potassium iodide solution is added to an aqueous solution of potassium chloride?

2 There is one video available on the Internet that shows both rubidium and caesium being thrown into a bath of water; the bath explodes spectacularly. If you look carefully, however, it can be seen that the video is a fake, because wires leading to explosive charges in the bath tub can be seen. Do you think that it can ever be justified to fake a reaction to make it more interesting and exciting to students, and hopefully increase their enjoyment of chemistry, or is it always ethically wrong?

The acid–base properties of the oxides are also linked to their type of bonding. The ionic oxides of the electropositive elements are very basic and form solutions that are alkaline. Sodium hydroxide is very soluble in water but magnesium hydroxide is only very sparingly soluble.

$$Na_2O(s) + H_2O(l) \rightarrow 2Na^+(aq) + 2OH^-(aq)$$
$$MgO(s) + H_2O(l) \rightarrow Mg(OH)_2(aq)/(s)$$

Aluminium oxide is amphoteric. It can behave either as an acid or as a base. It reacts as a base with an acid such as hydrochloric acid, and as an acid with alkalis such as sodium hydroxide.

Acting as a base:

$$Al_2O_3(s) + 6HCl(aq) \rightarrow 3AlCl_3(aq) + 3H_2O(l)$$

Acting as an acid:

$$Al_2O_3(s) + 2NaOH(aq) + 3H_2O(l) \rightarrow 2NaAl(OH)_4(aq)$$
$$\text{sodium aluminate}$$

Silicon dioxide is insoluble in water, and it does not react with water to form an acid. Nevertheless, it is still classified as an acidic oxide, because it does react with sodium hydroxide at temperatures above about 350 °C to form sodium silicate and water:

$$SiO_2(s) + 2NaOH(aq) \rightarrow Na_2SiO_3(aq) + H_2O(l)$$
$$\text{sodium silicate}$$

The oxides of phosphorus, sulfur and chlorine are all acidic, and react and dissolve in water to form strongly acidic solutions:

$$P_4O_6(s) + 6H_2O(l) \rightarrow 4H_3PO_3(aq) \quad \text{phosphoric(III) acid}$$
$$P_4O_{10}(s) + 6H_2O(l) \rightarrow 4H_3PO_4(aq) \quad \text{phosphoric(V) acid}$$

$$SO_2(g) + H_2O(l) \rightarrow H_2SO_3(aq) \quad \text{sulfuric(IV) acid (sulfurous acid)}$$
$$SO_3(s) + H_2O(l) \rightarrow H_2SO_4(aq) \quad \text{sulfuric(VI) acid}$$

$$Cl_2O(g) + H_2O(l) \rightarrow 2HClO(aq) \quad \text{chloric(I) acid (hypochlorous acid)}$$
$$Cl_2O_7(l) + H_2O(l) \rightarrow 2HClO_4(aq) \quad \text{chloric(VII) acid (perchloric acid)}$$

Table 6 Oxides of Period 3 elements

Property	Formula						
	Na_2O	MgO	Al_2O_3	SiO_2	P_4O_{10} (P_4O_6)	SO_3 (SO_2)	Cl_2O_7 (Cl_2O)
State at 25 °C	solid	solid	solid	solid	solid (solid)	liquid (gas)	liquid (gas)
Melting point / °C	1275	2852	2027	1610	24	17	−92
Boiling point / °C	–	3600	2980	2230	175	45	80
Electrical conductivity in molten state	good	good	good	very poor	none	none	none
Structure	ionic			covalent macro-molecular	simple covalent molecular		
Reaction with water	Forms Na(OH)(aq), an alkaline solution	Forms $Mg(OH)_2$, weakly alkaline solution	Does not react	Does not react	P_4O_{10} forms H_3PO_4, an acidic solution	SO_3 forms H_2SO_4, a strong acid	Cl_2O_7 forms $HClO_4$, an acidic solution
Nature of oxide	basic		amphoteric		acidic		

Chlorides

The physical properties of the chlorides are related to the structure in the same way as those of the oxides. Sodium chloride, Na^+Cl^-, and magnesium chloride, $Mg^{2+}(Cl^-)_2$, are ionic. Both conduct electricity when molten and have relatively high melting points. Compared with aluminium oxide, aluminium chloride possesses a much more covalent character. There is evidence that, in the solid state, it exists as aluminium and chloride ions, but as it melts there is a dramatic change in the bonding, and the covalent dimer Al_2Cl_6 is formed. Molten aluminium chloride is a poor conductor of electricity. In the gaseous state there is an equilibrium between aluminium chloride monomer and aluminium chloride dimer (Figure 12).

$$2AlCl_3(g) \rightleftharpoons Al_2Cl_6(g)$$

The remaining chlorides of Period 3 all have simple molecular structures (Figure 13). These molecules are held together by weak van der Waals' forces, which results in low melting and boiling points.

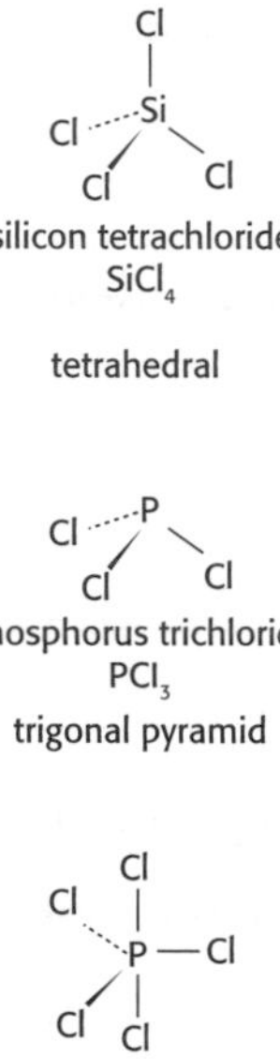

Sodium chloride is the only neutral chloride in Period 3. When it dissolves in distilled water it gives a solution with a pH of 7. Magnesium chloride gives a slightly acidic solution with distilled water. The magnesium ion is small, and has a charge of +2. This means that the density of the charge on the ion is much higher than that on a sodium ion. The smaller, more densely charged magnesium ion attracts water molecules and causes some of them to dissociate to form hydrogen ions. A full description of this process and the reaction of anhydrous aluminium chloride with water are given in Chapter 8. Aluminium has an even smaller ion, with a charge of +3, so the charge density is even higher. When anhydrous aluminium chloride is added to water, a very exothermic reaction takes place, and hydrochloric acid is formed.

$$2AlCl_3(s) + 3H_2O(l) \rightarrow Al_2O_3(s) + 6HCl(aq)$$

Strictly speaking, the aluminium ion becomes hexahydrated to form the hexaaquoaluminium(III) ion, $[Al(H_2O)_6]^{3+}$. The water molecules are strongly attracted to the high charge density on the small aluminium ion, and three of them successively dissociate to give hydrated aluminium hydroxide and hydrochloric acid:

$$[Al(H_2O)_6]^{3+}(aq) \rightleftharpoons Al(H_2O)_3(OH)_3(s) + 3H^+(aq)$$

All the other chlorides also react vigorously with water to produce acidic solutions of hydrochloric acid, together with fumes of hydrogen chloride:

$$SiCl_4(l) + 4H_2O(l) \rightarrow Si(OH)_4(aq) + 4HCl(aq)$$
$$PCl_3(l) + 3H_2O(l) \rightarrow H_3PO_3(aq) + 3HCl(aq)$$
$$PCl_5(s) + 4H_2O(l) \rightarrow H_3PO_4(aq) + 5HCl(aq)$$

Chlorine itself reacts with water to some extent to form an acidic solution:

$$Cl_2(aq) + H_2O(l) \rightleftharpoons HCl(aq) + HClO(aq)$$

Figure 12 The Lewis structure of aluminium chloride dimer. Each line represents a pair of electrons. A full explanation is given in Chapter 4.

silicon tetrachloride
$SiCl_4$

tetrahedral

phosphorus trichloride
PCl_3

trigonal pyramid

phosphorus pentachloride
PCl_5

trigonal bipyramid

Figure 13 Simple molecular structures of non-metallic chlorides

Table 7 Chlorides of period 3 elements

Property	Formula						
	NaCl	MgCl$_2$	AlCl$_3$ (Al$_2$Cl$_6$)	SiCl$_4$	PCl$_3$ (PCl$_5$)	(S$_2$Cl$_2$)	Cl$_2$
State at 25 °C	solid	solid	solid	liquid	liquid (solid)	liquid	gas
Melting point / °C	801	714	178 (sublimes)	−70	−112	−80	−101
Boiling point / °C	1413	1412	–	58	76	136	−35
Electrical conductivity in molten state	good	good	poor	none	none	none	none
Structure	ionic		simple covalent molecular				
Reaction with water	dissolves easily		fumes of HCl produced				some reaction with water
Nature of solution	neutral	weakly acidic	acidic				

Extension material

The reaction of the chlorides of carbon and silicon with water

Tetrachloromethane (also known as carbon tetrachloride), CCl$_4$, is a colourless liquid at room temperature, with a boiling point of 77 °C. Silicon tetrachloride, SiCl$_4$, is a colourless liquid at room temperature, with a boiling point of 58 °C. Both are the covalent chlorides of Group 4 elements, and both consist of non-polar tetrahedral molecules, so they obviously have many similarities, as we would expect. However, there is at least one very obvious difference in their chemical properties. Carbon tetrachloride is immiscible with water, and if the two are placed together they form two separate layers, and no reaction takes place. If water is added to silicon tetrachloride a very exothermic reaction occurs, and fumes of hydrogen chloride are seen.

$$SiCl_4(l) + 4H_2O(l) \rightarrow Si(OH)_4(aq) + 4HCl(aq)$$

Why is there such a difference in their chemical behaviour with water? The answer can be found if we look at the electronic configurations of the central atom. Both have four outer electrons and have the general outer level configuration ns^2np^2. Carbon's configuration is $1s^22s^22p^2$ and silicon's configuration is $1s^22s^22p^63s^23p^2$. Both atoms possess empty 3d orbitals. For carbon, these empty 3d orbitals are completely unavailable, as they are much higher in energy than the outer electrons and so cannot be utilized. With silicon, however, the 3d empty orbitals are only slightly higher in energy than the 3p sub-level, and hence they are readily available to be used. When a water molecule collides with a molecule of silicon tetrachloride one of the two non-bonding pairs of electrons on the oxygen atom of the water molecule can form a coordinate bond by utilizing one of the empty 3d orbitals on the silicon atom. See Figure 14.

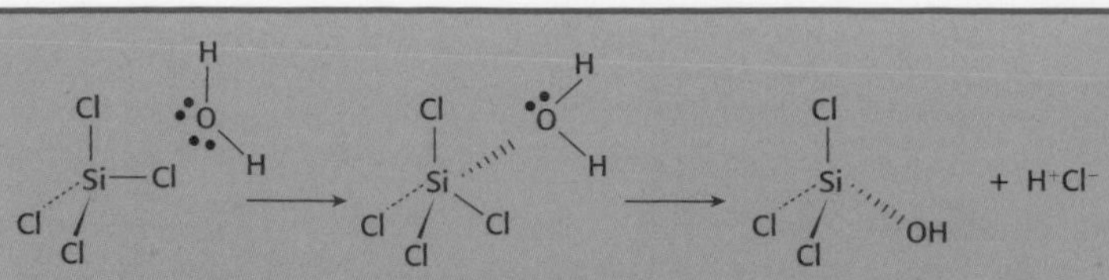

Figure 14 Silicon tetrachloride reacts with water, because it is able to expand its octet by using available empty d orbitals.

During this process an intermediate is formed in which the silicon atom is able to expand its octet and temporarily accommodate five pairs of electrons. Then one of the silicon-to-chlorine bonds breaks, and the water molecule substitutes in its place, releasing a hydrogen ion at the same time. Because carbon has no readily available empty 3d orbitals it is unable to expand its octet, and so tetrachloromethane remains completely unreactive with water.

HL : First row d-block elements

Definition of a transition element

Transition elements are defined as elements that contain an incomplete d level of electrons in one or more of their oxidation states. It is the 3d sub-level that is incomplete for the first row of transition elements (Table 8). Zinc, which has the configuration $[Ar]4s^23d^{10}$, is clearly not a transition element, because neither the metal nor the dipositive ion, Zn^{2+}, with the electron configuration $[Ar]3d^{10}$, contains an incomplete d sub-level. The situation is less clear with scandium. Scandium is usually also described as a non-transition element, because its common oxidation state is +3, and the tripositive ion, Sc^{3+}, has the electronic configuration of argon. However, the electron configuration of scandium metal is $[Ar]4s^23d^1$, and so in a sense the metal is a transition element but its compounds are not.

Table 8 The first row transition elements

Element	(Sc)	Ti	V	Cr	Mn	Fe	Co	Ni	Cu	(Zn)
Electron configuration [Ar]	$4s^23d^1$	$4s^23d^2$	$4s^23d^3$	$4s^13d^5$	$4s^23d^5$	$4s^23d^6$	$4s^23d^7$	$4s^23d^8$	$4s^13d^{10}$	$4s^23d^{10}$

As the atomic number increases, the 3d sub-level fills up regularly, except for chromium and copper. These have configurations of $[Ar]4s^13d^5$ and $[Ar]4s^13d^{10}$ respectively, because it is more energetically favourable to half-fill and completely fill the 3d sub-level rather than spin-pair two of the electrons in the 4s orbital.

Transition elements show characteristic properties, although they are not totally unique just to transition elements. These properties include variable oxidation states, complex ion formation, the existence of coloured compounds, and good catalytic properties.

Variable oxidation states

According to the *Aufbau* principle, when we are filling up the sub-levels of the neutral atoms the 4s level is lower in energy than the 3d sub-level and so fills up first. However, when positive ions (cations) are being formed, it is the 4s electrons that are lost first, so that, for example, the electron configuration of the manganese(II) ion, Mn^{2+}, is $[Ar]3d^5$. All the energy levels will be closer to the nucleus in the ion because there are more protons than electrons, and as they move down in energy the order of the 4s and 3d sub-levels reverses. See Figure 15.

All the first row transition elements (apart from chromium and copper) contain two 4s electrons, and so all of them show the +2 oxidation state. This includes chromium and copper, because they can also lose one of the 3d electrons along with the single 4s electron.

Some of the transition metals can form the +3 or +4 ion (e.g. Fe^{3+}, Mn^{4+}), because the ionization energies are such that up to two 3d electrons can also be lost. The M^{4+} ion is rare, because it is so small and so very highly charged that it polarizes surrounding molecules or ions, leading to covalent bonding. In the higher oxidation states the element is usually found not as the free metal ion but either covalently bonded or as the oxyanion, such as the manganate(VII) ion, MnO_4^-, and the dichromate(VI) ion, $Cr_2O_7^{2-}$. Some common examples of variable oxidation states in addition to +2 are

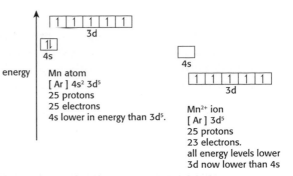

Figure 15 In a positive ion the energy levels are lower than in the free atom due to the stronger attraction for the remaining electrons by the nucleus.

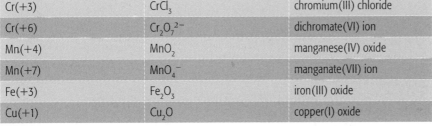

Cr(+3)	$CrCl_3$	chromium(III) chloride
Cr(+6)	$Cr_2O_7^{2-}$	dichromate(VI) ion
Mn(+4)	MnO_2	manganese(IV) oxide
Mn(+7)	MnO_4^-	manganate(VII) ion
Fe(+3)	Fe_2O_3	iron(III) oxide
Cu(+1)	Cu_2O	copper(I) oxide

Complex ion formation

Because of their small size, transition metal ions attract species that are rich in electrons. Such species are known as **ligands**. Ligands are neutral molecules or negative ions that contain a non-bonding pair of electrons. These electron pairs can form coordinate covalent bonds with the metal ion to form **complex ions**. As we shall see in Chapter 8, this is an example of an acid–base reaction in terms of Lewis theory. Transition metal ions are good Lewis acids, because they can accept a pair of electrons, and ligands are good Lewis bases, because they can donate a pair of electrons. Water is a common ligand, and most (but not all) transition metal ions exist as hexahydrated complex ions in aqueous solution, for example the hexaaquairon(III) ion, $[Fe(H_2O)_6]^{3+}$ (Figure 16).

Other examples of ligands include the neutral molecule ammonia, and the chloride and cyanide anions. A nice example to illustrate how one ligand can be replaced by another is the addition of ammonia to an aqueous solution of copper(II) sulfate. Initially, as the ammonia is added the solution gives a precipitate owing to the formation of copper(II) hydroxide, but as more ammonia is added this dissolves to give the deep blue colour of the tetraamminecopper(II) complex ion, $[Cu(NH_3)_4]^{2+}$. Similarly, if concentrated hydrochloric acid is added to a solution of copper(II) ions the yellow tetrachlorocuprate(II) complex anion is formed. Note that in this ion the overall charge on the ion is −2, as each of the

Figure 16 The hexaaquairon(III) ion, Fe^{3+}(aq).

four ligands has a charge of −1. The solution will actually appear green as there will still be some hydrated copper(II) ions present. This is a good example to illustrate Le Châtelier's principle, as the position of equilibrium can easily be altered by adding more of one of the ligands, with the corresponding colour change:

$$[CuCl_4]^{2-}(aq) \underset{H_2O}{\overset{Cl^-}{\rightleftharpoons}} [Cu(H_2O)_4]^{2+}(aq) \underset{H_2O}{\overset{NH_3}{\rightleftharpoons}} [Cu(NH_3)_4]^{2+}(aq)$$

  yellow/green   pale blue   deep blue

The number of lone pairs bonded to the metal ion is known as the **coordination number**. Compounds with a coordination number of 6 are octahedral in shape, those with a coordination number of 4 are tetrahedral or square planar, and those with a coordination number of 2 are usually linear (Figure 17).

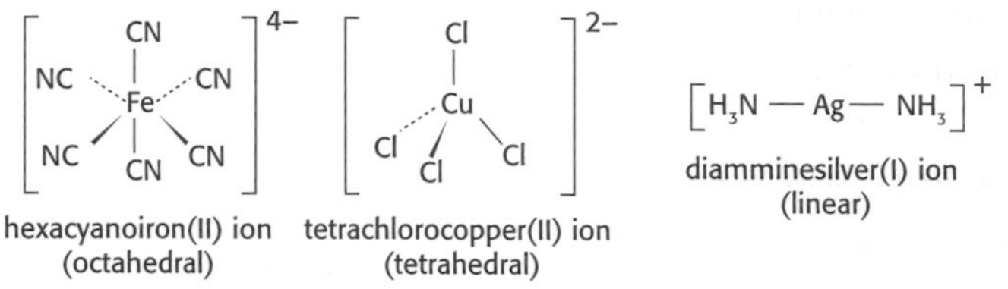

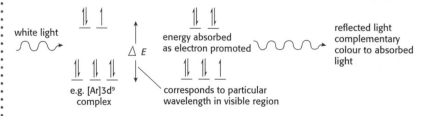

hexacyanoiron(II) ion   tetrachlorocopper(II) ion
(octahedral)   (tetrahedral)

Figure 17 Shapes of some transition metal complex ions.

Coordination number	6	4	2
Examples	$[Fe(CN)_6]^{3-}$	$[CuCl_4]^{2-}$	$[Ag(NH_3)_2]^+$
	$[Fe(CN)_6]^{4-}$	$[Cu(NH_3)_4]^{2+}$	

Coloured complexes

When white light falls on a solution containing a complex ion of a transition metal, some of the light corresponding to a particular wavelength or energy is absorbed. The solution transmits the remaining light, and this complementary colour is the colour that is observed. Copper(II) sulfate solution is blue. This is because the tetraaquacopper(II) ion, $[Cu(H_2O)_4]^{2+}$, absorbs red light and transmits the blue light. The wavelength of light that is absorbed depends on several different factors, but all of them are due to the fact that, when six ligands bond to the transition metal ion to form an octahedral complex, the 3d sub-level is split into two. The light that is absorbed corresponds to the energy required to promote a d electron from the lower split level to the higher split level (Figure 18).

Figure 18 The excitation of an electron in the split d orbitals

There are four main factors that affect the size of the d–d splitting and hence the colour. Three of them are the nature of the transition element, the oxidation state of the transition element, and the type of ligand (Table 9). The fourth factor is the shape of the complex ion.

Table 9

Nature of the transition element	$[Mn(H_2O)_6]^{2+}$	$[Fe(H_2O)_6]^{2+}$	
	pale pink	green	
Oxidation state	$[Fe(H_2O)_6]^{2+}$	$[Fe(H_2O)_6]^{3+}$	
	green	yellow/brown	
Type of ligand	$[Cu(H_2O)_4]^{2+}$	$[Cu(NH_3)_4]^{2+}$	$CuCl_4^{2-}$
	light blue	deep blue	yellow

Catalytic properties

Catalyst are substances that increase the rate of a chemical reaction without themselves being chemically changed at the end of the reaction. In Chapter 6 we shall see that, essentially, they work by providing an alternative pathway for the reaction—one with a lower activation energy. They do this by helping the two reacting species to come into closer contact with each other. If they are acting as **heterogeneous** catalysts, where the catalyst is in a different phase from the reactants and products, they may do this by adsorbing reactant molecules onto the surface of the metal. An example is the use of nickel or palladium as a hydrogenation catalyst. Compounds containing carbon-to-carbon double bonds (C=C) are said to be **unsaturated**. When hydrogen is added across the double bond they become **saturated**. The catalyst works by adsorbing hydrogen and the unsaturated compound (e.g. ethene) onto its surface and aligning them so that they are in the correct orientation to react (Figure 19).

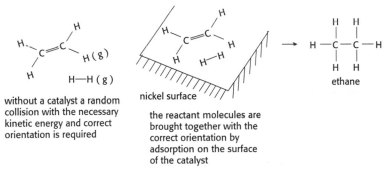

without a catalyst a random collision with the necessary kinetic energy and correct orientation is required

nickel surface

the reactant molecules are brought together with the correct orientation by adsorption on the surface of the catalyst

ethane

Figure 19 The reaction of ethene with hydrogen to form ethane.

Transition metals are particularly good at adsorbing small molecules, so the metals themselves make good heterogeneous catalysts. Other examples are the use of iron in the Haber process, where ammonia is manufactured from nitrogen and hydrogen:

$$3H_2(g) + N_2(g) \underset{}{\overset{Fe(s)}{\rightleftharpoons}} 2NH_3(g)$$

and the use of rhodium, platinum and palladium in catalytic converters in automobiles (Figure 20). These catalysts convert carbon monoxide, oxides of nitrogen and unburnt hydrocarbons into the less polluting gases carbon dioxide, nitrogen and water.

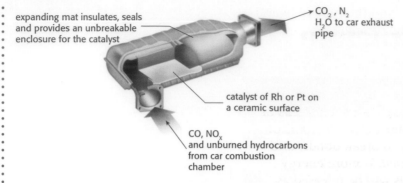

expanding mat insulates, seals and provides an unbreakable enclosure for the catalyst

CO_2, N_2 H_2O to car exhaust pipe

catalyst of Rh or Pt on a ceramic surface

CO, NO_x and unburned hydrocarbons from car combustion chamber

Figure 20 Cross-section of a catalytic converter.

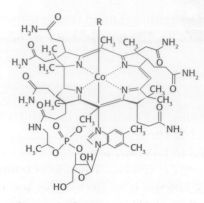

Figure 21 The structure of vitamin B_{12}, showing the central cobalt atom.

Homogeneous catalysts are in the same phase as the reactants and products. In homogeneous catalysis often the two reacting species bond chemically to the transition metal to form an intermediate, react together, and then leave. During the process the oxidation state of the central element in the catalyst will increase and then decrease. One of the reasons why transition metal compounds are such good catalysts is that they have variable oxidation states: that is, they can be relatively easily oxidized and reduced. Examples of compounds of transition metals that are important catalysts are:

- The use of manganese(IV) oxide to decompose hydrogen peroxide:

$$2H_2O_2(aq) \xrightarrow{MnO_2(s)} 2H_2O(l) + O_2(g)$$

- Vanadium(V) oxide in the conversion of sulfur dioxide into sulfur trioxide during the manufacture of sulfuric acid in the contact process:

$$2SO_2(g) + O_2(g) \underset{}{\overset{V_2O_5(s)}{\rightleftharpoons}} 2SO_3(g)$$

- Iron and cobalt in biological catalysts, for example cobalt in vitamin B_{12}. In Figure 21, R represents different groups (e.g. a methyl group, $-CH_3$) to which the cobalt atom can bond in different forms of the vitamin. Vitamin B_{12}, which is found in foods of animal origin (fish, meat, liver, eggs and milk), is essential for the production of red blood cells. It is also essential for the correct functioning of the central nervous system. A lack of vitamin B_{12} causes pernicious anaemia.

Catalysts play an important part in the world economy. A suitable catalyst will make a reaction more efficient. The product will be obtained more quickly, and less energy will be needed to produce it, so the process will be much more cost-effective and less damaging to the environment. There is a whole branch of chemistry called **green chemistry** that covers these issues. Essentially, green chemistry is the design of chemical products and processes that reduce or eliminate the use and generation of substances that are hazardous to the environment. Clearly this does also have significant cost implications. For example, the main reason why the price of some precious metals such as rhodium and platinum is so high is the heavy demand for catalysts, particularly as catalytic converters in automobiles.

❓

Transition elements are said to have characteristic properties. These include: variable oxidation states, complex ion formation, coloured compounds, and catalytic properties. One other characteristic property is that they readily form alloys. Consider the element lead in Group 4 of the Periodic Table and answer the following (you may need to consult a textbook or the Internet):

1 What colour is lead(II) iodide?

2 What are the two common oxidation states of lead? Illustrate your answer by giving the formulas of the three oxides of lead.

3 Give the formula of the complex ion formed when lead(II) chloride reacts and dissolves in concentrated hydrochloric acid.

4 Solder is a low melting point alloy used to join electrical wires together on a circuit board. What is the chemical composition of solder?

5 Based on your answers above, would you classify lead as having the characteristic properties of a transition metal?

6 Why is lead not classified as a transition element?

4 Bonding and structure

During any chemical reaction bonds are broken between atoms or ions in the reactants, and new bonds are made between atoms or ions to form the products. It requires energy to break bonds, and energy is released when new bonds are formed. If more energy is given out than needs to be put in then this will be released as heat, and the reaction is said to be **exothermic**. If more energy is required to break the reactant bonds than is given out when the products are formed then extra heat will need to be added, and the reaction is said to be **endothermic**. The bonds that are present between atoms *within* a molecule are due to **intramolecular** forces of attraction. These forces of attraction tend to be strong.

When a substance changes state from a solid to a liquid or from a liquid to a gas, then bonds are also broken. However, in this case no new chemical substance is formed, so the change is a physical change. The bonds that are broken in physical changes are the forces of attraction *between* the particles of the compound that hold it together in the solid or liquid state. These are known as **intermolecular** forces of attraction. For this reason a few ice cubes are much more effective than the same mass of cold water at cooling a drink, because heat is absorbed in breaking down the structure of the ice. When the change of state is in the other direction, and intermolecular forces of attraction are being formed, heat is given out. Thus when steam condenses to liquid water at 100 °C heat is evolved, even though the temperature remains constant. This explains why steam at 100 °C will cause a worse scald than water at 100 °C.

In this chapter we shall first look at the different ways in which atoms or ions can bond together in compounds. Then we shall consider how the properties of the molecules or ions formed determine the type of intermolecular forces of attraction between them, and hence their melting and boiling points and other physical properties.

Ionic (or electrovalent) bonding

Sodium chloride provides the standard example of ionic bonding. Sodium is a reactive metal, and if a piece of hot sodium metal is lowered into a gas jar of chlorine then the sodium burns in the chlorine to produce a large amount of heat, and sodium chloride (Figure 1).

Clearly the reaction is very exothermic. The equation for the reaction is

$$2Na(s) + Cl_2(g) \rightarrow 2NaCl(s)$$

To form 1 mol of sodium chloride from its elements the equation can be written

$$Na(s) + \tfrac{1}{2}Cl_2(g) \rightarrow NaCl(s)$$

Figure 1 Sodium burning in chlorine.

The electronic arrangement of a sodium atom is 2.8.1 ($1s^22s^22p^63s^1$ at HL). When sodium bonds with chlorine it loses its outer electron to gain the electronic configuration of the inert gas neon, 2.8. In the process a chlorine atom gains the electron from the sodium atom to achieve the electronic configuration of the inert gas argon, 2.8.8. What was the chlorine atom now has one more electron than it has protons, so it has become a negative chloride ion, Cl^-. Similarly what was the sodium atom now has one less electron than it has protons, so it has become a positive sodium ion, Na^+. The formula of sodium chloride is thus Na^+Cl^- although often the charges on the ions are left out and we simply write NaCl. This is illustrated in Figure 2.

Note that this diagram contains much that was learned in previous chapters.

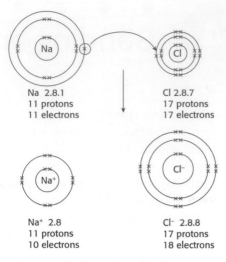

Na 2.8.1
11 protons
11 electrons

Cl 2.8.7
17 protons
17 electrons

Na$^+$ 2.8
11 protons
10 electrons

Cl$^-$ 2.8.8
17 protons
18 electrons

Figure 2 The formation of sodium chloride (NaCl) from atoms of sodium and chlorine.

- The electrons are shown in pairs, because each pair of electrons occupies an orbital.
- The successive energy levels in the atoms and ions are shown getting closer together.
- The radius of a sodium atom is approximately twice that of a chlorine atom.
- The radius of a sodium ion is approximately half that of a sodium atom.
- The radius of a chloride ion is approximately twice that of a chlorine atom.

Even so, the diagram still does not fit the equation given above, because it shows a sodium atom reacting with a chlorine atom, not half a chlorine molecule. Energy needs to be put in first to break the Cl–Cl bond in the chlorine molecule to form chlorine atoms. Energy also needs to be put in to ionize the sodium atom (the first ionization energy of sodium is $+494$ kJ mol^{-1}). As this involves the gaseous state, energy is also required to turn solid sodium into gaseous sodium atoms. Some energy is given out when a chlorine atom gains an electron (this is known as the **electron affinity** of chlorine, and is equal to -364 kJ mol^{-1}), but this is not enough to account for all the energy that has to be put in. So why is this reaction so exothermic, if so much energy has to be put in? The answer lies in the very strong ionic crystal lattice that is formed. In a crystal of sodium chloride each sodium ion is surrounded by six chloride ions, and each chloride ion is in turn surrounded by six sodium ions (Figure 3). The ionic "bond" is the sum of all these electrostatic attractions between the oppositely charged ions, and is very strong, so much energy is given out when it is formed. This energy is known as the **lattice enthalpy**.

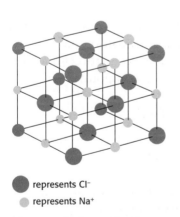

● represents Cl$^-$
● represents Na$^+$

Figure 3 Sodium chloride structure.

Ionic bonds are formed between metals on the left of the Periodic Table, which can lose one or more electrons to form an inert gas configuration, and non-metals on the right of the Periodic Table, which can gain one or more electrons to form an inert gas configuration. For example, in the reaction between magnesium and chlorine each magnesium atom loses two electrons and each chlorine atom gains one electron, so the ratio of magnesium ions to chloride ions in the product is 1 to 2, [$Mg^{2+}2Cl^-$], which is written more simply as $MgCl_2$ (Figure 4).

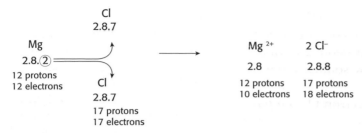

Figure 4 The formation of magnesium chloride ($MgCl_2$).

Similarly, when calcium combines with oxygen, each calcium atom loses two electrons to form the calcium ion, Ca^{2+}, and each oxygen atom gains two electrons to form the oxide ion, O^{2-}, so the formula of calcium oxide is $[Ca^{2+}O^{2-}]$ or CaO.

By knowing how many electrons an atom needs to lose or gain to form an inert gas configuration it is easy to determine the simple ion they will form, and hence the formula of the ionic compound, because the overall sum of all the charges must equal zero.

Positive ions (cations)			Negative ions (anions)		
Group 1	Group 2	Group 3	Group 5	Group 6	Group 7
+1	+2	+3	−3	−2	−1
Li^+, Na^+, K^+	Mg^{2+}, Ca^{2+}	Al^{3+}	N^{3-}, P^{3-}	O^{2-}, S^{2-}	F^-, Cl^-, Br^-, I^-

Formulas of simple ionic compounds		
sodium chloride NaCl	calcium bromide $CaBr_2$	aluminium iodide AlI_3
lithium oxide Li_2O	magnesium sulfide MgS	aluminium oxide Al_2O_3
potassium nitride K_3N	calcium phosphide Ca_3P_2	aluminium phosphide AlP

Transition metals can also form ionic compounds. A common property of transition metals is that they can form more than one ion. For example, iron can form the Fe^{2+} ion and the Fe^{3+} ion. The name of the compound will distinguish between them. Iron(II) chloride has the formula $FeCl_2$, and iron(III) chloride has the formula $FeCl_3$.

Positive and negative ions can also be formed when more than one element is joined to another. These are known as **polyatomic ions**, and the positive or negative charge is delocalized: that is, it is spread over the whole of the ion rather than localized on one particular atom. Examples include the ammonium ion NH_4^+ and the negative ions formed from acids (sometimes known as acid radicals) such as the nitrate ion, NO_3^-, which is formed when nitric acid, HNO_3, loses a proton.

Polyatomic negative ions (anions)		
nitrate NO_3^-	hydroxide OH^-	sulfate SO_4^{2-}
carbonate CO_3^{2-}	phosphate PO_4^{3-}	hydrogen carbonate HCO_3^-

The formulas of ionic compounds containing polyatomic ions are arrived at in exactly the same way, in that the sum of all the positive and negative charges must total zero. Brackets are used to show that the subscript covers all the elements in the polyatomic ion: for example, iron(III) hydroxide, $Fe(OH)_3$; calcium nitrate, $Ca(NO_3)_2$; and ammonium phosphate, $(NH_4)_3PO_4$.

Covalent bonding

When they form ionic bonds, atoms gain or lose electrons to form an inert gas configuration. In covalent bonding atoms achieve an inert gas configuration by sharing one or more pairs of electrons. The simplest covalent compound is hydrogen, H_2. Each hydrogen atom has one electron in its outer shell. The electrons from two hydrogen atoms combine together to form a pair of electrons. This electron pair is electrostatically attracted by both nuclei, so that it forms a directional bond between the two atoms, resulting in a molecule of hydrogen. When one pair of electrons is shared between two atoms it is known as a **single covalent bond**. Other examples of diatomic molecules containing a single covalent bond are fluorine and chlorine. See Figure 5.

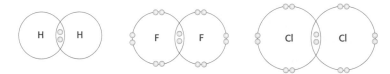

Figure 5 Covalent bonding in hydrogen, fluorine and chlorine (only outer electrons shown).

A diagram that shows all the pairs of outer electrons in a molecule is known as the **Lewis structure**. It is important that *all* the pairs of outer electrons are shown—not just the bonding pairs. There are several accepted ways of doing this. In some countries dots and crosses are used to represent individual electrons; in other countries a line is used to represent a pair of electrons. All the diagrams in Figure 6 are correct ways of showing the Lewis structures for a chlorine molecule.

Figure 6 Different representations of the Lewis structure of chlorine.

Single covalent bonds also occur in molecules made up from atoms of different elements. In both methane and tetrachloromethane the carbon atom has four of its own outer electrons, and so it needs a share in one each from the four hydrogen or chlorine atoms. In ammonia the nitrogen atom needs only to share three electrons to achieve an inert gas configuration, so it has one non-bonded pair of electrons. This non-bonded pair is also sometimes known as a **lone pair**. The oxygen atom in water has two non-bonded (or lone) pairs of electrons, and the chlorine atom in hydrogen chloride has three non-bonded pairs of electrons. See Figure 7.

Figure 7 Lewis structures of methane (CH_4), tetrachloromethane (CCl_4), ammonia (NH_3), water (H_2O) and hydrogen chloride (HCl).

1 Give the formulas for the following compounds: potassium hydroxide, aluminium hydroxide, copper(II) chloride, ammonium carbonate, manganese(II) sulfate, calcium phosphate, magnesium hydrogen carbonate, iron(III) sulfate.

2 Name the following compounds: K_2SO_4, $Ba(OH)_2$, $CuSO_4$, Cu_2SO_4, Mn_2O_3, MnO_2, FeS, $Fe(NO_3)_3$

Oxygen atoms have six electrons in their outer shell. For them to gain a noble gas configuration with another oxygen atom, each oxygen atom needs to share two of its electrons with the other oxygen atom. This results in a **double covalent bond**. Similarly, in nitrogen the three shared pairs of electrons between the two nitrogen atoms result in a **triple covalent bond**. See Figure 8.

Other compounds containing multiple covalent bonds include: carbon dioxide, CO_2; ethene, C_2H_4; ethyne, C_2H_2; and hydrogen cyanide, HCN (Figure 9).

Figure 8 Lewis structures of oxygen (O_2) and nitrogen (N_2).

Figure 9 Lewis structures of carbon dioxide (CO_2), ethene (C_2H_4), ethyne (C_2H_2) and hydrogen cyanide (HCN).

Coordinate or dative bonds

In the above examples the shared pair of electrons is made up using one electron each from the two bonding atoms. In some cases both of the electrons in the shared pair can originate from the same atom. This is known as a **coordinate** or **dative covalent bond** and is formed when a non-bonding pair of electrons on one of the atoms donates its electrons to an atom or ion that is deficient in electrons. For example, when ammonia reacts with an acid the ammonium ion is formed by the non-bonding pair of electrons on the nitrogen atom combining with the hydrogen ion, H^+, released by the acid (Figure 10).

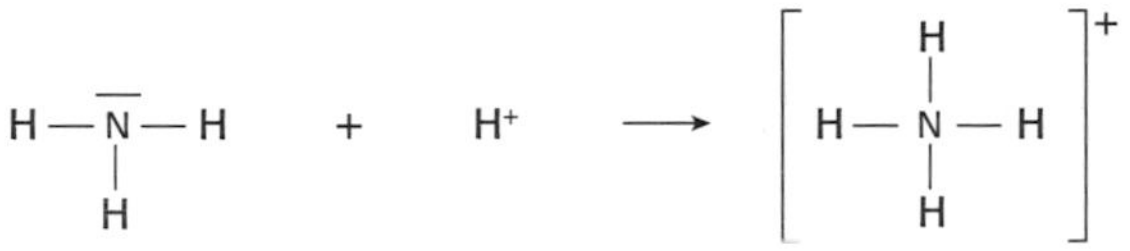

Figure 10 Ammonia reacting with hydrogen ions to form the ammonium ion, NH_4^+.

Once the bond has been formed it is no different from a normal covalent bond. There are now four equal N–H single covalent bonds. However, the ion formed now contains one extra proton with the positive charge spread out or **delocalized** over the whole ion. This is shown by enclosing the ion within brackets and writing the charge outside the bracket. Although all the electron pairs are now equal, sometimes a coordinate bond is shown with an arrowhead to show where the pair of electrons originated.

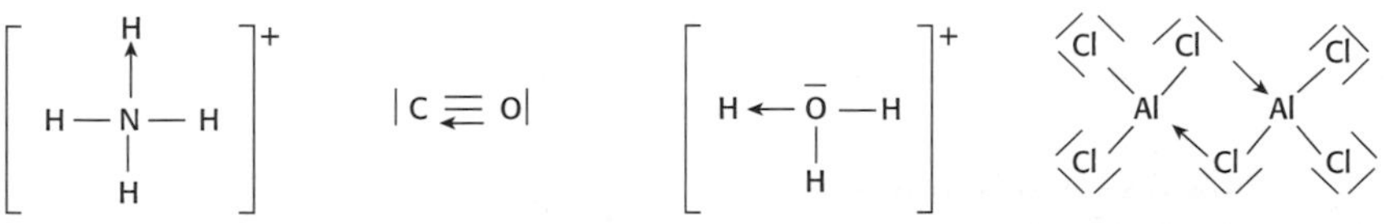

ammonium ion   carbon monoxide   hydronium ion   aluminium chloride dimer

Figure 11 Examples of compounds and ions containing coordinate bonds.

Bond length and bond strength

The more pairs of electrons that are shared between two atoms in a bond then the stronger the bond and the shorter the bond length. In almost all cases single bonds are weaker and longer than double bonds, and triple bonds are even stronger and shorter than double bonds. See Table 1 and Figure 12.

Table 1 Bond length and bond strength

		Length / nm	Strength / kJ mol⁻¹
Single bonds	O–O	0.148	146
	Cl–Cl	0.199	242
	N–N	0.146	163
	C–C	0.154	348
	H–H	0.074	436
Double bonds	O=O	0.121	496
	N=N	0.120	409
	C=C	0.134	612
Triple bonds	C≡C	0.120	837
	N≡N	0.110	944

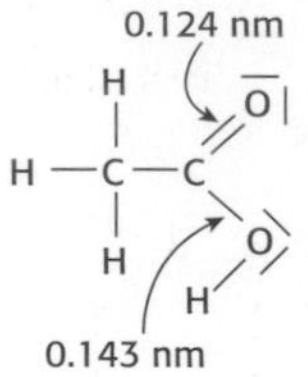

Figure 12 The different carbon-to-oxygen bond lengths in ethanoic acid.

Resonance hybrids

Consider the Lewis structure for the carbonate ion, CO_3^{2-}. The carbon atom has four outer electrons of its own, and each oxygen has six outer electrons. Together with the two extra electrons this gives a total of 24 valence electrons: that is, 12 pairs. The structure shows that one of the carbon-to-oxygen bonds is a double bond, whereas the other two are single bonds.

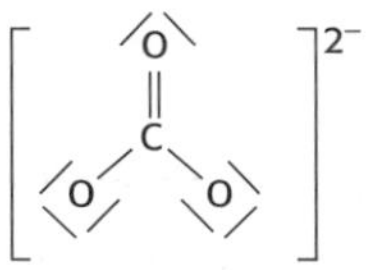

If this structure is correct we would expect one of the carbon-to-oxygen bonds to be shorter than the other two. In fact all the carbon-to-oxygen bonds have the same length, which is intermediate between a carbon-to-oxygen single bond and a carbon-to-oxygen double bond. This can be explained by stating that the true structure of the carbonate ion lies somewhere in between the three possible extreme structures, which are called **resonance hybrids** (Figure 14).

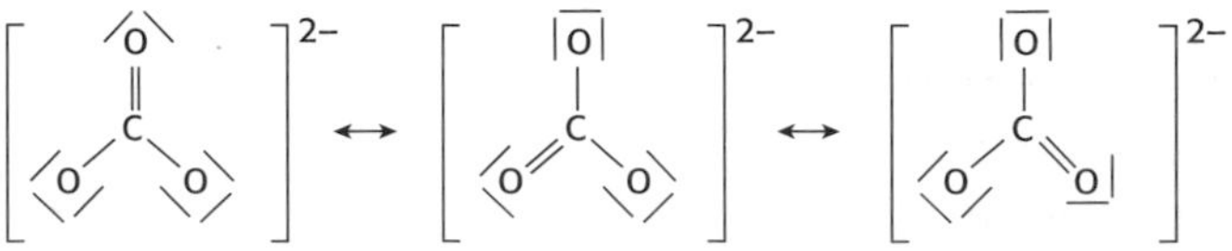

Figure 13 A Lewis structure for the carbonate ion.

Figure 14 The resonance hybrid structures of the carbonate ion.

Other compounds or ions for which resonance hybrid structures can be drawn include: sulfur dioxide, SO_2; ozone, O_3; the nitrate ion, NO_3^-; and benzene, C_6H_6 (Figure 15).

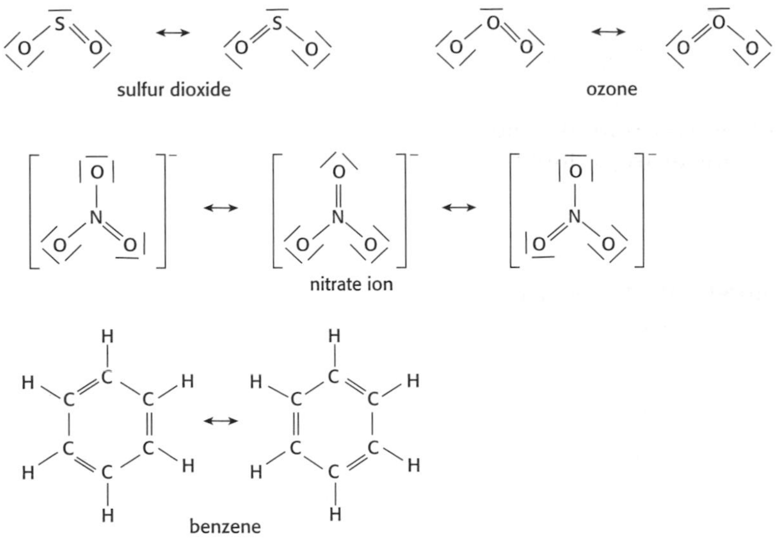

Figure 15 The resonance hybrid structures of sulfur dioxide (SO_2), ozone (O_3), the nitrate ion (NO_3^-), and benzene (C_6H_6).

Ionic or covalent?

We can predict whether the bond between two elements will be ionic or covalent by looking at the difference in electronegativity values. Elements in Groups 1, 2 and 3 tend to have low electronegativity values, whereas elements in Groups 5, 6 and 7 tend to have high electronegativity values. The greater the difference in electronegativity values, then the more likely it is that the bond will be ionic. Generally, the difference in electronegativity value between the two elements needs to be greater than about 1.8 for ionic bonding to occur (Table 2).

Table 2 Electronegativity and bonding

	Mg	Cl	Be	Cl	Al	O	Al	Cl
Electronegativity	1.2	3.0	1.5	3.0	1.5	3.5	1.5	3.0
Difference	1.8		1.5		2.0		1.5	
Formula	$MgCl_2$		$BeCl_2$		Al_2O_3		Al_2Cl_6	
Type of bonding	ionic		covalent		ionic		covalent	

However, this is only a general rule, and there are many exceptions. For example, lead(II) bromide, $PbBr_2$, is ionic, and yet the difference in electronegativity values is only 1.0, whereas boron trifluoride, BF_3, is not ionic, and yet the difference in electronegativity values is 2.0. In fact many ionic compounds show some covalent character, and many covalent compounds show some ionic character, so often it is not a case of ionic or covalent but somewhere in between. Generally we say that ionic bonding predominates if the compound conducts electricity when molten or in aqueous solution, and covalent bonding predominates if the compound is a poor conductor of electricity when molten or in aqueous solution.

Polarity of bonds

When a single covalent bond is formed between two atoms of the same element, for example a chlorine molecule, Cl_2, the electron pair will on average be shared equally between the two atoms, and the bond will be non-polar.

However, if two different atoms are bonded together covalently, for example a hydrogen chloride molecule, HCl, then the nuclei of the different atoms will exert different attractive forces on the electron pair, and it will not be shared equally. The atom that attracts the electron pair more strongly will then be slightly negatively charged compared with the other atom, which will be slightly positively charged. This results in a polar bond, with the two small charges represented by δ^- and δ^+ (Figure 16). If the molecule is placed between two electrically charged plates, then it is said to have a **dipole moment**, because the δ^- end of the molecule will be attracted to the positive plate and the δ^+ end of the molecule will be attracted to the negative plate. The bigger the difference in electronegativities, then the more polar the bond and the greater the dipole moment.

non polar

$H \overset{x}{\underset{•}{}} H$

pair of bonding electrons
shared equally

polar

$\overset{\delta^+}{H} \overset{x}{\underset{•}{}} \overset{\delta^-}{Cl}$

pair of bonding electrons
closer to chlorine atom

Figure 16 Non-polar and polar bonds.

Shapes of simple molecules and ions

The bonds in a molecule may be polar, but it does not follow that the molecule itself will be polar, because the resulting dipoles may cancel each other out. This can be illustrated by carbon dioxide and water. The two carbon-to-oxygen bonds in carbon dioxide are both polar, but as the molecule is linear the resulting effect is zero. In water the two hydrogen-to-oxygen bonds are both polar, and as the angle between them is 105° there will be a resultant dipole (Figure 17).

$\overset{\delta^-}{O} = \overset{2\delta^+}{C} = \overset{\delta^-}{O}$

resultant dipole is zero
non-polar

$2\delta^-$
O
$\overset{\delta+}{H} \qquad \overset{\delta+}{H}$

resultant dipole polar

Figure 17 The resultant dipoles in carbon dioxide and water.

In order to determine whether or not a molecule will be polar we therefore need to know the shape of the molecule as well as the difference in electronegativities between the atoms in the molecule. The shapes of simple molecules and ions can be obtained by using the **valence shell electron pair repulsion** (VSEPR) theory. This states very simply that because of the repulsion between pairs of electrons around the central atom in the molecule or ion, the pairs of electrons are arranged so that they are as far away from each other as possible.

If there are two pairs of electrons the angle between them will be 180° and the shape will be linear. For three pairs the angle will be 120° and the shape will be trigonal planar. Four pairs will be arranged in a tetrahedral shape with a bond angle of 109.5°.

two pairs–linear three pairs–trigonal planar four pairs–tetrahedral

Figure 18 The three basic shapes for two, three and four pairs of electrons around a central atom or ion.

In order to be able to work out the shapes of actual molecules or ions two other factors need to be taken into account. Clearly, if there are double or triple bonds between two atoms then the two or three pairs of electrons involved must all point in the same direction. Therefore, for the purposes of VSEPR, multiple bonds count as if they are just one pair of electrons. For this reason the theory is sometimes said to be concerned with the repulsion between negative charge centres rather than between pairs of electrons. The second factor is that non-bonding pairs exert a greater repulsion than bonding pairs. This is because bonding pairs are held in a fixed position by the two nuclei, whereas a non-bonding pair is more diffused or spread out, because it is being attracted by only one nucleus.

To work out the shape of a molecule or ion it is first necessary to calculate the number of pairs of electrons around the central atom. This can be done by taking the number of valence electrons contributed by the central atom (that is, its group number in the Periodic Table) and then adding the number shared by each surrounding atom. If it is an ion, the charge on the ion must also be taken into account. Finally, we need to determine whether there are any multiple bonds, and also how many of the electron pairs are non-bonding pairs.

Two negative charge centres

There are very few molecules that genuinely have only two pairs of electrons around the central atom. The best example of this is beryllium chloride, $BeCl_2$. The other examples tend to have four actual pairs but involve multiple bonds so that they count as just two pairs. All the examples are linear (Figure 19).

Cl—Be—Cl   O=C=O   H—C≡C—H   H—C≡N

Figure 19 Two negative charge centres.

Three negative charge centres

There are also not many examples of compounds containing three genuine electron pairs, because this means that the central atom does not have a share in a noble gas configuration. Examples that do exist include compounds from Group 3, such as boron trifluoride or aluminium chloride monomer. Other examples actually have four pairs of electrons but one of them is a double bond so they count as three pairs. See Figure 20.

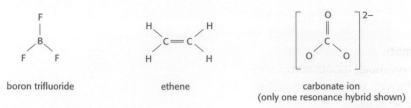

boron trifluoride ethene carbonate ion
(only one resonance hybrid shown)

Figure 20 Examples of molecules or ions with three bonding pairs of electrons.

When one of the pairs of electrons is a non-bonding pair then, although the "parent" shape (that is, the basic shape it is derived from) will be trigonal planar, the actual shape of the molecule will be bent, with a bond angle in the region of 120°.

sulfur dioxide   ozone

Figure 21 Examples of molecules with two bonding pairs and one non-bonding pair of electrons.

Four negative charge centres

This provides the most numerous examples, because it involves a noble gas configuration around the central atom. Straightforward examples of four bonded pairs include the hydrides and chlorides of Group 4 elements such as: methane, CH_4; tetrachloromethane, CCl_4; and silicon tetrachloride, $SiCl_4$. Other examples of four bonded pairs include ions such as the ammonium ion, NH_4^+, and the tetrafluoroborate ion, BF_4^- (Figure 22).

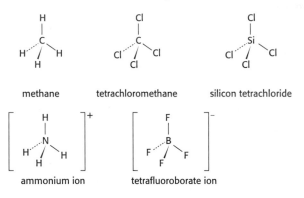

methane   tetrachloromethane   silicon tetrachloride

ammonium ion   tetrafluoroborate ion

Figure 22 Examples of molecules or ions with four bonding pairs of electrons.

When there are three bonding pairs and one non-bonding pair, the non-bonding pair exerts a greater repulsion, so the bond angle becomes slightly less than the regular tetrahedral angle of 109.5°: in ammonia, for example, it is 107°. Although it is derived from a tetrahedron, the actual shape of ammonia is trigonal pyramid. In water there are two non-bonding pairs, so there is even greater repulsion and the H–O–H bond angle is 105°, giving a bent shape (Figure 23).

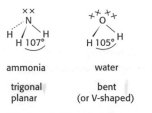

ammonia   water

trigonal   bent
planar   (or V-shaped)

Figure 23 Shapes of ammonia and water molecules.

Predict the shape and the bond angles for the following molecules or ions:
H_2S, $PbCl_4$, H_2CO, NO_3^-, PH_3, NO_2^-, NH_2^-, $POCl_3$

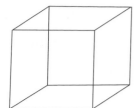

Thinking about science

The use of models in chemistry and perception

Much of the time when we use models—for example, model aeroplanes or models of people, houses or towns—the models are smaller than the objects they represent. In chemistry the reverse is usually true. There is an additional problem in that we often have to use two dimensions, for example the page of this book or a computer screen, to depict three-dimensional models. There are many good computer simulations (e.g. ChemDraw) to show molecular modelling, and you are advised to look at some of these, because the models can be shown in different formats and can be made to rotate. It is also possible to use them to calculate bond angles and bond lengths. For more complicated molecules, such as enzymes and the receptors for drug action, much use of modelling is made in chemistry to see whether the substrate or potential drug can fit the site where the interaction occurs.

Figure 24

In order to be able to "see" three-dimensional shapes in two dimensions we need to learn to alter our perception. This can be easily demonstrated by looking at the two diagrams in Figure 24. Everyone will recognise the first drawing as a cube, although if they are young they may call it a brick or a box. However, normally only a chemist will recognise the second one as a tetrahedron (if you don't believe this, try it with a friend who does not study chemistry).

In reality the first drawing is not a cube. It is some straight lines joined together in two dimensions, which we have learned to recognise as a cube. In fact your brain has not been given enough information to interpret which is the back and which is the front, and when you look hard at the "cube" it will continually invert as your brain considers the possibilities. To "see" the second diagram as a tetrahedron you need to have been taught that the two normal lines are in the plane of the paper, the dotted line is going behind the paper, and the wedge-shaped line is sticking out in front of the paper. Once this has been learned, chemists see the second shape in three dimensions just as much as they see a cube when they look at the first shape.

Because most molecules are three-dimensional, and some of their properties can depend on the way in which the atoms are orientated in space (known as their **stereochemistry**), it is important to be able to visualize them in three dimensions. You should practise making actual models of molecules in three dimensions using special modelling kits (e.g. Molymod®; see Figure 25) or, if these are unavailable, using balls and sticks. It is worth remembering, though, that however useful models may be, they are only representations of what we assume to be reality, and all models do have their limitations. For example, ball-and-stick models tend to give the idea that bond lengths are fixed and rigid, whereas in reality bonds vibrate.

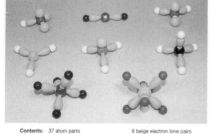

Figure 25 Molecule models.

Intermolecular forces

There is a gravitational force of attraction between any two bodies that have mass. However, the gravitational attraction between two molecules is incredibly weak, and cannot account for the magnitude of the attraction that does exist. We know there must be attractive forces between molecules, otherwise gases would not condense into liquids at very low temperatures. Even helium (where the attractive forces between the individual atoms are extremely weak) condenses to form a liquid at 4 K (−269 °C). For non-polar molecules there are no permanent electrostatic forces of attraction between them. The forces that do exist are called **van der Waals' forces**. The stronger the van der Waals' forces, then the higher is the boiling point, as more energy is required to overcome the attraction between the molecules and separate them. See Figure 26.

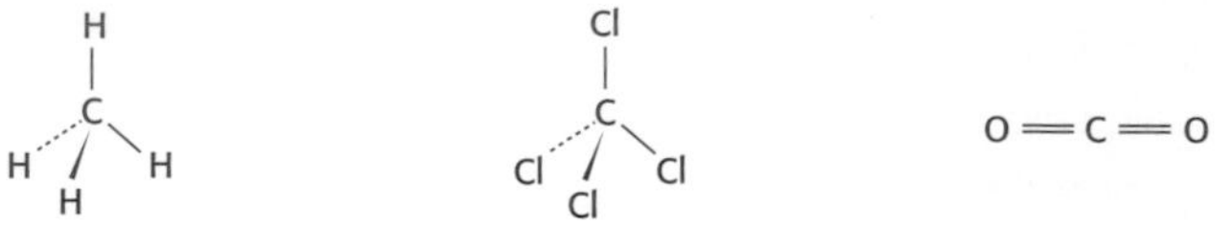

methane
b.p. −182 °C

tetrachloromethane
b.p. −23.0 °C

carbon dioxide
b.p. −78 °C

Figure 26 Examples of non-polar molecules and their boiling points.

Factors affecting the strength of van der Waals' intermolecular forces

Van der Waals' forces increase down Group 7, and also along a homologous series. In both of the examples in Table 3 the number of electrons and the molar mass both increase along the series.

Table 3

Halogen	Fluorine, F_2	Chlorine, Cl_2	Bromine, Br_2	Iodine, I_2
M_r	38	71	160	254
Boiling point / °C	−188	−34.7	+58.8	+84
Alkane	Methane, CH_4	Ethane, C_2H_6	Propane, C_3H_8	Butane, C_4H_{10}
M_r	16	30	44	58
Boiling point / °C	−162	−88.6	−42.2	−0.5

Critical considerations

Some chemists explain van der Waals' forces by assuming that as the molecules approach each other they cause temporary induced dipoles, because at any moment in time the electron pairs are not shared equally. They then go on to state that the attractive forces increase according to the number of electrons present. Other chemists tend to emphasise that the attractive forces increase as the molar mass of the molecules increases. Rather than simply accept hypotheses like this it is good practice to try to test them with suitable examples.

If we find examples of non-polar molecules with the same number of electrons but different molar masses we can see that their boiling points (b.p.) differ: hence it cannot depend simply on the number of electrons.

Non-polar molecule	M_r	No. of electrons	b.p. / °C
C_2H_6	30	18	−89
SiH_4	32	18	−111
Ar	40	18	−186

Equally, if we find examples of non-polar molecules with the same mass but a different number of electrons, then their boiling points also differ.

Non-polar molecule	M_r	No. of electrons	bp / °C
CO_2	44	22	−78
C_3H_8	44	26	−42

Clearly, it is not as simple as some chemists make out, and perhaps no one has yet provided a thorough explanation of van der Waals' forces, even though we do know they exist.

❓

Consider the following three sets of data:

Non-polar molecule	M_r	No. of electrons	bp / °C
H_2	2	2	−252.8
D_2	4	2	−249.6
$CH_3-CH_2-CH_2-CH_2-CH_3$ pentane	72	42	36.3
$CH_3-CH_2-\overset{\displaystyle CH_3}{\overset{\displaystyle \vert}{CH}}-CH_3$ 2-methylbutane	72	42	27.9
$CH_3-\overset{\displaystyle CH_3}{\underset{\displaystyle \underset{\textstyle CH_3}{\vert}}{\overset{\displaystyle \vert}{C}}}-CH_3$ 2,2-dimethylpropane	72	42	9.5
Br_2	160	70	58.8
UF_6	352	146	56.5
SF_6	146	70	−63.9

Can you think of any other factors apart from molar mass and number of electrons that might help to explain the size of van der Waals' forces of attraction?

Dipole–dipole forces

When one of the hydrogen atoms in methane is replaced by a chlorine atom to form chloromethane, the resultant bond polarities now do not equal zero, and the molecule has a **dipole**. The dipole moment is shown by an arrow, with the arrowhead pointing to the resultant negative pole (Figure 27).

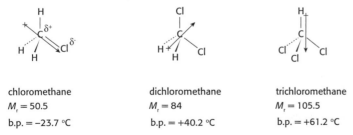

chloromethane
$M_r = 50.5$
b.p. = −23.7 °C

dichloromethane
$M_r = 84$
b.p. = +40.2 °C

trichloromethane
$M_r = 105.5$
b.p. = +61.2 °C

Figure 27 Polarity of substituted methane molecules.

The negative pole of one polar molecule is attracted to the positive pole of another polar molecule, and the resulting electrostatic attraction is stronger than the Van der Waals' attraction that is still present, resulting in higher boiling points. This can clearly be seen if we compare the boiling points of two substances with the same, or a very similar, relative molar mass but where one is polar and the other non-polar (Figure 28).

propane, non-polar
$M_r = 44$
b.p. = − 42.2 °C

ethanal, polar
$M_r = 44$
b.p. = +20.8 °C

Figure 28 The difference in boiling points between a non-polar molecule and a polar molecule with the same relative molecular mass.

Distinguishing between polar and non-polar molecules

In addition to comparing boiling points, the traditional way to see whether the molecules of a liquid are polar is to let a stream of the liquid run from a burette and then bring a charged rod near the liquid stream. If the liquid is polar then the stream of liquid will be attracted to the rod (Figure 29).

A more modern way is to use a microwave oven. Microwave radiation involves oscillating electric and magnetic fields. In a microwave oven the electric field changes direction in the order of 5×10^9 times each second. The water molecules in food try to line up their dipoles in line with the field, and so rotate rapidly as the field changes and try to orientate with it. These rotations cause the water molecules to collide with other water molecules, increasing their kinetic energy and thus their temperature. However, to do this the molecules must be polar and free to move. In ice the polar water molecules are held in fixed positions in the crystal structure, so ice is unaffected by microwave radiation, provided no liquid water is present. Similarly, non-polar molecules are unaffected.

In a simple experiment in a home microwave oven 20 cm³ of water in a glass beaker increased in temperature from 18 °C to 81 °C after a 30-second burst. When 20 cm³ of the non-polar liquid tetrachloromethane (CCl_4) was placed in a similar beaker and subjected to the same amount of microwave radiation, the temperature was unaltered. (NB: tetrachloromethane is a chlorinated organic compound, and you should not try this at home.)

Hydrogen bonding

When hydrogen is bonded directly to one of the small, highly electronegative atoms fluorine, oxygen or nitrogen, then the polarity of the covalent bond is very high. In addition, as the electron pair is drawn away from the hydrogen atom, all that remains is the proton in the nucleus, as there are no inner electrons. The negative electronegative atom of another molecule is thus attracted by a very strong dipole–dipole attraction. This type of very strong dipole–dipole attraction is given its own name—**hydrogen bonding**. The strength of this attraction is clearly demonstrated by the hydrides of the elements of Groups 5, 6 and 7. Each of the first members of their respective series—ammonia (NH_3), water (H_2O), and hydrogen

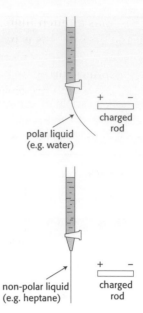

polar liquid
(e.g. water)

charged rod

non-polar liquid
(e.g. heptane)

charged rod

Figure 29 The effect of a charged rod on a polar and a non-polar liquid.

fluoride (HF)—has a much higher boiling point than the other members of the group. This is particularly noticeable with water, which is a liquid at room temperature with a boiling point of 100 °C at atmospheric pressure compared with all the other Group 6 hydrides, which are gases at room temperature and atmospheric pressure (Figure 30).

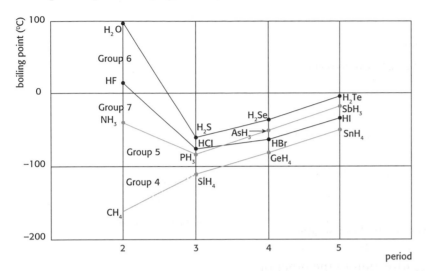

Figure 30 Graph of the boiling points of the hydrides of Groups 4, 5, 6 and 7.

Although water is a substance we are all familiar with, it is in fact an almost unique liquid, because when it freezes it expands. Nearly all other liquids contract in volume when they freeze. The structure of ice is very open (see Figure 31). Each oxygen atom is bonded to four hydrogen atoms in a giant tetrahedral arrangement. Two of these bonds are the strong covalent O–H bonds in the water molecule; the other two are weaker and longer hydrogen bonds between the $2\delta^-$ charge on each oxygen atom and the δ^+ charge on each of the two hydrogen atoms from other water molecules. When ice melts, the molecules can move closer together, and water has its maximum density at 4 °C.

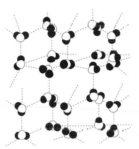

Figure 31 The structure of ice.

1 Butane ($CH_3CH_2CH_2CH_3$), propanone (CH_3COCH_3) and propan-1-ol ($CH_3CH_2CH_2OH$) all have very similar molecular masses. State the type of attractive forces present between the molecules in each of the substances, and predict the order of their increasing boiling points.
2 Water, with a relative molecular mass of 18, has a boiling point of 100 °C. Ethanol, with a relative molecular mass of 46, has a boiling point of 78 °C. Explain why water has the higher boiling point, even though both water and ethanol exhibit hydrogen bonding.
3 In which of the following substances would hydrogen bonding be expected to occur: hydrogen peroxide (H_2O_2); ethylamine ($C_2H_5NH_2$); difluoromethane (CH_2F_2); methoxymethane (CH_3OCH_3); ethanoic acid (CH_3COOH)?

An experiment to demonstrate the types of attraction between molecules

The different types of bonding present between organic molecules can be elegantly shown by the following simple experiment. 10 cm³ of propan-1-ol is placed in a round-bottom flask and the apparatus is set up for reflux, as shown in Figure 32. (Refluxing involves using a condenser placed vertically above the volatile organic compound being heated so that all the vapour condenses back into the flask and none is lost.)

Heat is applied to the bottom of the flask, and when the liquid is refluxing gently the temperature of the boiling point is recorded. 2 cm³ of propan-2-ol is now added though the top of the condenser, and once the reading has steadied the boiling point of the mixture is recorded. Further readings of the boiling point are taken after successive additions of 2 cm³ of propan-2-ol until a total of 10 cm³ has been added and a 50 : 50 mixture by volume obtained. The apparatus is now dismantled and the experiment repeated, but this time starting with 10 cm³ of propan-2-ol and then adding successive 2 cm³ amounts of propan-1-ol. A graph of boiling point against percentage by volume of the mixture is then plotted (Figure 33).

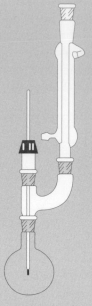

Figure 32 Reflux apparatus.

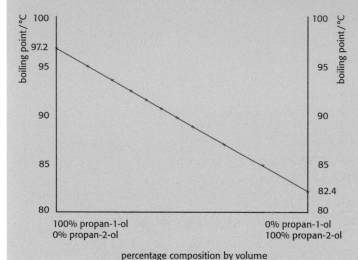

Figure 33 Graph of boiling point against percentage composition by volume of propan-1-ol and propan-2-ol.

The accuracy of the experiment can be checked by comparing the boiling points of both pure propan-1-ol and propan-2-ol with the Data Booklet values. The graph plots as a straight line, which shows that there is no change in the type or strength of bonding taking place. There is hydrogen bonding in both pure propan-1-ol and pure propan-2-ol. When the two substances are mixed, hydrogen bonding also then occurs between propan-1-ol and propan-2-ol (Figure 34).

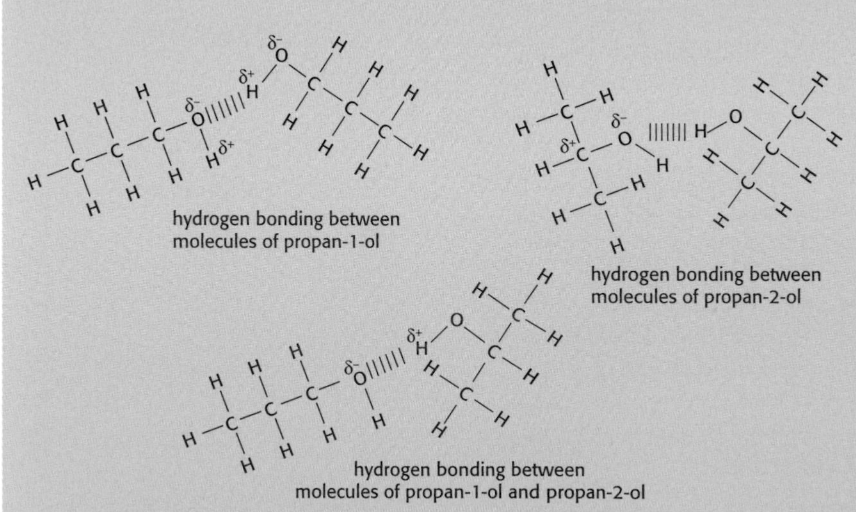

hydrogen bonding between molecules of propan-1-ol

hydrogen bonding between molecules of propan-2-ol

hydrogen bonding between molecules of propan-1-ol and propan-2-ol

Figure 34 Hydrogen bonding in pure propan-1-ol, in pure propan-2-ol, and in the mixture.

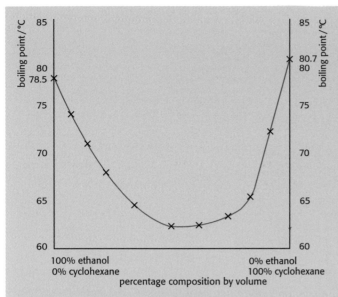

Figure 35 Graph of boiling point against percentage composition by volume for ethanol/cyclohexane mixture.

The experiment can then be repeated using ethanol and cyclohexane instead of propan-1-ol and propan-2-ol. Now a very different-shaped graph is obtained (Figure 35). The boiling point of the mixture is lower than for either of the two pure substances, which indicates that the forces of attraction between the substances in the mixture are lower. In pure ethanol the main attractive forces are hydrogen bonding, whereas cyclohexane is non-polar so the forces of attraction in pure cyclohexane are van der Waals' forces (Figure 36). When the two compounds mix, the attractive forces between ethanol and cyclohexane will be van der Waals' forces, but as ethanol molecules have less mass and there is less surface area interaction they will be weaker than in pure cyclohexane (Figure 37).

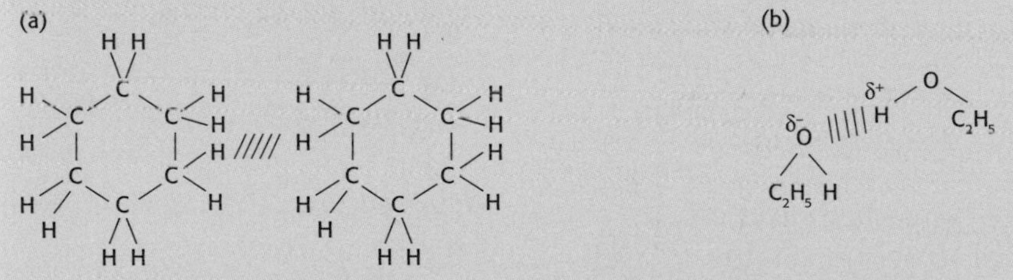

(a)

(b)

Figure 36 (a) Weak van der Waals' forces of attraction between cyclohexane molecules; (b) hydrogen bonding between ethanol molecules.

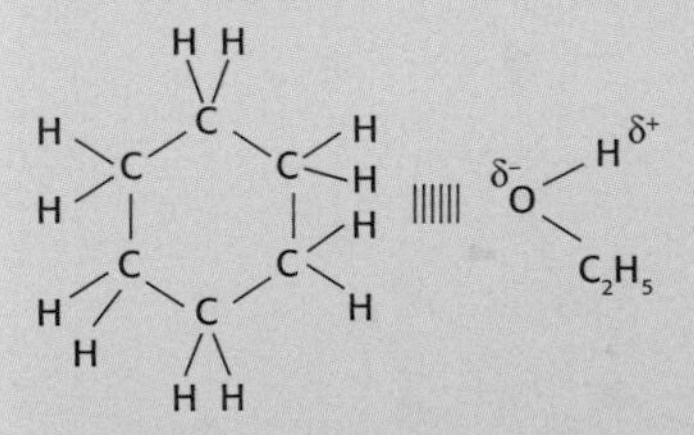

Figure 37 Weak van der Waals' forces of attraction between non-polar cyclohexane molecules and polar ethanol molecules.

If the experiment is repeated using trichloromethane and propanone, then the graph shows that the mixture has a higher boiling point than either of the two pure components (Figure 38). (You are advised not to use this mixture in a school laboratory, as it was once reported that an explosion occurred. A similar-shaped graph can be obtained using a mixture of trichloromethane and ethyl ethanoate.)

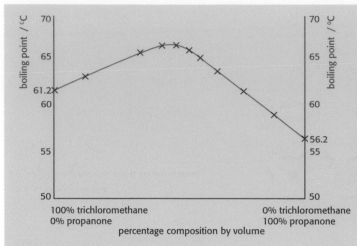

Figure 38 Graph of boiling point against percentage composition by volume for trichloromethane/propanone mixture.

Both trichloromethane and propanone are polar substances, so the attractive forces between their molecules in the pure substances will be dipole–dipole interactions. When they are mixed, the δ^+ on the hydrogen atom from the trichloromethane molecule will be attracted to the δ^- on the oxygen atom from the propanone molecule, resulting in a type of hydrogen bonding that will be stronger than the dipole–dipole interactions in the pure substances (Figure 39).

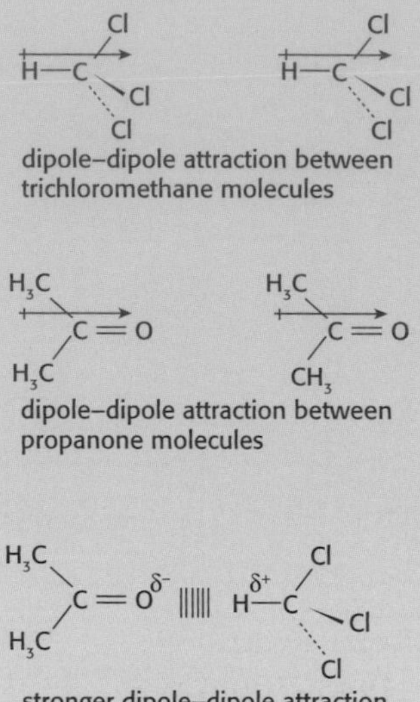

Figure 39 Stronger dipole–dipole attraction between propanone and trichloromethane.

Giant covalent structures

Carbon exhibits **allotropy**. This means that it can exist in more than one physical form, or allotrope. The three main allotropes of pure carbon are diamond, graphite, and fullerene (C_{60}) (Figure 40). In all three allotropes the carbon atoms are bonded covalently, but in diamond and graphite, instead of small simple molecules, the covalent bonds link across the carbon atoms to form a single large (or giant) molecule. In order to break the structure down, the bonds that have to be broken are strong covalent bonds, so diamond and graphite have extremely high melting points—in the region of 4000 °C.

In diamond, each carbon atom is bonded equally to four other carbon atoms to form a giant tetrahedral structure (Figure 41). All the C–C bond lengths are equal, and there is no plane of weakness through the structure, so diamond is an extremely hard substance. All the outer electrons around each carbon atom are localized to form the four bonds to other carbon atoms, so diamond does not conduct electricity, because there are no delocalized electrons. Both silicon and silicon dioxide (silica) also have a diamond structure.

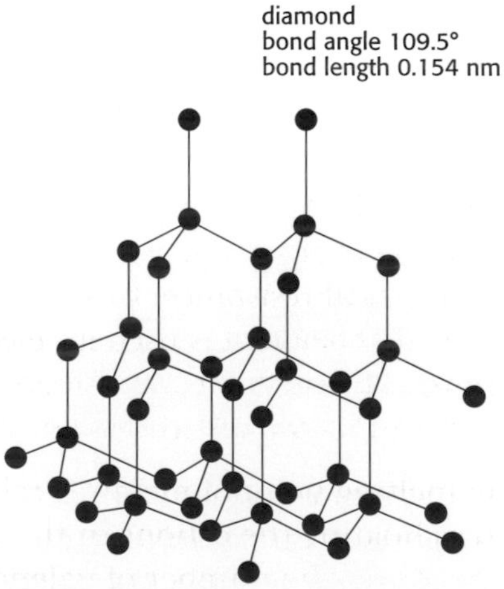

In graphite each carbon atom forms strong covalent bonds to three other carbon atoms in a trigonal planar structure so that the carbon atoms link up to form layers of hexagonal rings (Figure 42). The C–C bonds in the rings are actually stronger and shorter than the C–C bond in diamond. However, the forces of attraction between the layers are very weak, because they are formed by delocalized electrons that can move between the layers. This explains why graphite is one of the few non-metals to be a good conductor of electricity. The fact that the attractive forces between the layers are weak gives graphite another useful physical property. The layers can easily slide over each other, so graphite feels waxy to the touch and is a good lubricant.

Figure 40 Models of graphite, C_{60} and diamond.

diamond
bond angle 109.5°
bond length 0.154 nm

Figure 41 Structure of diamond, with carbon-to-carbon bond length and bond angle.

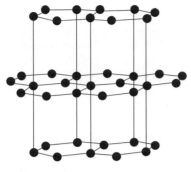

graphite
bond angle 120°
bond lengths
- in layers 0.141 nm
- between layers 0.335 nm

Figure 42 Structure of graphite.

Figure 43 Structure of C$_{60}$.

1 The dioxides of the first two elements in Group 4 have the formulas CO_2 and SiO_2. Explain why carbon dioxide is a gas at room temperature and pressure, whereas silicon dioxide is a solid with a high melting point.
2 Both diamond and ice have a giant tetrahedral structure. Explain why the melting point of ice (0 °C) is much lower than the melting point of diamond (approximately 4000 °C).

In 1996 Robert Curl (1933–), Harold Kroto (1939–) and Richard Smalley (1943–2005) were jointly awarded the Nobel Prize in Chemistry for their discovery of **fullerenes**. The basic fullerene is C$_{60}$ in which 60 carbon atoms are joined in a combination of hexagonal and pentagonal rings to form a sphere (Figure 43).

Since their initial discovery, more than 1000 new compounds involving fullerenes have been made. Some contain metals (e.g. lanthanum) trapped inside a fullerene cage; others consist of long tubes that can be closed or open at one end. These are called **nanotubes**, because they have an extremely small diameter, in the order of one nanometre (1×10^{-9} m) (Figure 44).

Metallic bonding

When the atoms of metals bond together in the solid state, one or more of their valence electrons becomes detached from each atom to become **delocalized**. These valence electrons are no longer associated with a particular atom, but are free to move throughout the metallic structure. The bonding in metals thus consists of the attraction between these delocalized valence electrons and the remaining positive metal ions (cations). It is sometimes said that metals are made up of an array of cations in a "sea" of mobile electrons.

Physical properties of metals

It is because the valence electrons are no longer located on a particular atom, but are free to move throughout the structure, that metals are such excellent conductors of electricity. If there are impurities in the metal, then this can hinder the movement of electrons and increase the electrical resistance: this explains why copper needs to be refined or purified before it is used for electrical wiring. The movement of electrons through metals also enables the transmission of kinetic energy, so metals are also good conductors of heat.

The melting point of metals is related to the strength of the attractive forces holding the cations in the "sea" of delocalized electrons. This depends on the number of valence electrons delocalized from each atom, the size of the cations, and the way in which the cations are packed together. In general, the melting point decreases as the size of the cation increases, which explains why melting points decrease down Group 1 (alkali metals) (see Table 4).

Figure 44 Scientists working at NASA's Ames Research Center in Moffet Field, California, have helped find a way to create transistors much smaller than can be created using silicon chips, by building them from carbon nanotubes.

Table 4

Alkali metal	Ionic radius / nm	Melting point / °C
Lithium, Li	0.060	180
Sodium, Na	0.095	97.8
Potassium, K	0.133	63.7
Rubidium, Rb	0.148	38.9
Caesium, Cs	0.169	28.7

Only one common metal, mercury, is a liquid at room temperature and pressure. Many metals, such as copper, have melting points in the region of about 1000 °C, but some metals, such as tungsten, have very high melting points (Table 5).

Many metals are **malleable** and **ductile**. Malleable means that the metal can be beaten into shape without breaking. Ductile means that the metal can be drawn into a wire. Both of these properties can be explained by the cations being able to slide past each other to rearrange the overall shape of the solid. Because the electrons are delocalized this can happen without a significant change in the bonding forces.

Table 5

Metal	Melting point / °C
Mercury, Hg	−38.9
Sodium, Na	97.8
Magnesium, Mg	650
Aluminium, Al	660
Copper, Cu	1083
Iron, Fe	1535
Platinum, Pt	1769
Tungsten, W	3407

Extension material

The importance of metals in the world economy

The particular properties of metals mean that they are an essential part of the world economy. Their multitude of uses range from the construction of buildings and reinforcement of concrete to the manufacture of cars, ships and aeroplanes, to "white" goods (such as refrigerators and washing machines), to water pipes and electrical wiring and, increasingly, in packaging. Because of their economic importance, metals are traded as commodities.

The most common metal in use is still iron, but aluminium is rapidly challenging this position. In recent years the price of metals and their ores has increased rapidly. This is due partly to the emergence of China and India as industrialized nations. For instance, during the last ten years the Chinese economy has been growing rapidly. The annual production of aluminium in China has increased more than fivefold during that time (Figure 45), and the price of alumina—the raw material used to make aluminium—has increased considerably because of demand.

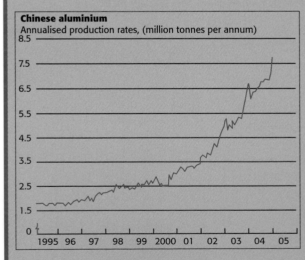

Figure 45 Annual production of aluminium in China, 1995–2005.

If everyone on the planet aspires to a similar high standard of living as that currently enjoyed by the heavily industrialised nations, then this may result in considerable tensions, as the resources are limited.

Physical properties

Throughout this chapter reference has been made to the way in which the type of bonding determines the physical properties of the substance in question. Substances made up of simple covalent molecules that have weak forces of attraction between the molecules will be volatile, with low melting and boiling points. Substances that contain non-polar molecules will tend to be insoluble in water but soluble in non-polar solvents such as heptane. Substances composed of polar molecules will have higher melting and boiling points, will be less volatile, and will tend to be more soluble in polar solvents. Ionic compounds will conduct electricity (and be decomposed in the process) when molten or in aqueous solution, because their ions are free to move to the electrodes, whereas in the solid state they are non-conductors as the ions are held in fixed positions. Some ionic substances such as sodium chloride are very soluble in water, but others such as silver chloride are almost insoluble in water. Metals and graphite are good conductors because the delocalized electrons are able to move freely through them. The substances with the highest melting and boiling points of all are giant molecular solids such as diamond and silica, because the atoms are all held together by strong covalent bonds.

Although there are always some exceptions, if the type of bonding is known then the physical properties can be predicted (Table 6).

Table 6 Bonding type and physical properties

Type of bonding	Example	Properties
van der Waals'	Butane, C_4H_{10} $M_r = 58$	Low melting point (−138 °C); low boiling point (−0.5 °C); insoluble in water, soluble in non-polar solvents. Non-conductor of electricity.
Dipole : dipole	Propanone, C_3H_6O $M_r = 58$	Melting point (−95.4 °C) and boiling point (56.2 °C) higher than for non-polar substances with the same M_r. Miscible with water and non-polar solvents. Non-conductor of electricity.
Hydrogen bonding	Propan-1-ol, C_3H_8O $M_r = 60$	Boiling point (97.2 °C) higher than dipole : dipole bonded substances with a similar M_r. Miscible with water. Less soluble in non-polar solvents. Non-conductor of electricity.
Ionic bonding	Sodium chloride, NaCl $M = 58.5$ g mol^{-1}	High melting point (801 °C) and boiling point (1467 °C). Soluble in water (although some ionic compounds are insoluble in water). Non-conductor of electricity when solid. Good conductor of electricity when molten or in aqueous solution – decomposed chemically in the process.
Metallic bonding	Nickel, Ni $A_r = 58.7$	High melting point (1453 °C) and boiling point (2730 °C), although some metals have lower values. Insoluble in water (but may react to form ions) and non-polar solvents. Good conductor of electricity – not chemically decomposed in the process.
Giant covalent	Carbon, C	High melting point (graphite sublimes at 3730 °C) and boiling point (4830 °C). Insoluble in water and non-polar solvents. Non-conductor of electricity (except graphite, which is a good conductor of electricity).

1 Since about 600 BCE, when gold coins were first introduced, unreactive metals have been used for coins. Gradually over time, as the face value of a coin became less than its intrinsic value because of inflation, the amount of precious metals in the coins has been reduced. For example, the old English one penny was originally made of pure silver. In 1971, when decimalization of the UK's coinage began, the new penny was made out of bronze (an alloy of copper and tin), but since 1992 they have been made from copper-plated steel, and so pennies dated after 1992 are magnetic. Find out the composition of the coins in your country. Are any of them worth more for the metals they contain than the face value of their currency?

2 The metals commonly listed on the commodities market include aluminium, copper, lead, nickel, tin, zinc, gold, silver and platinum. Apart from coinage, list one major use for each of these metals.

not simply appear out of nowhere. The background work has to be put in beforehand. Thomas Alva Edison (1847–1931) famously said that genius is 99% perspiration and 1% inspiration.

It is significant that before he turned to chemistry Friedrich Kekulé (1829–1896), a German national, originally studied architecture. More of a theoretical chemist than a practical chemist, Kekulé pondered for a long time on the possible structure of benzene, until the answer apparently came to him through a dream. He dreamed that he was watching atoms moving rapidly, twisting and twining in a snakelike motion. One of the snakes caught hold of its own tail, and the form rotated in front of his eyes. When he woke with a sudden flash of inspiration he realised that the structure of benzene was a closed hexagonal ring. With hindsight we can see that his architectural training probably gave him the ability to think of the structure in a different way because the true leap of the imagination was in realising that carbon compounds can form rings as well as chains.

Data response question

The following extract is taken from a very well-respected British chemistry textbook for 16–19 year olds (specifically: G. Hill and J. Holman, *Chemistry in Context*, 5th edn, Nelson, 2000). It introduces the topic of ionic bonding using lithium oxide as the example. Many other books use sodium chloride as the example, set out in a similar way.

8.2 Transfer of electrons – electrovalent (ionic) bonding

Typical ionic compounds are formed when metals in Group I or Group II react with non-metals in Group VI or Group VII. When the reaction occurs, electrons are transferred from the metal to the non-metal unit the outer electron shells of the resulting ions are identical to those of a noble gas.

Figure 8.1 shows how the transfer of electrons from lithium to oxygen forms ions in lithium oxide. In figure 8.1, the nucleus of each atom is represented by its symbol and the electrons in each shell are represented by circled dots of crosses around the symbol. Ions are shown in square brackets with the charge at the top right-hand corner.

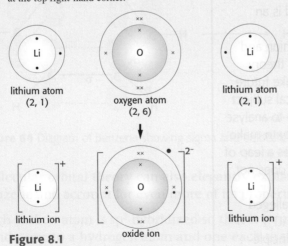

Figure 8.1

Although the electrons of the different atoms in figure 8.1 are shown by dots and crosses, you must not think that electrons of lithium are any different from those of oxygen. All electrons are identical. They are shown differently in the diagram so that you can follow transfer more easily.

Figure 67 An extract on ionic bonding from Chemistry in Context.

Read the extract carefully then answer the four questions that appeared after the extract in the book:

1   What is the electron structure of: (a) the lithium ion; (b) the oxide ion?
2   Which noble gas has an electron structure like that of Li^+?
3   Which noble gas has an electron structure like that of O^{2-}?
4   Why is it that two lithium atoms react with only one oxygen atom?

Let us now ask some different questions that were not asked in the book. (You will need the IB Data Booklet to answer some of the questions.)

1   Why are some of the electrons shown singly and others in pairs?
2   Comment on the size of the lithium atom compared with that of the oxygen atom. (Support your answer by finding out the values for their atomic radii.)
3   Is the process of a lithium atom losing an electron exothermic or endothermic? Find its value.
4   Removal of an electron from lithium involves the first ionization energy of lithium. What must be done to the lithium atom before the value for the first ionization energy can be used? Is this process exothermic or endothermic? Find its value.
5   What name is given to the enthalpy change for adding two electrons to an oxygen atom? Is it an exothermic or endothermic process? Find its value.
6   Comment on the size of a lithium ion compared with that of a lithium atom.
7   Comment on the size of an oxide ion compared with that of an oxygen atom.
8   Write the equation for the complete combustion of lithium in oxygen. Include the physical states in your balanced equation.
9   Does the diagram represent the equation you have written? If not, what extra enthalpy change is necessary for oxygen? Is this change exothermic or endothermic? Find its value (remember this is for the formation of one atom of gaseous oxygen).
10 Look at all the values for the changes taking place that have been discussed so far and add them up (remember there are two atoms of lithium reacting with one atom of oxygen).
11 The total enthalpy change for the reaction is -596 kJ mol^{-1}. Suggest which energy factor you have not yet taken into account, and calculate its value.
12 If you were asked to write a chemistry textbook, could you now come up with a better diagram to illustrate ionic bonding using lithium oxide?
13 The enthalpy term arrived at in question 11 is known as the lattice enthalpy change. Would you expect the lattice enthalpy change for sodium chloride (NaCl) to be larger or smaller in absolute terms than that for lithium oxide (Li_2O)? Explain your answer.
14 Finally, a tough question! The lattice enthalpy change can be calculated theoretically and obtained experimentally. For sodium chloride there is excellent agreement between the two values obtained. However, the experimental value for lithium oxide (Li_2O) is -2906 kJ mol^{-1}, whereas the theoretical value is -2799 kJ mol^{-1}. Suggest a reason why the two values are not the same.

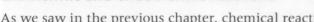

5 Energetics

Exothermic and endothermic reactions

As we saw in the previous chapter, chemical reactions involve the breaking and making of bonds. It requires energy to break bonds, and energy is given out when new bonds are formed. If more energy is given out than needs to be put in, then this will be released as heat to the surroundings, and the reaction is said to be **exothermic**. If more energy is required to break the reactant bonds than is given out when the products are formed, then extra heat will need to be added, and the reaction is said to be **endothermic**.

The energy contained in chemical bonds that can be converted into heat is known as **enthalpy** and is given the symbol H. It is impossible to measure the actual heat content or enthalpy of a particular substance, but what can be readily measured is the enthalpy change for a reaction, ΔH. For exothermic reactions the heat content (enthalpy) of the products is less than that of the reactants. By convention the enthalpy change for the reaction, ΔH, is said to be negative $(-)$. For endothermic reactions, where the enthalpy of the products is more than the enthalpy of the reactants, the enthalpy change, ΔH, will have a positive value $(+)$. Note that the plus and minus symbols indicate the direction of heat flow; they do not imply positive or negative energy.

Examples of exothermic reactions include the combustion of fuels, such as hydrogen or gasoline (petrol), and the neutralization of an acid by a base:

$$C_8H_{18}(l) + 12\tfrac{1}{2}O_2(g) \rightarrow 8CO_2(g) + 9H_2O(l)\ \Delta H = -5512\ \text{kJ mol}^{-1}$$

$$HCl(aq) + NaOH(aq) \rightarrow NaCl(aq) + H_2O(l)\ \Delta H = -57.3\ \text{kJ mol}^{-1}$$

An example of an endothermic reaction is the combination of nitrogen and oxygen to form nitrogen dioxide:

$$\tfrac{1}{2}N_2(g) + O_2(g) \rightarrow NO_2(g) \qquad\qquad \Delta H = +33.9\ \text{kJ mol}^{-1}$$

Another interesting example of an endothermic reaction is the reaction between solid hydrated barium hydroxide and solid ammonium chloride:

$$Ba(OH)_2.8H_2O(s) + 2NH_4Cl(s) \rightarrow BaCl_2.2H_2O(s) + 2NH_3(g/aq)$$
$$+ 8H_2O(l)\ \Delta H = +96.0\ \text{kJ}$$

This is interesting not only because it is a reaction between two solids, but also because the temperature decrease is such that the flask will often stick to the bench after the reaction.

When there is a change of state from a solid to a liquid, a solid to a gas, or a liquid to a gas at a constant temperature, the process is also endothermic. For example:

$$Na(s) \rightarrow Na(g) \qquad\qquad \Delta H = +108\ \text{kJ mol}^{-1}$$

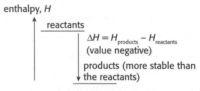

enthalpy, H

Representation of an exothermic reaction using an enthalpy diagram.

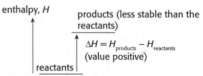

enthalpy, H

Representation of an endothermic reaction using an enthalpy diagram.

Figure 1 Use of enthalpy diagrams to represent: (a) an exothermic reaction; (b) an endothermic reaction.

Standard enthalpy change of reaction

The actual amount of heat given out or taken in during a chemical reaction will depend on several factors:

- The nature of the reactants and products: different reactions have different values of enthalpy changes, ΔH, as different bonds with different strengths are being broken and formed.
- The amount (or concentration) of reactants: the greater the amount that reacts, the greater the heat change.
- The states of the reactants and products—changing state involves an enthalpy change, and so will affect the total amount of heat given out or taken in.
- The temperature of the reaction.
- The pressure at which the reaction is carried out.

All these factors are taken into account by referring to standard enthalpy changes. The **standard enthalpy change of reaction**, $\Delta H^{\ominus}$, is defined as the enthalpy change when molar quantities of reactants in their normal states react to form products in their normal states under standard conditions of temperature and pressure. Standard pressure is 101.3 kPa (1 atmosphere) and standard temperature is any specified temperature, but usually this is taken as 298 K (25 °C). The symbol $^{\ominus}$ signifies that the enthalpy change has been measured under standard conditions.

It is worth noting that enthalpy changes are normally carried out under **constant pressure**. If the volume is kept constant then the value for the heat change will be slightly different from that at constant pressure, because some work is involved in expanding or contracting a gas. Strictly speaking the enthalpy change, ΔH, is the heat change if the pressure is kept constant; if the volume is kept constant then the total heat change is designated by ΔU.

Calculation of enthalpy changes

Energy is defined as the ability to do work: that is, to move a force through a distance.

$$\text{Energy} = \text{force} \times \text{distance}$$
$$\text{(J)} \qquad \text{(N} \times \text{m)}$$

$\Delta H^{\ominus}$ values are normally given in kJ mol^{-1}. However, the actual amount of heat evolved in an exothermic reaction is normally measured in practice by using the heat given out to the surroundings to increase the temperature of a known mass of water. Water has a high specific heat capacity and is able to absorb considerable quantities of heat without its temperature increasing significantly. One gram of water requires 4.18 J of heat energy for the temperature of the water to increase by 1 °C. This is known as the **specific heat capacity of water**, and is often quoted in SI units as 4180 J kg^{-1} K^{-1}.

An experiment to determine the enthalpy change when 1 mol of ethanol is burned completely

The mass of a spirit lamp containing ethanol was determined accurately. The lamp was then lit and placed under a beaker containing 150 ± 1 g of pure water (Figure 2). The water was continually stirred. After the temperature of the water had increased by 12.0 ± 0.5 °C the flame was extinguished and the mass of the spirit lamp containing the unburnt ethanol was determined.

Data collection and processing
Observations:
The ethanol burned with a yellow flame. Some black soot was formed on the bottom of the beaker. It was noted that the spirit lamp was warm when it was weighed at the end of the experiment.

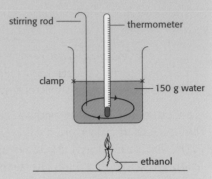

Figure 2 Determining the enthalpy change of combustion of ethanol.

Quantitative results:
Mass of spirit lamp + ethanol before burning = 42.597 ± 0.001 g
Mass of spirit lamp + ethanol after burning = 42.200 ± 0.001 g
Mass of ethanol burned = 0.397 ± 0.002 g

Calculation:
Molar mass of ethanol, $C_2H_5OH = [(2 \times 12.01) + (6 \times 1.01) + 16.00] = 46.1$ g mol^{-1}

Amount of ethanol burned $= \dfrac{0.397 \text{ g}}{46.1 \text{ g mol}^{-1}} = 0.00861$ mol

Heat energy required to raise the temperature of 150 g of water by 12.0 °C
$= 150 \times 4.18 \times 12.0 = 7524$ J
Thus heat energy evolved by burning 0.0861 mol of ethanol = 7524 J

Heat energy evolved by burning 1 mol of ethanol $= \dfrac{7524}{0.00861} = 874\,000$ J = 874 kJ

Since the reaction is exothermic, $\Delta H^{\ominus} = -874$ kJ mol^{-1}

Total uncertainties:

$$\text{Mass of water} = \frac{1}{150} \times 100 = 0.67\%$$

$$\text{Temperature} = \frac{0.5}{12} \times 100 = 4.17\%$$

$$\text{Mass of ethanol} = \frac{0.002}{0.397} \times 100 = 0.50\%$$

Total percentage uncertainties = 0.67% + 4.17% + 0.50% = 5.34%
5.34% of 874 kJ mol^{-1} = 47 kJ mol^{-1}

Summary:
$C_2H_5OH(l) + 3O_2(g) \rightarrow 2CO_2(g) + 3H_2(l)$ $\Delta H^{\ominus} = -874 \pm 47$ kJ mol^{-1}

Conclusion and evaluation
The Data Booklet value for the standard enthalpy of combustion of ethanol is

-1371 kJ mol^{-1}. The percentage error for this experiment $= \dfrac{1371 - 874}{1371} \times 100$

$= 36.2\%$. This error is large, and cannot be accounted for by the uncertainties in the experimental measurements, although it is worth noting that the largest uncertainty will be due to the thermometer, and a more accurate thermometer could have been used. The major uncertainties lie in the underlying assumptions used in the experiment. These are many. For example:
- Complete combustion of the ethanol occurred. In fact the ethanol was being burned in air, and the yellow flame and the black soot on the flask indicated that carbon, one of the products of incomplete combustion, was also formed.

- All the heat from the reaction was transferred to the water. Some of the heat would have been lost round the sides of the beaker, and from the water to the atmosphere.
- Only the water was being heated. The glass beaker, the stirrer and the thermometer were also being heated.
- None of the ethanol escaped through evaporation during the time the flame was extinguished and the spirit lamp and its contents were reweighed. Ethanol is a volatile liquid, and the fact that the spirit lamp was hot indicates that some of it would have evaporated.

The result from this experiment emphasises that it is good practice when considering uncertainties in an experiment to consider *all* the underlying assumptions made, not just simply list and quantify the experimental uncertainties due to the apparatus used to perform the measurements. If these assumptions are not true, then they may have a large effect on the reliability of the result.

1 Can you now devise a more accurate method to determine the enthalpy of combustion of ethanol in a school laboratory?
2 The Data Booklet values for enthalpies of combustion have been obtained using a bomb calorimeter. Use the Internet to discover how a bomb calorimeter works, and why it will give a more accurate value.
3 Ethanol is part of a series of alcohols, each having the general formula $C_nH_{(2n+1)}OH$. Use the Data Booklet to plot a graph of the enthalpy of combustion of the alcohol against the corresponding value of n for methanol, ethanol, propan-1-ol and butan-1-ol. Extrapolate the graph to predict the value for the enthalpy of combustion of pentan-1-ol.

Reactions carried out in solution

In the above experiment the heat evolved from an exothermic reaction was used to heat up water in a separate container. When the reaction takes place in aqueous solution the heat change of the solution can be measured directly.

For example, when 50.0 cm³ of 1.00 mol dm⁻³ sodium hydroxide solution (NaOH(aq)) at 25.0 °C was added with stirring to 50.0 cm³ of 1.00 mol dm⁻³ hydrochloric acid solution (HCl(aq)), also at 25.0 °C, in a polystyrene "coffee cup" (acting as a simple calorimeter; Figure 3) the temperature of the resulting solution rose very quickly to 31.8 °C.

The equation for the reaction is

$$HCl(aq) + NaOH(aq) \rightarrow NaCl(aq) + H_2O(l)$$

The temperature rise = 31.8 − 25.0 = 6.8 °C

The total volume of final solution = 100 cm³

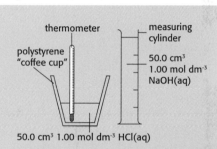

Figure 3 Experiment to measure the heat change of a solution.

To calculate the heat evolved in this reaction two common assumptions are made. The first is that the resulting solution has the same specific heat capacity as pure water, whereas in reality it is a solution of sodium chloride in water. The second is that the density of the solution is 1.00 g cm⁻³ (i.e. the same as that of pure water). Assuming that the two volumes add cumulatively to give 100 cm³ this means that we can assume that the mass of the resulting solution is 100 g. In fact, upon reflection we know that the density of the salt solution will be greater than 1.00 g cm⁻³, because icebergs, which are basically pure water, float in the sea. Neither of these assumptions is actually true; however, within the confines of a school laboratory they are reasonable assumptions to make and do not affect the result greatly.

Heat evolved = mass of water × specific heat capacity of water × temperature rise
= 100 g × 4.18 J °C⁻¹ g⁻¹ × 6.8 °C = 2842 J = 2.842 kJ
Amount of hydrochloric acid = amount of sodium hydroxide = $\frac{50}{1000}$ × 1 = 0.050 mol

Thus enthalpy of neutralisation for this reaction = $\frac{-2.842}{0.050}$ = −56.8 kJ mol⁻¹

That is:

$$HCl(aq) + NaOH(aq) \rightarrow NaCl(aq) + H_2O(l) \qquad \Delta H = -56.8 \text{ kJ mol}^{-1}$$

This compares quite favourably with the Data Booklet value of -57.3 kJ mol^{-1}. The error is less than 1%, which shows that very simple apparatus (in this case a "coffee cup") can often be used effectively in a school laboratory.

Neutralization reactions in solution are fast, because they involve oppositely charged ions, and so we can safely ignore any heat loss due to cooling during the reaction. The same is true of most precipitation reactions. However, if the reaction occurs at a slower rate then all the time the solution is being heated it is also losing heat to the surroundings. We can compensate for this heat loss to the surroundings by plotting a graph of temperature against time and extrapolating. A good example of this is the reaction between solid zinc metal and copper(II) sulfate solution, $CuSO_4(aq)$:

$$Zn(s) + CuSO_4(aq) \rightarrow ZnSO_4(aq) + Cu(s)$$

50.0 cm³ of 0.200 mol copper(II) sulfate solution, $CuSO_4$, was placed in a polystyrene "coffee cup" and the temperature was taken every 30 seconds for 2 minutes. This was to ensure that the solution was at the same temperature as the surroundings (room temperature), and that no heat was being lost or gained before the reaction started. 1.20 g of powdered zinc metal was then added, and the solution was continuously stirred. The temperature was recorded every 30 seconds until it was clear that the reaction had finished and the temperature had dropped back by one or two degrees as the solution started to cool down to room temperature. The graph shown in Figure 4 was obtained.

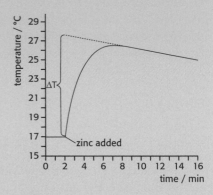

Figure 4

Answer the following questions:

1. Suggest a reason why the reaction is quite slow.
2. The highest temperature was reached 5 minutes after the zinc had been added. What *exactly* was happening at this point?
3. How high would the temperature have risen had there been no heat loss to the surroundings?
4. Calculate the amounts of zinc and copper(II) sulfate involved, and state which is the limiting reagent, and which is present in excess.
5. Copper(II) sulfate solution is blue; zinc sulfate solution is colourless. Zinc is a silver-grey metal, copper is a reddish-brown metal. State what you would expect to observe as the reaction proceeds.
6. Use the temperature increase from the extrapolation shown on the graph to calculate how much heat was given out in this reaction.
7. State the enthalpy change for the reaction in kilojoules per mole.
8. State *three* assumptions you have made in arriving at your answer.
9. The literature value for this reaction is -218 kJ mol^{-1}. Calculate the percentage error, and comment on the validity of the assumptions you have stated.

Hess's law

Consider a chemical reaction in which reactant A is converted into product B. Let us suppose it can either be converted directly into B with an enthalpy change of ΔH_x, or can be converted into B via an intermediate product C with a total enthalpy change of $\Delta H_y + \Delta H_z$. In other words, the reaction can proceed by two different pathways (Figure 5).

If all the reactions are exothermic and ΔH_x is larger than $\Delta H_y + \Delta H_z$, we could turn A directly into B and use some of that energy to convert B back to A via C. The remaining energy we could use to do what we like with. By repeating the cycle we could get an

Figure 5 Energy cycle to illustrate Hess's law.

unlimited supply of free energy and solve many of the world's problems (and probably get very rich in the process!). Unfortunately this cycle, which is an example of what is known as a perpetual motion machine, cannot happen, because it contravenes the **law of conservation of energy**. Germain Henri Hess (1802–1850), a Swiss-born Russian chemist, realised this, and restated the law of conservation of energy to apply to chemical systems. Hess's law states that the total enthalpy change of a reaction depends only on the difference between the enthalpy of the products and the enthalpy of the reactants. It is independent of the reaction pathway. This is a useful law in chemistry, because it enables us to work out some enthalpy change values indirectly. Consider the standard enthalpy of formation of methane. The equation for this reaction, in which methane is formed from its elements in their normal states under standard conditions, is

$$C(s) + 2H_2(g) \xrightarrow{\Delta H_f^\circ} CH_4(g)$$

Superficially this looks a simple enough reaction. We could react carbon with hydrogen and determine the heat change. However, it is difficult to get carbon to react with hydrogen, and when it does there are many possible compounds of carbon and hydrogen (called **hydrocarbons**) that could be formed, so the reaction is actually impossible to achieve experimentally. However, what we can do easily is burn carbon, hydrogen and methane separately and determine the standard enthalpy change of combustion, ΔH_c°, directly for each of the three substances.

The three equations and their enthalpy values are

(a) $C(s) + O_2(g) \rightarrow CO_2(g)$ $\Delta H_c^\circ = -393$ kJ mol^{-1}

(b) $H_2(g) + \frac{1}{2}O_2(g) \rightarrow H_2O(l)$ $\Delta H_c^\circ = -286$ kJ mol^{-1}

(c) $CH_4(g) + 2O_2(g) \rightarrow CO_2(g) + 2H_2O(l)$ $\Delta H_c^\circ = -890$ kJ mol^{-1}

If we multiply equation (b) by 2 and then add equation (a) we get

$$C(s) + 2H_2(g) + 2O_2(g) \rightarrow CO_2(g) + 2H_2O(l)$$
$$\Delta H^\circ = [-393 + (2 \times -286)] = -965 \text{ kJ mol}^{-1}$$

Now subtract equation (c) to give

$$C(s) + 2H_2(g) - CH_4(g) \rightarrow \text{zero}$$
$$\Delta H^\circ = [-965 - (-890)] = -75 \text{ kJ mol}^{-1}$$

By rearranging the equation we arrive at the enthalpy change of formation of methane:

$$C(s) + 2H_2(g) \rightarrow CH_4(g)$$ $\Delta H_f^\circ = -75$ kJ mol^{-1}

We have solved this problem mathematically simply by manipulating equations. A more elegant solution based on chemistry is to draw the corresponding energy cycle (Figure 6). In this cycle the reactants are carbon and hydrogen, and the products are carbon dioxide and water. The intermediate product is methane.

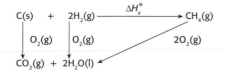

Figure 6

By Hess's law the enthalpy change going by the direct route from reactants to products is the same as the enthalpy change going by the indirect route via methane:

$$[(-393) + (2 \times -286)] = \Delta H_f^\circ + (-890)$$
$$\Delta H_f^\circ = [(-393) + (2 \times -286) - (-890)] = -75 \text{ kJ mol}^{-1}$$

Solve the following problems using Hess's law. It might be good practice to solve each question by drawing an appropriate energy cycle and also by simply manipulating the equations mathematically.

1 The standard enthalpy changes of combustion $\Delta H_c^{\ominus}$ in kJ mol^{-1} for carbon, hydrogen and propane (C_3H_8) are -393, -286 and -2220 respectively. Use these values to determine the standard enthalpy change of formation of propane. (The equation for the formation of propane is $3C(s) + 4H_2(g) \rightarrow C_3H_8(g)$.)

2 Consider the following table of standard enthalpy changes of formation, $\Delta H_f^{\ominus}$:
   (a) Why is the standard enthalpy change of formation of fluorine zero?

Compound	Formula	$\Delta H_f^{\ominus}$/ kJ mol^{-1}
Fluorine	$F_2(g)$	0
Ammonia	$NH_3(g)$	-46
Hydrogen fluoride	$HF(g)$	-269
Nitrogen trifluoride	$NF_3(g)$	-114

   (b) What will be the standard enthalpy change of reaction for the following reaction between fluorine and ammonia?

$$3F_2(g) + NH_3(g) \rightarrow 3HF(g) + NF_3(g)$$

3 The standard enthalpy change of combustion $\Delta H_c^{\ominus}$ of carbon is -393 kJ mol^{-1}, and the standard enthalpy change of combustion of carbon monoxide is -283 kJ mol^{-1}. Use these values to determine the enthalpy change when 1 mol of carbon (graphite) burns to form carbon monoxide, CO, under standard conditions.

4 In the thermite reaction, aluminium powder is mixed with iron(III) oxide and ignited. Aluminium oxide and iron metal are produced in a highly exothermic reaction. Given the following information, calculate the enthalpy change of reaction for the thermite reaction.

$$2Al(s) + 1\tfrac{1}{2} O_2(g) \rightarrow Al_2O_3(s) \qquad \Delta H^{\ominus} = -1669 \text{ kJ mol}^{-1}$$
$$2Fe(s) + 1\tfrac{1}{2} O_2(g) \rightarrow Fe_2O_3(s) \qquad \Delta H^{\ominus} = -822 \text{ kJ mol}^{-1}$$

Extension material

Carbon trading and carbon footprints

Since the onset of the Industrial Revolution and, later, the use of the internal combustion engine in motor vehicles and the jet engine in aircraft, the combustion of fossil fuels has provided an important source of energy. All fossil fuels (coal, natural gas, oil, and oil products such as diesel and gasoline) produce carbon dioxide as one of the major products when they are burned. Currently the world is emitting into the atmosphere nearly 7×10^9 (seven thousand million) tonnes of carbon in the form of carbon dioxide each year (Figure 7).

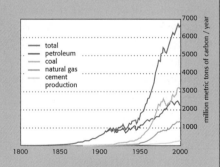

Figure 7 Global fossil carbon emissions.

Despite the fact that plants absorb carbon dioxide, regular readings taken since 1957 from Mauna Loa in Hawaii show that the amount of carbon dioxide in the atmosphere has been steadily increasing for the past 50 years (Figure 8).

Carbon dioxide is a greenhouse gas. High-energy radiation from the Sun passes through our atmosphere including the greenhouse gases and is radiated back from the Earth as lower-energy infra-red radiation. The vibrations of the bonds in carbon dioxide molecules and other greenhouse gases absorb much of this lower-energy infra-red radiation, preventing it from leaving the atmosphere and instead re-radiating it back to the Earth's surface. This

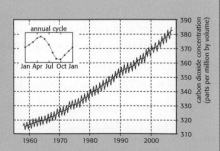

Figure 8 Atmospheric carbon dioxide, measured at Mauna Loa, Hawaii.

results in **global warming**. During the past 100 years the mean temperature of the Earth has increased by about 1 °C.

Evidence from ice core samples in Greenland shows that there have also been large fluctuations in global temperatures in the past; however, for about the last 10–15 years it has been generally accepted by scientists that the current global warming is a direct consequence of the increased emission of greenhouse gases. The Intergovernmental Panel on Climate Change (IPCC) has predicted mean global temperature rises of between 1.4 °C and 5.8 °C between the years 1990 and 2100. Many people now think that it is imperative that greenhouse gas emissions be reduced before a global catastrophe occurs.

In the late 1990s, 163 countries throughout the world ratified the **Kyoto Protocol** negotiated in Japan in 1997. This agreement, which came into force on 16 February 2005, set objectives to lower the amount of greenhouse gas emissions within set time frames. One way in which this is to be achieved has been to set up a trading scheme whereby countries (and industries within them) can buy and sell allocated carbon credits. If a country or business emits more carbon than it has been allocated, it must purchase extra credits from "cleaner" countries or businesses that have not filled their quota. Countries that have developing economies are being encouraged to develop cleaner production techniques and so gain **certified emission reduction** (CER) credits, which they can sell to developed countries that fall short of their targets. Meanwhile, these developed countries are being urged to do all they can to reduce their carbon dioxide emissions in as many ways as possible.

Individuals too are becoming increasingly aware that they are also contributing to global warming through the lifestyle they lead. There are now many websites (e.g. www.carbonneutral.com and www.climatecare.org) that enable you to calculate your **carbon footprint**. For example, a return trip by airplane from Lima in Peru to Amsterdam in the Netherlands produces 2.3 tonnes of carbon dioxide for each passenger. By using the calculator on these websites it is possible to donate a sum of money to a woodland project (one tree growing to maturity absorbs about one tonne of carbon dioxide during its lifespan) or an alternative energy project such as a wind farm to offset this amount of carbon and leave no carbon footprint.

1 Find out whether your own country has ratified the Kyoto agreement. What is being done in your country to reduce greenhouse gas emissions?

2 Suggest a reason why the amount of carbon dioxide in the atmosphere at Mauna Loa fluctuates annually, with the measurement being at its highest during May each year.

3 Should airlines reward frequent flyers with airmiles and access to executive lounges, or should they encourage or even force frequent flyers to contribute to carbon-neutralizing projects?

4 Try to list at least ten different practical ways in which you personally can reduce your dependence on fossil fuels and so reduce your own carbon footprint.

Bond enthalpies

Enthalpy changes of reactions are the result of bonds breaking and new bonds being formed. By knowing the bond enthalpies it is possible to deduce the overall enthalpy change of a reaction.

Bond enthalpies are defined as the energy change when 1 mol of a covalent bond in the gaseous state is formed from its gaseous atoms:

$$X(g) + Y(g) \rightarrow X\text{–}Y(g)$$

Because bonds are being formed, the enthalpy change will be exothermic, so it will have a negative value. It is also perfectly reasonable to define bond enthalpies in terms of bond breaking:

$$X\text{–}Y(g) \rightarrow X(g) + Y(g)$$

in which case the enthalpy change will be endothermic and the value will be positive, but the quantitative value in kilojoules will be the same.

Some bond enthalpies are known precisely, as the bond occurs in only one element or compound. For example:

Bond	Bond enthalpy / kJ mol^{-1}
H–H(g)	436
Cl–Cl(g)	242
F–F(g)	158
H–Cl(g)	431
H–F(g)	562

However, for other bonds the precise bond enthalpy will be affected by surrounding bonds and will be slightly different in different compounds. In these cases we use **average bond enthalpies**. For example:

Bond	Average bond enthalpy / kJ mol^{-1}
C–C(g)	348
C=C(g)	612
C⋯C(g) (in benzene)	518
C–H(g)	412
C=O(g)	743
O–H(g)	463
N–H(g)	388
O=O(g)	496

To calculate the overall enthalpy change for a reaction, calculate the quantity of energy that has to be put in to break all the bonds. Then calculate all the energy that is released as all the new bonds are formed. From these two totals it is easy to see whether the reaction is exothermic or endothermic: that is, whether the overall value will be negative or positive. The difference between the two values will be the overall enthalpy change.

In question 2 on page 98, enthalpy changes of formation were given to find the enthalpy change for the reaction between fluorine and ammonia. This can also be solved using bond enthalpies (bond enthalpy for the N–F bond = 272 kJ mol^{-1}).

$$3 \text{ F-F(g)} + \underset{\displaystyle |}{\underset{\displaystyle H}{\text{H-N-H(g)}}} \longrightarrow 3 \text{ H-F(g)} + \underset{\displaystyle |}{\underset{\displaystyle F}{\text{F-N-F(g)}}}$$

Energy in / kJ mol^{-1}	Energy out / kJ mol^{-1}
3F–F = 3 × 158 = 474	3 × H–F = 3 × 562
3N–H = 3 × 388 = 1164	3 × N–F = 3 × 272
Total energy in: 1638 kJ mol^{-1}	Total energy out = 2502 kJ mol^{-1}

The reaction is exothermic, as more energy is given out. This is due principally to the very weak fluorine-to-fluorine bond (F–F) in the reactants and the very strong hydrogen-to-fluorine bond (H–F) in the products. The difference between the two values is 864 kJ mol^{-1}. The value of ΔH for this reaction is therefore -864 kJ mol^{-1}.

It is worth noting that the value obtained for the above reaction using average bond enthalpies is slightly different (by about 1.3%) from the value obtained using the standard enthalpies of formation. This is because average bond enthalpies have been used, whereas the enthalpies of formation are specific to the compounds in question.

1 In the 1960s the American space agency, NASA, considered the relative merits of using hydrogen and oxygen compared with hydrogen and fluorine as rocket fuels. Assuming all the reactants and products are in the gaseous state, use bond enthalpies to calculate the enthalpy change of reaction (in kJ mol^{-1}) for both fuels. As mass is more important than amount in the choice of rocket fuels, which reaction would give more energy per kilogram of fuel?

2 You are given the following data:
$I_2(s) \rightarrow I_2(g)$ $\Delta H = +62.3$ kJ mol^{-1}
$Br_2(l) \rightarrow Br_2(g)$ $\Delta H = +30.5$ kJ mol^{-1}
Bond enthalpies (in kJ mol^{-1}):
H–I   297
H–Br 368
Br–Br 192
I–I   151
H–H 436
(a) Calculate the enthalpy change for the reaction
$2HI(g) + Br_2(g) \rightarrow 2HBr(g) + I_2(g)$
(b) Use your result in (a) and the above data to calculate the enthalpy change for
$2HI(g) + Br_2(l) \rightarrow 2HBr(g) + I_2(s)$
(c) Discuss reasons for the difference in your answers for parts (a) and (b).

HL ⋮ ## Use of standard enthalpy changes of reaction

It is important that the states of all reactants and products are known, otherwise there may be additional enthalpy changes involving changes of state. For this reason reactants and products are taken as being in their normal states under standard conditions.

The standard enthalpy change of formation of a substance, $\Delta H_f^{\ominus}$, is defined as the enthalpy change when 1 mol of the substance in its normal state is formed from its elements in their normal states under standard conditions of temperature and pressure (298 K, 101.3 kPa). From this it follows that the enthalpy change of an element in its normal state is always zero.

The standard enthalpy change of combustion of a substance, $\Delta H_c^{\ominus}$, is defined as the enthalpy change when 1 mol of the substance in its normal state is completely burned in oxygen to form products in their normal states under standard conditions of temperature and pressure (298 K, 101.3 kPa).

A good example to illustrate why states are important is to calculate the standard enthalpy change of combustion of methane using average bond enthalpies and then compare the result with the value given in the Data Booklet.

The equation for the reaction using bond enthalpies is

H
|
H-C-H(g) + 2O=O(g) ⟶ O=C=O(g) + 2O (g)
| / \
H   H   H

Energy in / kJ mol⁻¹	Energy out / kJ mol⁻¹
4 × C–H = 4 × 412 = 1648	2 × C=O = 2 × 743 = 1486
2 × O=O = 2 × 496 = 992	4 × O–H = 4 × 463 = 1852
Total energy in = 2640 kJ mol⁻¹	Total energy out = 3338 kJ mol⁻¹

The calculated enthalpy change for this reaction using bond enthalpies is therefore -698 kJ mol⁻¹.

The Data Booklet value is -890 kJ mol⁻¹, so clearly there is a large difference, equal to 21.6%. However, the value we have calculated involves water in the gaseous state, whereas enthalpy changes of combustion refer to reactants and products in their normal states under standard conditions, so the water will be a liquid product not a gaseous product.

One mole of water releases 44 kJ of heat energy when it turns from gas to liquid. This is due to the formation of hydrogen bonds between the water molecules, and is known as the **enthalpy change of vaporization of water**. As 2 mol of water are involved, the extra energy given out will be 88 kJ, so the total energy evolved will now be -786 kJ mol⁻¹. This is a better value, but still not very close to -890 kJ mol⁻¹. In the calculation it has been assumed that, in carbon dioxide, the bond enthalpy of the two C=O bonds is the same as twice the average bond enthalpy for the C=O bond. In carbon dioxide the C=O bond enthalpy is in fact very different, and this has not been taken into account.

Born–Haber cycle

In Chapter 4 we saw that, when an ionic compound is formed from its elements, the overall reaction can be broken down into separate steps. Most of these separate steps are endothermic processes, but the overall process is still exothermic, owing to the high lattice enthalpy. This can be set out formally in an energy cycle that is known as a **Born–Haber cycle**. See Figure 9.

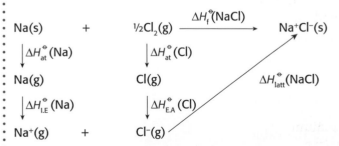

Figure 9 Born–Haber cycle for formation of sodium chloride.

1 Calculate the standard enthalpy change of hydrogenation of ethene ($C_2H_4(g)$) to ethane ($C_2H_6(g)$), given that the standard enthalpy changes of combustion, in kJ mol⁻¹, for hydrogen, ethene and ethane are -286, -1409, and -1560 respectively.

2 The boiling point of hex-1-ene is 63.9 °C, and the boiling point of hexane is 68.7 °C. Both hex-1-ene and hexane are liquids at room temperature. Apart from the bond enthalpy for hydrogen (H–H) and the average bond enthalpies for the C–C, C=C and C–H bonds, what additional information is required in order to work out a value for the standard enthalpy change of hydrogenation of hex-1-ene to hexane using average bond enthalpies?

The overall reaction involving the standard enthalpy of formation of sodium chloride, $\Delta H_f^{\ominus}$(NaCl), is given by the equation

$$Na(s) + \tfrac{1}{2}Cl_2(g) \rightarrow Na^+Cl^-(s)$$

The first step involves changing 1 mol of sodium atoms in the solid state to 1 mol of sodium atoms in the gaseous state. The enthalpy change for this step is known as the **standard enthalpy of atomization** (or the standard enthalpy of vaporization), and the reaction is endothermic. The value for sodium is +108 kJ mol^{-1}.

$$Na(s) \rightarrow Na(g) \quad \Delta H_{at}^{\ominus}(Na) = +108 \text{ kJ mol}^{-1}$$

The sodium atoms are then ionized to form sodium ions. The value for the first ionization energy of sodium atoms is +494 kJ mol^{-1}.

$$Na(g) \rightarrow Na^+(g) + e^- \quad \Delta H_{IE}^{\ominus}(Na) = +494 \text{ kJ mol}^{-1}$$

During the cycle ½ mol of gaseous chlorine molecules are converted into 1 mol of gaseous chlorine atoms. This involves breaking the chlorine-to-chlorine bond, Cl–Cl(g). The energy is equivalent to half the bond enthalpy. This is more usually known as the standard enthalpy of atomization, because 1 mol of gaseous atoms is being formed from the element in its standard state under standard conditions. Enthalpies of atomization always have positive values, and the value for chlorine is +121 kJ mol^{-1}.

$$\tfrac{1}{2}Cl_2(g) \rightarrow Cl(g) \quad \Delta H_{at}^{\ominus}(Cl) = +121 \text{ kJ mol}^{-1}$$

The gaseous chlorine atoms now each gain one electron. This enthalpy change is known as the **first electron affinity**. In the case of chlorine this process is exothermic, because energy is always evolved when gaseous atoms gain one electron. For example:

Element	H	Li	B	C	N	O	F	Cl	Br	I
Electron affinity/ kJ mol^{-1}	−72	−52	−29	−120	−3	−142	−348	−364	−342	−314

Hence

$$Cl(g) + e^- \rightarrow Cl^-(g) \quad \Delta H_{at}^{\circ}(Cl) = -364 \text{ kJ mol}^{-1}$$

However, if we had been discussing the formation of sodium oxide rather than sodium chloride, it is worth noting that when an oxide ion is formed a second electron must be added to the $O^-(g)$ ion to form the $O^{2-}(g)$ ion. The addition of the second electron is very endothermic, with a value of +844 kJ mol^{-1}. Thus the formation of 1 mol of oxide ions from 1 mol of oxygen atoms is an endothermic process overall:

$$\begin{aligned}
O(g) + e^- &\rightarrow O^-(g) & \Delta H^{\ominus} &= -142 \text{ kJ mol}^{-1} \\
\underline{O^-(g) + e^- \rightarrow O^{2-}(g)} & & \underline{\Delta H^{\ominus}} &= \underline{+844 \text{ kJ mol}^{-1}} \\
O(g) + 2e^- &\rightarrow O^{2-}(g) & \Delta H^{\ominus} &= +702 \text{ kJ mol}^{-1}
\end{aligned}$$

The final step is the formation of the crystalline solid sodium chloride from its gaseous ions. This is known as the **lattice enthalpy**. The standard lattice enthalpy can be defined either as the enthalpy change when 1 mol of a solid ionic salt is formed from its gaseous ions under standard conditions, in which case it has a negative value, or as the enthalpy change when 1 mol of a solid ionic salt is decomposed into its gaseous ions under standard conditions, in which case it has a positive value. In the way the cycle has been drawn in Figure 9 it has a negative value of -771 kJ mol^{-1}.

$$Na^+(g) + Cl^-(g) \rightarrow Na^+Cl^-(s) \qquad \Delta H_{latt}^{\ominus}(NaCl) = -771 \text{ kJ mol}^{-1}$$

Using Hess's law it can be seen that

$$\Delta H_f^{\ominus}(NaCl) = \Delta H_{at}^{\ominus}(Na) + \Delta H_{IE}^{\ominus}(Na) + \Delta H_{at}^{\ominus}(Cl)$$
$$+ \Delta H_{EA}^{\ominus}(Cl) + \Delta H_{latt}^{\ominus}(NaCl)$$

Substituting the values:

$$\Delta H_f^{\ominus}(NaCl) = [(+108) + (+494) + (+121) + (-364) + (-771)]$$
$$= -412 \text{ kJ mol}^{-1}$$

From the cycle it can clearly be seen that the main reason why the formation of sodium chloride is overall such an exothermic process is the large amount of heat given out when the ionic lattice is formed from its gaseous ions.

Lattice enthalpies

Lattice enthalpy is a measure of the strength of the electrostatic attraction between all the ions in the crystal structure. Its value will depend principally upon two factors:

- The charge on the ions: the greater the charge on the ions, the stronger the electrostatic attraction between them.
- The size of the ions: the smaller the ionic radius, the shorter the distance between the ions and hence the stronger the electrostatic attraction, as it is proportional to the reciprocal of the distance squared (i.e. $1/d^2$).

This explains why the lattice enthalpy for magnesium oxide (-3889 kJ mol^{-1}) is so much larger than the lattice enthalpy for sodium chloride (-771 kJ mol^{-1}). Not only are the charges larger in magnesium oxide (Mg^{2+} compared with Na^+, and O^{2-} compared with Cl^-) but the ions are smaller (0.065 nm for Mg^{2+} compared with 0.098 nm for Na^+, and 0.146 nm for O^{2-} compared with 0.181 nm for Cl^-; Figure 10).

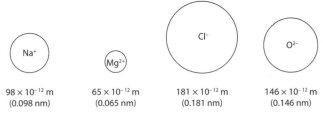

98×10^{-12} m (0.098 nm)	65×10^{-12} m (0.065 nm)	181×10^{-12} m (0.181 nm)	146×10^{-12} m (0.146 nm)

Figure 10

As well as the charge and size of the ions, which are the two most important factors, the lattice enthalpy will also depend to a lesser extent on the way in which the ions are packed. For example, in sodium chloride each sodium ion is surrounded by six chloride ions and vice versa, whereas in caesium chloride each caesium ion is surrounded by eight chloride ions and each chloride ion by eight caesium ions (Figure 11).

1 Construct two Born–Haber cycles—one for the formation of [Ca^+Cl^-] and one for the formation of [$Ca^{2+}2Cl^-$]. The second ionization energy of calcium is $+1150$ kJ mol^{-1}. Explain why even though this large amount of extra energy is required, calcium chloride has the formula $CaCl_2$ and not CaCl.

2 Draw a Born–Haber cycle for the formation of magnesium oxide. The lattice enthalpy of magnesium oxide is -3889 kJ mol^{-1}, the enthalpy of atomization of magnesium is $+150$ kJ mol^{-1}, and the second ionization energy of magnesium is $+1450$ kJ mol^{-1}. Use the Data Booklet to find the values of any other relevant steps, and then calculate the standard enthalpy of formation of magnesium oxide.

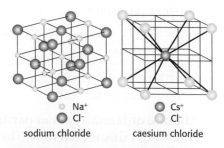

Figure 11 Sodium chloride (left) and caesium chloride (right).

Use of Born–Haber cycles

There are two main uses for Born–Haber cycles. Like any energy cycle they can be used to find an unknown value provided that all the other values are known. The main use, however, is to determine the degree of ionic character of a salt. The cycle assumes that the salt has 100% ionic character. Lattice enthalpies can be calculated theoretically from the size, charge and packing of the constituent ions. They can also be determined directly from the cycle. If there is good agreement between the two values obtained it is reasonable to assume that the salt has a high degree of ionic character. This is usually the case if there is a large difference in electronegativity values between the two constituent elements: sodium and chlorine, for example. When the difference in electronegativity values is lower, then the salt will contain an appreciable amount of covalent bonding, and this will result in a higher experimental value for the lattice enthalpy than that obtained theoretically. This is clearly the case with both silver chloride (AgCl) and silver iodide (AgI):

	NaCl	AgCl	AgI
Difference in electronegativity	2.1	1.1	0.6
Theoretical value for lattice enthalpy / kJ mol^{-1}	766	770	736
Experimental value for lattice enthalpy / kJ mol^{-1}	771	905	876

Entropy and free energy

Imagine two rooms connected by a single door. One room contains oxygen gas, and the other room contains nitrogen gas at the same temperature and pressure. What happens when the door is opened? Clearly the two gases will diffuse and mix, so that both rooms will quite quickly contain equal mixtures of oxygen and nitrogen. This is known as a **spontaneous process**. However, there is no chemical reaction taking place, and no heat is taken in or given out during the process. The enthalpy change for the mixing, ΔH, is zero. So why do the gases mix, if there is no enthalpy change? A related question might be to ask why the air in the room you are in at the moment does not "unmix" on its own accord, and all the oxygen diffuse to one end of the room and all the nitrogen diffuse to the other. Such a process is extremely improbable and hence would be non-spontaneous.

The answer is that systems naturally tend towards disorder. **Entropy** can be thought of as a measure of disorder. Unlike enthalpy, absolute entropy values can actually be measured, but usually it is the change in entropy, ΔS, that is more important.

$$\Delta S^{\ominus} = \sum S^{\ominus} \text{ (products)} - \sum S^{\ominus} \text{ (reactants)}$$

When a system becomes more disordered, then the change in entropy has a positive value. Examples of changes with increased disorder include:

- mixing different types of particles, e.g. dissolving a salt in water (although this is generally true there are a few cases—usually involving the hydration of small ions with charges of +2 or −2 (or +3 or −3)—when this is not so)

1 The experimental lattice enthalpy for silver fluoride, AgF, is 955 kJ mol^{-1}. Explain why this value is greater than the experimental value for the lattice enthalpy of silver chloride.

2 The theoretical value for the lattice enthalpy of silver fluoride is 870 kJ mol^{-1}. The difference between the experimental value and the theoretical value is 8.9%.
   (a) Explain why there is a difference between the two values.
   (b) Calculate the percentage difference between the experimental and theoretical values for the lattice enthalpy of silver chloride, and comment on the value you obtain.

- change of state where the distance between the particles increases, e.g. a liquid boiling or evaporating and forming a gas
- increasing the movement of particles, e.g. heating a solid, liquid or gas
- increasing the number of particles, e.g. $N_2O_4(g) \rightarrow 2NO_2(g)$.

Systems that become less disordered (more ordered) will have negative ΔS values: for example, the formation of solid ammonium chloride from hydrogen chloride gas and ammonia gas.

The standard entropy values, $S^\ominus$, for ammonia, hydrogen chloride and ammonium chloride are 193, 187 and 96 J K^{-1} mol^{-1} respectively. When 1 mol of gaseous ammonia reacts with 1 mol of gaseous hydrogen chloride, the product is 1 mol of solid ammonium chloride. Not only have 2 mol of reactants formed 1 mol of product, but two gases have reacted to form a solid. The system is considerably more ordered. $\Delta S^\ominus$ will equal $[96 - (193 + 187)] = -284$ J K^{-1} mol^{-1}.

$$NH_3(g) + HCl(g) \rightarrow NH_4Cl(s) \quad \Delta S^\ominus = -284 \text{ J K}^{-1} \text{ mol}^{-1}$$

Free energy (more properly known as the Gibbs free energy) has the symbol ΔG. When its value is negative it is equal to the total amount of energy given out in a reaction that is available to do work. It is made up of a combination of the enthalpy change and the entropy change multiplied by the absolute temperature, and is governed by the equation

$$\Delta G^\ominus = \Delta H^\ominus - T\Delta S^\ominus$$

where T represents the absolute temperature in kelvin.

A reaction is said to be **spontaneous** when $\Delta G^\ominus$ has a negative value. Once started, a spontaneous reaction will continue on its own accord without any extra energy having to be added. However, it is important to realize that a spontaneous reaction may proceed so slowly that it appears not to be reacting at all. For example, the combustion of diamond is spontaneous at all temperatures, but diamond is virtually completely stable in air at room temperature. This is because the reaction has an extremely high activation energy that has to be overcome first, and only a minute fraction of the reactants will have this necessary energy at room temperature.

$$C(s) + O_2(g) \rightarrow CO_2(g) \quad \Delta H^\ominus = -395.4 \text{ kJ mol}^{-1}$$
$$\Delta S^\ominus = +6.6 \text{ JK}^{-1} \text{ mol}^{-1}$$

$$\Delta G^\ominus = -395.4 - \frac{298 \times 6.6}{1000} = -397.4 \text{ kJ mol}^{-1} \text{ at 298 K}$$

Similarly, to light a coal fire it is quite common to light paper first, and then wood, to provide the necessary activation energy for the spontaneous combustion of the coal.

Some reactions are either spontaneous or non-spontaneous at all temperatures. For example, if, as in the reaction above, $\Delta H^\ominus$ is negative and $\Delta S^\ominus$ is positive, then $\Delta G^\ominus$ will always be negative and the reaction will always be spontaneous. Similarly, if $\Delta H^\ominus$ is positive and $\Delta S^\ominus$ is negative, then $\Delta G^\ominus$ will always be positive and the reaction can never be spontaneous.

HL However, for other reactions, whether or not they are spontaneous depends on the temperature. If $\Delta H^{\ominus}$ is positive and $\Delta S^{\ominus}$ is positive or if $\Delta H^{\ominus}$ is negative and $\Delta S^{\ominus}$ is negative then $\Delta G^{\ominus}$ will be either positive or negative depending on the value of the absolute temperature T.

For example, is the decomposition of calcium carbonate into calcium oxide and carbon dioxide spontaneous at 298 K?

$$CaCO_3(s) \rightarrow CaO(s) + CO_2(g) \quad \Delta H^{\ominus} = +178 \text{ kJ mol}^{-1}$$
$$\Delta S^{\ominus} = +165 \text{ JK}^{-1} \text{ mol}^{-1}$$

At 298 K, $\Delta G^{\ominus} = \Delta H^{\ominus} - T\Delta S^{\ominus} = +178 - \dfrac{298 \times 165}{1000}$
$= +129 \text{ kJ mol}^{-1}$

The answer is clearly no, as $\Delta G^{\ominus}$ has a positive value. This is reassuring, as many types of building material are essentially made of calcium carbonate. However, calcium carbonate is decomposed in a blast furnace during the industrial production of iron from its ore, where temperatures of up to 1400 °C are used. Assuming that the value of the entropy change $\Delta S^{\ominus}$ is independent of temperature (which is always assumed, but is not strictly true), we can calculate the temperature at which the reaction does become spontaneous. $\Delta G^{\ominus}$ will have the value of zero when $\Delta H^{\ominus} = T\Delta S^{\ominus}$. Therefore the temperature at which $\Delta G^{\ominus} = 0$ will be equal to 178 kJ mol^{-1} divided by (165.3/1000 kJ K^{-1} mol^{-1}), which equals 1077 K or 804 °C. Above this temperature, $T\Delta S^{\ominus} > \Delta H^{\ominus}$, which means that $\Delta G^{\ominus}$ will have a negative value and the reaction will be spontaneous.

When doing calculations involving enthalpy and entropy changes to find the Gibbs free energy $\Delta G^{\ominus}$, remember that enthalpy changes are normally given in kJ mol^{-1} whereas entropy changes are in normally given in J K^{-1} mol^{-1}, so it is necessary to first convert the entropy change into kJ K^{-1} mol^{-1} by dividing by 1000.

1 Write the equation for the combustion of hydrogen to form water under standard conditions, and state whether you would expect the entropy change to have a positive or negative value. The standard enthalpy of combustion of hydrogen is −286 kJ mol^{-1}. The standard entropy values for hydrogen, oxygen and water are 131, 205 and 69.9 J K^{-1} mol^{-1} respectively. Calculate the value for the Gibbs free energy at 298 K to show that this reaction is spontaneous at this temperature. Above what temperature will this reaction become non-spontaneous?

2 Consider the following two reactions:

$2Fe(s) + O_2(g) \rightarrow 2FeO(s)$ $\Delta G^{\ominus} = -450$ kJ mol^{-1} at 500 K
$2CO(g) + O_2(g) \rightarrow 2CO_2(g)$ $\Delta G^{\ominus} = -480$ kJ mol^{-1} at 500 K

Use Hess's law to determine $\Delta G^{\ominus}$ at 500 K for the reaction

$2FeO(s) + 2CO(g) \rightarrow 2Fe(s) + 2CO_2(g)$

Is the reduction of iron(II) oxide, FeO, by carbon monoxide spontaneous or non-spontaneous at this temperature?

3 Earlier in the chapter the reaction between hydrated barium hydroxide and ammonium chloride was given as an example of an endothermic reaction:

$Ba(OH)_2.8H_2O(s) + 2NH_4Cl(s) \rightarrow BaCl_2.2H_2O(s) + 2NH_3(g/aq) + 8H_2O(l)$

Use the data given in Table 1 to calculate a value for the Gibbs free energy $\Delta G^{\ominus}$ for this reaction at 298 K. (You will need to assume that one of the moles of ammonia formed is in the gaseous state and the other is in aqueous solution.)

Table 1

Substance	ΔH_f°/ kJ mol^{-1}	S°(system)/ J mol K^{-1}	ΔG_f°/ kJ mol^{-1}
Ba(OH)$_2$.8H$_2$O(s)	−3342	427	−2793
NH$_4$Cl(s)	−314	95	−203
BaCl$_2$.2H$_2$O(s)	−1460	203	−1297
NH$_3$(g)	−46	192	−16
NH$_3$(aq)	−80	111	−27
H$_2$O(l)	−286	70	−237

Extension material
The second law of thermodynamics

It is convenient to think of entropy in terms of disorder, and this will suffice for IB Diploma chemistry, but it is an oversimplification, and leads to questions that are difficult to answer. For example, if systems naturally tend towards disorder, why do we exist? Humans (and other life forms) are incredibly organized and ordered. Similarly, when water freezes it forms very beautiful and highly ordered crystals of ice (Figure 12).

Figure 12 Ice crystals

The second law of thermodynamics concerns the directional flow of heat in relation to work. Put simply, it states that heat cannot spontaneously pass from a colder body to a hotter body. Rather like the first law of thermodynamics, which is concerned with the conservation of energy, the second law can be stated in several different ways. Another way of stating the second law is: The entropy of an isolated system not at equilibrium will tend to increase over time, approaching a maximum value.

So what exactly *is* entropy, if it is not simply "disorder"? Perhaps a better way to understand entropy is to see it as the dispersal of energy. An isolated system can be thought of as consisting of two parts: a subsystem of particular interest, and the surroundings around the subsystem.

Consider, as an example, a piece of ice taken out of the refrigerator and placed in a warm room. The ice is the subsystem and the warm room is the surroundings. The air in the room consists of molecules in rapid random motion. As energy from the air is transferred to the ice the movement of the air molecules will slow slightly. The water molecules in the ice have vibrational energy but are not free to move around. Once the ice melts and the water molecules are free to move, there are many more possible arrangements in which the water molecules can exist. In addition, as the water molecules become free to move, the energy states within the molecules due to vibration, rotation and translation increase.

Figure 13 "Discipline 2002" by Roger Hiorns. Roger Hiorns is a British artist who fashions works of art by growing highly ordered copper sulfate crystals.

Thus, as well as depending upon the number of ways in which the energy can be distributed over all the different possible arrangements of the particles relative to each other, the entropy changes also depend upon the ways in which the energy can be distributed within the particles themselves. Increasing the temperature increases the energy available, and increases the number of ways in which it can be distributed or dispersed. In the example with the ice, the entropy change for the ice will be positive, but the surroundings—the air in the room—have become cooler, so the entropy change for the surroundings is negative. However, because the temperature of the surroundings is initially higher, and there are more possible ways in which the energy can be distributed, the entropy loss to the surroundings is less than the entropy gain to the subsystem, so the overall effect for the isolated system is positive:

$$\Delta S_{total} = \Delta S_{subsystem} + \Delta S_{surroundings}$$

This is in accordance with the second law of thermodynamics, which states that the total entropy change for an isolated system must be positive, so:

$$\Delta S_{subsystem} + \Delta S_{surroundings} > 0$$

However, the heat lost by the surroundings will be the same as the heat gained by the subsystem, because by definition no heat can be lost or gained from the isolated system. If the value for the heat gained by the subsystem is $\Delta H_{subsystem}$ then the heat lost by the surroundings will be $-\Delta H_{subsystem}$. The entropy change depends on the absolute temperature T, so

$$\Delta S_{surroundings} = -\frac{\Delta H_{subsystem}}{T}$$

Hence

$$\Delta S_{subsystem} - \frac{\Delta H_{subsystem}}{T} > 0$$

So for the subsystem in question we can rearrange and shorten this to

$$\Delta H - T\Delta S < 0$$

Thus for a spontaneous reaction the expression $(\Delta H - T\Delta S)$ will always have a negative value. As we have seen, the actual value is known as the Gibbs free energy ΔG: hence the expression

$$\Delta G = \Delta H - T\Delta S$$

One of the important words in the second law of thermodynamics is the word "isolated" when referring to isolated systems. Complex and highly ordered forms of life can evolve with negative entropy changes, because they are not isolated systems. The Sun provides a large amount of energy to the Earth, and this flow of heat results in huge positive changes in entropy, when compared with the negative changes associated with the formation of biological systems. Perhaps the only truly isolated system is the universe itself. If this is the case, then the second law of thermodynamics predicts the eventual end of the universe—the so-called "heat death"—since eventually the universe will reach maximum entropy and energy can no longer be dispersed. But—and it is a big but—is the universe really an isolated system? Astronomers disagree on this point, and the role of entropy in cosmology remains a controversial subject.

Within chemistry, and science in general, the second law of thermodynamics is one of the most fundamental of all concepts. The novelist and scientist C.P. Snow (1905–1980), in a lecture given in 1951, said:

> A good many times I have been present at gatherings of people, who by the standards of the traditional culture, are thought to be highly educated and who have with considerable gusto been expressing their incredulity at the illiteracy of scientists. Once or twice I have been provoked and have asked the company how many of them could describe the second law of thermodynamics. The response was cold: it was also negative. Yet I was asking something which is about the scientific equivalent of "Have you read a work of Shakespeare's?".

1 State three different ways in which the *first* law of thermodynamics can be expressed.
2 The alchemists searched for the "elixir of life". Could human beings ever be immortal?

6 Kinetics

Rate of reaction

Some reactions proceed very fast, whereas others can proceed so slowly that it can take millions of years before any appreciable amount of products are formed. An example of a very fast reaction that we have come to rely on is the rapid inflation of airbags used in automobiles. Typical reactants are a mixture of sodium azide (NaN_3), potassium nitrate (KNO_3) and silicon dioxide. The sodium azide and the potassium nitrate react to produce inert nitrogen gas, which inflates the bag:

$$10NaN_3(s) + 2KNO_3(s) \rightarrow K_2O(s) + 5Na_2O(s) + 16N_2(g)$$

The silicon dioxide is present to convert the oxides of potassium and sodium that are produced into harmless silicates. When a crash occurs, a sensor in the car's airbag will trigger the reaction if the driver (and passengers) continues to move forward with a momentum greater than the equivalent of hitting a brick wall at 16–24 km h^{-1}. The nitrogen has to be produced to inflate the bag more quickly than the time taken for the driver to hit the steering wheel. The whole reaction from triggering to complete inflation of the bag is complete in less than one 25th of a second (Figure 1).

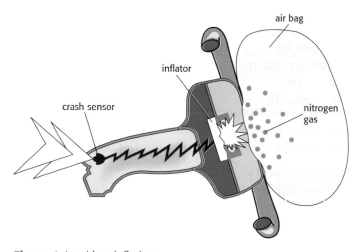

Figure 1 An airbag inflating.

Other reactions are much slower. It typically takes several days for equilibrium to be reached for the reaction between ethanoic acid and ethanol in the presence of a catalyst, H$^+$(aq), to form ethyl ethanoate and water:

$$CH_3COOH(l) + C_2H_5OH(l) \rightleftharpoons CH_3COOC_2H_5(l) + H_2O(l)$$
$$\text{ethanoic acid} \quad \text{ethanol} \qquad \text{ethyl ethanoate} \quad \text{water}$$

Many famous paintings have needed cleaning and restoring after several hundred years, because of the slow reaction of the

111

step was the slowest step—that is, the rate-determining step—then the overall order of the reaction would be second order, and hydrogen would not appear in the rate equation.

If the second step is the rate-determining step, then the rate depends on the concentration of dinitrogen dioxide and hydrogen. But from the first step the concentration of dinitrogen dioxide depends on the square of the concentration of nitrogen monoxide. The overall rate equation will therefore be

$$\text{rate} = k[\text{H}_2(\text{g})][\text{NO}(\text{g})]^2$$

This is the same as the experimentally determined expression, so this three-step mechanism, with the second step being the slowest step, is consistent with the rate equation. However, it is important to realize that this is not proof that this is the mechanism. Experimental results must support the theory, but they do not prove it. Another possible mechanism is

$$\text{H}_2(\text{g}) + \text{NO}(\text{g}) \rightarrow \text{H}_2\text{O}(\text{g}) + \text{N}(\text{g}) \qquad \text{Fast}$$
$$\text{N}(\text{g}) + \text{NO}(\text{g}) \rightarrow \text{N}_2\text{O}(\text{g}) \qquad \text{Slow}$$
$$\text{N}_2\text{O}(\text{g}) + \text{H}_2(\text{g}) \rightarrow \text{N}_2(\text{g}) + \text{H}_2\text{O}(\text{g}) \qquad \text{Fast}$$

Provided that the second step is the rate-determining step, this mechanism is consistent with the rate equation.

What chemists have to do is use their chemical intuition to decide which is the most likely mechanism and then look for supplementary evidence to support their hypothesis. In this case the first mechanism described is more probable, because dinitrogen dioxide is more likely to be formed as an intermediate than a very unstable nitrogen atom. In industry it can be especially helpful to understand the mechanism of a reaction, because it may be possible to alter the reaction to enable it to proceed more efficiently and cheaply. Ultimately, however, a proposed mechanism can never be proved from the rate equation, but it can be disproved if it is not consistent with the rate equation. This fits in neatly with Popper's approach to falsification in science, which asserts that a theory that is incapable of being falsified or of being tested is not scientific.

1 Evidence suggests that the reaction between nitrogen dioxide and fluorine in the gaseous state is a two-step process:
$2\text{NO}_2(\text{g}) + \text{F}_2(\text{g}) \rightarrow 2\text{NO}_2\text{F}(\text{g})$
Step 1: $\text{NO}_2(\text{g}) + \text{F}_2(\text{g}) \rightarrow \text{NO}_2\text{F}(\text{g}) + \text{F}(\text{g})$ (Slow)
Step 2: $\text{F}(\text{g}) + \text{NO}_2(\text{g}) \rightarrow \text{NO}_2\text{F}(\text{g})$ (Fast)
(a) State which is the rate-determining step, and explain why.
(b) State which of the two steps is expected to have the higher activation energy, and explain why.
(c) Give the rate expression of the reaction based on your answer.
(d) What feature of the fluorine molecule makes it reasonable to propose that a fluorine atom may be formed as an intermediate in this reaction?

2 The experimental data obtained for the acid-catalysed iodination of propanone showed that the rate equation is

$\text{rate} = k[\text{H}_3\text{CCOCH}_3(\text{aq})][\text{H}^+(\text{aq})]$

Propose a mechanism for the reaction between iodine and propanone in the presence of an acid catalyst to form iodopropanone that is consistent with this rate equation.

Activation energy

The rate constant k for a reaction is a constant only at a fixed temperature. As the temperature increases, the value of the rate constant increases exponentially. The relationship between the rate constant and the absolute temperature is given by the **Arrhenius equation**:

$$k = A \exp \frac{-E_a}{RT}$$

where E_a is the activation energy for the reaction and R is the gas constant. A is known as the **Arrhenius constant** and is related to the orientation of the reactants at the point of collision. The Arrhenius equation is often expressed in logarithmic form:

$$\ln k = \frac{-E_a}{RT} + \ln A$$

This expression is in the form $y = mx + c$, so that a plot of $\ln k$ against $1/T$ will give a straight line. The gradient of the line will be equal to $-E_a/R$, and hence the value of the activation energy for the reaction can be determined. Extrapolating the straight line back to the $\ln k$ axis will give the intercept equal to $\ln A$ (Figure 17).

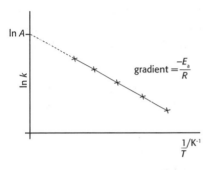

Figure 17 Activation energy against reciprocal of absolute temperature.

The Arrhenius equation is named after the Swedish physical chemist Svante Arrhenius (1859–1927). Arrhenius is best known for his work on electrolytes. In 1884, at the age of 25, he submitted his thesis for his doctorate at the University of Uppsala, in which he concluded that electrolytes dissociate in solution into positive and negative ions, and that the degree of dissociation depends on the dilution: the more diluted the electrolyte, the greater the degree of dissociation. This was a revolutionary idea, and his work was not well received by the university authorities, who awarded him the lowest possible passing mark for his thesis. In 1903, after the brilliance of his work on electrolytes had been fully recognized, he was awarded the Nobel Prize in Chemistry.

Critical considerations

It should be stressed that the Arrhenius equation cannot be mathematically proven. It is a good example of an empirical relationship—that is, one that has been determined by experiment. Even so, we can use it to test a "rule of thumb" that is often stated in chemistry textbooks.

It is often claimed that increasing the temperature by 10 °C will double the initial rate of the reaction. For the general reaction A + B → products, the rate = $k[A]^x[B]^y$, but if the concentrations are kept constant then the rate is directly proportional to the rate constant. Doubling the rate constant k will double the initial rate. A typical value for the activation energy of a reaction is 50 kJ mol^{-1}. As the value for the molar gas constant, R, is given in J K^{-1} mol^{-1} we shall need the value for the activation energy in J mol^{-1} —that is, 50 000 J mol^{-1}.

Critical considerations (continued)

At 298 K:

$$k_{298} = Ae^{-E_a/RT} = Ae^{-50000/(8.314 \times 298)} = A \times 1.72 \times 10^{-9}$$

At 308 K:

$$k_{308} = Ae^{-E_a/RT} = Ae^{-50000/(8.314 \times 308)} = A \times 3.31 \times 10^{-9}$$

$$\frac{k_{308}}{k_{298}} = \frac{A \times 3.31 \times 10^{-9}}{A \times 1.72 \times 10^{-9}} = 1.92 \simeq 2$$

So increasing the temperature by 10 °C (or 10 K) does virtually double the rate, which supports the "rule of thumb".

By now, though, you should be able to look critically at an argument like this. At least two assumptions have been made. The first is that all activation energies are about 50 kJ mol^{-1}. What happens if the activation energy is 100 kJ mol^{-1} instead?

At 298 K:

$$k_{298} = Ae^{-E_a/RT} = Ae^{-100000/(8.314 \times 298)} = A \times 2.96 \times 10^{-18}$$

At 308 K:

$$k_{308} = Ae^{-E_a/RT} = Ae^{-100000/(8.314 \times 308)} = A \times 1.10 \times 10^{-17}$$

$$\frac{k_{308}}{k_{298}} = \frac{A \times 1.10 \times 10^{-17}}{A \times 2.96 \times 10^{-18}} = 3.71$$

So if the activation energy for a reaction is high, increasing the temperature by 10 °C has a much greater effect than merely doubling the rate of the reaction. You can confirm that the converse is also true by using a value of 25 kJ mol^{-1} for the activation energy.

The second assumption is that the value of the Arrhenius constant, A, is independent of the temperature. This is not strictly true, although it is a reasonable approximation.

1 Can you think of a third assumption that has been made in the argument above? Test the validity of this assumption.

2 The calculations above give the value for the rate constant at 298 K for two different values of activation energy. In one case the activation energy is 100 kJ mol^{-1} and in the second case it has been halved to 50 kJ mol^{-1}. By how many times has the rate increased when the activation energy has been halved? What is the relevance of this to the way a catalyst works?

Extension material

The Belousov–Zhabotinsky (BZ) reaction

We began this chapter by talking about the predictability of chemistry. The sodium azide and potassium nitrate contained within the airbag in your car have never reacted before, yet we can predict with almost absolute certainty that in the event of a crash of the required magnitude the chemicals will come together and react to fill the airbag with nitrogen gas within one 25th of a second. This ability to accurately predict the outcome of an event is one of the key achievements of science, and is one of the main reasons why scientific knowledge is sometimes seen as so powerful.

In the 1950s an unknown Russian scientist Boris Belousov (1893–1970) submitted a paper for publication. In this paper he claimed that an acidified solution of propanedioic acid (malonic acid), potassium(V) bromate, KBrO$_3$, and cerium(IV) sulfate, Ce(SO$_4$)$_2$, oscillated periodically between yellow and clear, and that it was impossible to predict the precise time interval between the oscillations. Even though Belousov gave the details of the relatively simple experiment, the editor of the journal did not bother to check the experiment. He rejected the paper outright, on the grounds that oscillating reactions were impossible in chemistry, because all reactions moved towards the position of equilibrium. The work was published several years later in an obscure scientific journal, but went unrecognized until a graduate student, Anatol Zhabotinsky, rediscovered the work in the 1960s and proposed a mechanism for the reaction.

There are now several examples of reactions that show similar behaviour. Oscillations occur in systems that are *approaching* equilibrium rather than at equilibrium. What they have in common is that the reaction that drives the process can follow at least two different pathways, and the reaction can switch between these pathways. In the BZ reaction the colour changes are due to two different oxidation states exhibited by the cerium ions. The cerium(IV) ions are reduced by propanedioic acid to cerium(III) ions and oxidized back to cerium(IV) ions by bromate(V) ions. The proposed reaction mechanism consists of many separate steps involving many intermediate substances. Some of these intermediates cause positive feedback whereas others cause inhibitions, resulting in oscillations between the various species. Because the reactions are non-linear—that is, depend on the concentrations of reacting species raised to a power greater than 1—a very small change in one step can have a large effect in subsequent steps, so that it is difficult to predict the exact time between the oscillations.

This has been compared to weather forecasting. In theory, the weather follows the laws of physics, but there are so many variables that completely accurate predictions are virtually impossible. In chaos theory this sensitive dependence on a small change in the initial conditions is known as the **butterfly effect**, because one small flap of a butterfly's wings could ultimately result in a tornado thousands of miles away.

Practical details for the BZ reaction

A tried and tested recipe (taken from an article by Stephen Scott in the *New Scientist*, 2 December 1989) for the BZ reaction is to stir a mixture containing:
500 cm^3 of 1.0 mol dm^{-3} sulfuric acid, H_2SO_4(aq)
14.30 g of propanedioic acid, $HOOCCH_2COOH$
5.22 g of potassium bromate(V), $KBrO_3$
0.548 g of ammonium cerium(IV) nitrate, $(NH_4)_2Ce(NO_3)_6$
1−2 cm^3 of 0.025 mol dm^{-3} ferroin

(The ferroin solution is made by dissolving 1.485 g of 1,10-phenanthroline and 0.685 g of heptahydrated iron(II) sulfate, $FeSO_4.7H_2O$, in 100 cm^3 of water.)

The ferroin, which is magenta when the iron is in the iron(II) state and blue when it is in the oxidized iron(III) state, makes a more visual indicator than the cerium ions alone. Once the mixture has been stirred you should see the resulting oscillations between blue and magenta with a period of about 1 minute that last for several hours.

Dynamic equilibrium

When wet clothes are hung on a washing line to dry, the water in them does not boil, because the temperature is still far below the normal boiling point of water. However, some of the water molecules will be travelling much faster than the average, and they will have enough kinetic energy to escape the surface of the water and enter the atmosphere. The pressure they produce is known as the **vapour pressure** of water. The vapour pressure is very dependent upon the temperature (Figure 1).

As the vapour leaves the wet clothes it will be blown away by the wind, and more water will evaporate to replace it, until eventually the clothes become dry. However, what will happen if the water is held in a closed container (Figure 2)?

Initially water will still vaporize, but once the air above the water becomes saturated with water vapour it can hold no more water. A state of equilibrium is established between the liquid water and the water vapour. However, some water molecules from the liquid water will still be travelling fast enough to have enough kinetic energy to escape the surface and enter the vapour phase; at the same time some of the vapour water molecules will slow down and condense back into the liquid state. Equilibrium is established when the rate of vaporization is equal to the rate of condensation:

$$H_2O(l) \rightleftharpoons H_2O(g)$$

This is an example of **dynamic equilibrium**. Another example of dynamic equilibrium is a mixture of ice and water at the freezing point of water:

$$H_2O(s) \rightleftharpoons H_2O(l)$$

At equilibrium the process taking place has not stopped, but the amounts on both sides of the process remain constant, as the rate of the forward process is equal to the rate of the reverse process. The opposite of dynamic equilibrium is **static equilibrium**: this occurs in many physical systems when all movement ceases once equilibrium is established. For example, a balanced see-saw will remain static until the position of equilibrium is disturbed in some way.

Closed system

Strictly speaking, the conical flask shown above is not a closed system (because heat can be slowly transferred through the glass walls), although it approximates to one. In a completely **closed (or isolated)** system neither matter nor energy can be lost or gained. When we use the term "equilibrium" we are usually referring to a closed system, because if the system is open some of the chemical

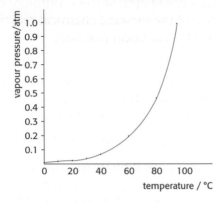

Figure 1 Vapour pressure of water at different temperatures. The water boils when its vapour pressure is equal to the external pressure.

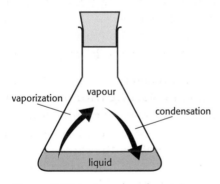

Figure 2 Water in a closed container.

species involved may escape from it, and the position of equilibrium will never be established.

Chemical equilibrium

Some chemical reactions do go virtually to completion; however, many do not. Those that do not go to completion are said to be **reversible**. Once some products have been formed they start to react to re-form the reactants.

In a closed system the position of equilibrium is reached when the rate of the forward chemical reaction is equal to the rate of the reverse chemical reaction.

We can use isotopes to show that the equilibrium is dynamic and the reaction has not stopped when the position of equilibrium has been reached. For example, consider the reaction of hydrogen with nitrogen to form ammonia:

$$3H_2(g) + N_2(g) \rightleftharpoons 2NH_3(g)$$

Suppose we now introduce some deuterium, D_2 (heavy hydrogen, $^2_1H_2$) into the system. A small amount will react to restore the position of equilibrium and will produce some deuterated ammonia: NH_2D, NHD_2 and ND_3. If the equilibrium was then static we would still have only a mixture of hydrogen (H_2) and deuterium (D_2) present, together with normal ammonia and the deuterated ammonia. However, if the equilibrium is dynamic then as well as hydrogen (H_2) and heavy hydrogen (D_2) we shall also find some mixed hydrogen (HD) in the mixture as the deuterated ammonia breaks down. By analysing the mixture using mass spectrometry the presence of mixed hydrogen can be confirmed.

For a reversible reaction the position of equilibrium can be reached either by starting with the reactants or by starting with the products. In each case, once equilibrium has been reached the relative concentrations of the reactants and products in the equilibrium mixture will be the same. This can be illustrated by looking at the dissociation of dinitrogen tetroxide, $N_2O_4(g)$. Whether we start with dinitrogen tetroxide or with nitrogen dioxide, once equilibrium has been attained the composition of the mixture is constant (Figure 3).

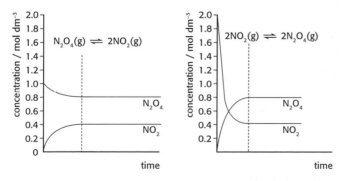

Figure 3 Dissociation of dinitrogen tetroxide, $N_2O_4(g)$.

Homogeneous chemical equilibria

A **homogeneous reaction** is one in which all the reactants and products are in the same phase. A **phase** is similar to a state such as solid, liquid or gas, but there is an important difference. There is a physically distinct boundary between two phases. Hence it is possible to have two phases but only one state. An example is two immiscible liquids such as oil and water. Both are in the liquid state, but there are two different phases, as there is a clear boundary where the two liquids meet (Figure 4).

If all the reactants and products are in the same phase, such as all in the gaseous state, all in the liquid phase or all in aqueous solution, we can write an equilibrium expression for the reaction.

Consider the general reaction in which a moles of A react with b moles of B to produce c moles of C and d moles of D:

$$a\text{A} + b\text{B} \rightleftharpoons c\text{C} + d\text{D}$$

At equilibrium the concentrations of A, B, C and D are written as $[\text{A}]_{eqm}$, $[\text{B}]_{eqm}$, $[\text{C}]_{eqm}$ and $[\text{D}]_{eqm}$ respectively. The equilibrium expression is written

$$K_c = \frac{[\text{C}]^c_{eqm} \times [\text{D}]^d_{eqm}}{[\text{A}]^a_{eqm} \times [\text{B}]^b_{eqm}}$$

K_c is known as the equilibrium constant, and it is a constant at a stated temperature.

Examples of specific homogeneous equilibria include the following:

- Esterification: the formation of ethyl ethanoate and water from ethanol and ethanoic acid

$$\text{C}_2\text{H}_5\text{OH(l)} + \text{CH}_3\text{COOH(l)} \rightleftharpoons \text{CH}_3\text{COOC}_2\text{H}_5\text{(l)} + \text{H}_2\text{O(l)}$$

$$K_c = \frac{[\text{CH}_3\text{COOC}_2\text{H}_5\text{(l)}]_{eqm} \times [\text{H}_2\text{O(l)}]_{eqm}}{[\text{C}_2\text{H}_5\text{OH(l)}]_{eqm} \times [\text{CH}_3\text{COOH(l)}]_{eqm}}$$

- The Haber process to manufacture ammonia (see page 133)

$$3\text{H}_2\text{(g)} + \text{N}_2\text{(g)} \rightleftharpoons 2\text{NH}_3\text{(g)}$$

$$K_c = \frac{[\text{NH}_3\text{(g)}]^2}{[\text{H}_2\text{(g)}]^3 \times [\text{N}_2\text{(g)}]}$$

- The dissociation of hydrogen iodide

$$2\text{HI(g)} \rightleftharpoons \text{H}_2\text{(g)} + \text{I}_2\text{(g)}$$

$$K_c = \frac{[\text{H}_2\text{(g)}] \times [\text{I}_2\text{(g)}]}{[\text{HI(g)}]^2}$$

- The contact process to manufacture sulfuric acid (see page 134)

$$2\text{SO}_2\text{(g)} + \text{O}_2\text{(g)} \rightleftharpoons 2\text{SO}_3\text{(g)}$$

$$K_c = \frac{[\text{SO}_3\text{(g)}]^2}{[\text{SO}_2\text{(g)}]^2 \times [\text{O}_2\text{(g)}]}$$

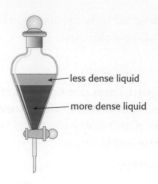

less dense liquid

more dense liquid

Figure 4 A separating funnel can be used to separate two immiscible liquids.

The magnitude of the equilibrium constant

The equilibrium constant K_c is, essentially, a measure of the amount of products in an equilibrium mixture compared with the amount of reactants. If the value of the equilibrium constant is very much greater than 1 then the position of equilibrium lies very much on the products side, and the mixture contains virtually all products: that is, the reaction has almost gone to completion. An example of this is the oxidation of zinc when it is added to a solution of copper(II) ions:

$$Zn(S) + Cu^{2+}(aq) \rightleftharpoons Zn^{2+}(aq) + Cu(s) \quad K_c = 1 \times 10^{37} \text{ at 298 K}$$

The value is so extremely large that this reaction is said to go to completion.

If the value of the equilibrium constant is much less than 1, then very little product will be formed. An example is a weak acid dissociating in water. For example:

$$CH_3COOH(aq) \rightleftharpoons CH_3COO^-(aq) + H^+(aq)$$
$$K_c = 1.8 \times 10^{-5} \text{ mol dm}^{-3} \text{ at 298 K}$$

From the value of the equilibrium constant we can deduce that in aqueous solution ethanoic acid exists mainly in the undissociated molecular form, CH_3COOH.

If the value of the equilibrium constant is in the order of 1 then both reactants and products will be present in the equilibrium mixture in appreciable amounts. This is the case with the esterification reaction between ethanoic acid and ethanol to form ethyl ethanoate and water. The equilibrium constant has a value of 4 at 298 K.

Le Châtelier's principle

Henri Le Châtelier (1850–1936) was a French chemist who described what happens to the position of equilibrium if a change to the conditions is made. **Le Châtelier's principle** states that if a dynamic equilibrium is disturbed by changing the conditions, the position of equilibrium moves to counteract the change. Le Châtelier's principle is often quoted, but it is important to realize that this statement does not explain why the position of equilibrium changes; it simply enables you to predict what will happen. For example, consider the reaction between ethanoic acid and ethanol to form ethyl ethanoate and water:

$$C_2H_5OH(l) + CH_3COOH(l) \rightleftharpoons CH_3COOC_2H_5(l) + H_2O(l)$$

If we remove some of the water from the equilibrium mixture, Le Châtelier's principle enables us to predict that more of the acid and alcohol will react to replace it. This concept of removing one of the products is a useful way to increase the amount of a desired product in an equilibrium reaction.

We can use the equilibrium expression to explain why this happens, rather than simply state it using Le Châtelier's principle.

The equilibrium expression for the esterification reaction is

$$K_c = \frac{[CH_3COOC_2H_5(l)]_{eqm} \times [H_2O(l)]_{eqm}}{[C_2H_5OH(l)]_{eqm} \times [CH_3COOH(l)]_{eqm}}$$

If we remove some of the water from the equilibrium mixture, the concentration of the water in the mixture decreases. This means that the value of the upper part of the expression becomes less, and one might expect that the value of the equilibrium constant, K_c, would therefore decrease. However, K_c is a constant. In order to keep the value of K_c the same the value of the lower part of the expression must also decrease, and the concentration of the ester will increase.

The equilibrium expression can always be used to explain changes in concentration and pressure. For example, in the above reaction the concentration of ethyl ethanoate in the equilibrium mixture could also be increased by increasing the concentration of the ethanoic acid or that of the ethanol, or both. In all three cases more ester will be formed to keep the value of K_c constant.

The effect of changing the pressure on gaseous reactions

If there is an overall volume change in a gaseous reaction then increasing the pressure will move the position of equilibrium towards the side with less volume. This is because the total number of molecules in the equilibrium mixture will be reduced, and hence the change caused by the increase in pressure will tend to be minimized. An example is the decomposition of dinitrogen tetroxide. This is a colourless gas, and it is in equilibrium with brown nitrogen dioxide.

$$N_2O_4(g) \rightleftharpoons 2NO_2(g)$$
colourless   brown

If we started with exactly 1 mol of dinitrogen tetroxide and it all dissociated, we would have 2 mol of nitrogen dioxide product. Avogadro's hypothesis states that equal amounts (moles) of different gases at the same temperature and pressure occupy the same volume: hence the volume of the nitrogen dioxide is twice that of the dinitrogen tetroxide. Often chemists convert molar quantities directly into volumes and write them under the gaseous reactants and products. They then treat volumes as if they were moles.

$$N_2O_4(g) \rightleftharpoons 2NO_2(g)$$
1 vol   2 vol

Increasing the pressure on the mixture will therefore increase the amount of dinitrogen tetroxide in the mixture by moving the position of equilibrium to the side with less volume. This can be seen in practice by filling a gas syringe with the mixture. If the end is sealed and the plunger is pushed in, the mixture will initially go a darker brown as the concentration of the nitrogen dioxide in the mixture increases; then as the position of equilibrium is re-established the mixture will quite quickly go lighter again. See Figure 5.

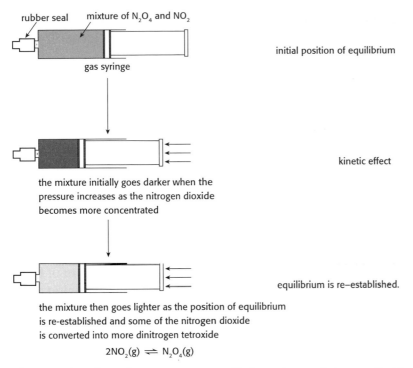

the mixture initially goes darker when the
pressure increases as the nitrogen dioxide
becomes more concentrated

the mixture then goes lighter as the position of equilibrium
is re-established and some of the nitrogen dioxide
is converted into more dinitrogen tetroxide

$$2NO_2(g) \rightleftharpoons N_2O_4(g)$$

Figure 5 The effect of pressure on an equilibrium mixture of nitrogen dioxide
and dinitrogen tetroxide.

The same conclusion can be reached using the equilibrium
expression:

$$K_c = \frac{[NO_2(g)]^2}{[N_2O_4(g)]}$$

Increasing the pressure will increase the concentration both of the
nitrogen dioxide (NO_2) and of the dinitrogen tetroxide (N_2O_4).
As the concentration of the NO_2 is squared in the equilibrium
expression, this will have the greater effect, and to keep K_c constant
the concentration of the NO_2 must decrease and that of the N_2O_4
must increase.

If there is no volume change in a gaseous reaction, then changing
the pressure will have no effect on the equilibrium constant,
because the concentrations of both the reactants and the products
will be affected in the same way, and there will be no change in the
total number of molecules present. This can be illustrated by the
reaction of hydrogen and iodine to form hydrogen iodide:

$$H_2(g) + I_2(g) \rightleftharpoons 2HI(g)$$

2 vol   2 vol

Increasing or decreasing the pressure has no change on the position
of equilibrium.

The effect of changing the temperature

The effect on the position of equilibrium of changing the
temperature will vary depending upon whether the reaction is
exothermic or endothermic.

Summary of the Haber process conditions:

$$\underbrace{3H_2(g) + N_2(g)}_{4 \text{ vol}} \xrightleftharpoons[]{\text{Fe(s), 450 °C, 200 atm}} \underbrace{2NH_3(g)}_{2 \text{ vol}} \qquad \Delta H^{\ominus} = -92 \text{ kJ mol}^{-1}$$

The contact process to manufacture sulfuric acid

The manufacture of sulfuric acid in the developed world has fallen in recent years. This is partly because the manufacture of phosphate fertilizers, which uses sulfuric acid, has declined. It is also because much of the industry has relocated to the developing world, where the plants are cheaper to build and operate, and where environmental restrictions can be less stringent.

The essential reaction is the conversion of sulfur dioxide to sulfur trioxide:

$$2SO_2(g) + O_2(g) \rightleftharpoons 2SO_3(g) \quad \Delta H^{\ominus} = -192 \text{ kJ mol}^{-1}$$

The sources of sulfur dioxide include the combustion of the element sulfur itself and the roasting (combustion) of sulfide ores in air.

As the reaction is exothermic, and involves a reduction in volume, the same arguments as used for the Haber process can be applied. A high yield will be favoured by a high pressure and a low temperature. The rate at which equilibrium is reached will be favoured by a high pressure, a high temperature, and the use of a suitable catalyst. As before, an optimum temperature must be used. In practice, plants operate at a temperature of about 450 °C, and the catalyst is vanadium(V) oxide. Although in theory a high pressure is required, in practice a pressure of only 2 atmospheres is used as this will give a yield of over 98%. Operating the plant at higher pressure than this would not be cost-effective.

Summary of contact process conditions:

$$\underbrace{2SO_2(g) + O_2(g)}_{3 \text{ vol}} \xrightleftharpoons[]{\text{V}_2\text{O}_5\text{(s), 450 °C, 2 atm}} \underbrace{2SO_3(g)}_{2 \text{ vol}} \qquad \Delta H^{\ominus} = -192 \text{ kJ mol}^{-1}$$

The reaction between sulfur trioxide and water then produces sulfuric acid:

$$SO_3(g) + H_2O(l) \rightarrow H_2SO_4(l)$$

The sulfur trioxide produced cannot be added directly to water, because the heat evolved is too great. Instead it is absorbed by a concentrated solution of sulfuric acid to which small quantities of water are then added to keep the acid concentrated. The sulfuric acid produced has many uses, ranging from a chemical feedstock to paints and pigments, fertilizers, detergents and soaps, and as the electrolyte in car batteries.

Figure 7 Construction in progress of a new plant to manufacture sulfuric acid at Richard's Bay, South Africa. The construction was completed in 2002 and the plant has the capacity to produce 3,000 tons of sulfuric acid per day.

1 Explain why the best conditions for drying clothes outside are when (a) it is hot, (b) it is windy, and (c) there is low humidity.
2 Explain why it is difficult to boil an egg on the summit of Chomolungma (Mount Everest). Would it be difficult to fry an egg there?

3 When hydrochloric acid is added to an aqueous solution of cobalt(II) chloride the following equilibrium is established:

$$[Co(H_2O)_6]^{2+} (aq) + 4Cl^- (aq) \rightleftharpoons CoCl_4^{2-} (aq) + 6H_2O(l)$$
pink   blue

(a) When the solution is heated, the blue colour increases. From this information, what can be deduced about the forward reaction?

(b) If more concentrated hydrochloric acid is added to the equilibrium mixture, what will be observed?

(c) If silver nitrate solution is added, a precipitate of silver chloride will be formed. What else will be observed?

4 If you answered "no" to the second part of question 2, think again—this time as a mountaineer, not a chemist!

Thinking about science

Fritz Haber

Fritz Haber (1868–1934) (Figure 8) was a German chemist who won the Nobel Prize in Chemistry in 1918 for his synthesis of ammonia from nitrogen and hydrogen.

Haber was a complex and highly intelligent man, whose life in many ways provides a good example of the political and social responsibilities facing eminent scientists. His discovery of a method to manufacture ammonia from its elements meant that food production could be increased through the use of artificial fertilizers, but his discovery also undoubtedly prolonged the First World War.

Ammonia can be oxidized to nitric acid, from which explosives (containing nitrates and nitro compounds) are made. Prior to Haber's discovery in 1911, nitrates came principally from natural deposits in South America. This meant that the country with the most powerful navy could control the supply of nitrates and hence the production of explosives. Haber's discovery enabled Germany to become independent of external sources of nitrates.

During the First World War Haber served his country by using his chemical knowledge to develop gas masks, and also to develop the use of chlorine as the first war gas. Haber himself personally supervised the release of chlorine in the trenches (Figure 9). Wilfred Owen's poem "Dulce et Decorum Est", which is reproduced in Chapter 3, eloquently refers to the "green sea" of chlorine faced by the soldiers at the time. Haber's wife, Clara, who was also a chemist, vehemently opposed his developing war gases because she saw them as morally wrong. She took her own life in protest by shooting herself at a dinner held in tribute to Haber for the successful use of the gas in 1917 at the Second Battle of Ypres in Belgium.

Figure 8 Fritz Haber in 1918.

Figure 9 Dispersion of chlorine gas in the First World War trenches.

From this we can see that the volume, V, *does* appear in the rate expression. If the volume is increased, which is equivalent to the pressure being decreased, the value of x must increase to ensure that K_c remains constant. This again confirms the prediction using Le Châtelier's principle that a decrease in pressure will favour the products.

An overall decrease in volume

The classic example used to illustrate this is the Haber process. We already know that an increase in pressure favours an increase in the amount of ammonia produced. This is the same effect as decreasing the volume, so now we would expect the volume to appear in the upper part of the equilibrium expression.

To simplify the expression we shall start with 3 mol of hydrogen and 1 mol of nitrogen, and assume that, once the position of equilibrium has been reached, x mol of nitrogen have reacted. The equilibrium volume is V dm^3.

	$3H_2(g)$	$+$	$N_2(g)$	$\rightleftharpoons$	$2NH_3(g)$
Initial amount/ mol	3		1		0
Equilibrium amount/ mol	$3 - 3x$		$1 - x$		$2x$
Equilibrium concentration/ mol dm^{-3}	$\dfrac{3(1-x)}{V}$		$\dfrac{1-x}{V}$		$\dfrac{2x}{V}$

$$K_c = \frac{[NH_3(g)]^2}{[H_2(g)]^3 \times [N_2(g)]}$$

$$= \frac{(2x/V)^2}{[3(1-x)/V]^3 \times [(1-x)/V]}$$

$$= \frac{4x^2\,V^2}{27(1-x)^4}$$

As can be seen the volume V (actually V^2) is on the top, and if the volume decreases (pressure increases) then x must increase in order for the value of the equilibrium constant K_c to remain the same.

1 Some pure phosphorus pentachloride, $PCl_5(g)$, was added to a one litre (1.00 dm^3) evacuated flask at 500 K. After the position of equilibrium had been reached, the concentration of the phosphorus pentachloride was 4×10^{-2} mol dm^{-3}. Given that the numerical value of the equilibrium constant for the dissociation of phosphorus pentachloride into chlorine and phosphorus trichloride is 1.00×10^{-2} at 500 K, calculate the concentration of chlorine in the equilibrium mixture.

2 (a) The concentrations of hydrogen iodide, hydrogen and iodine in an equilibrium mixture were found to be 1.0, 0.20 and 0.10 mol dm^{-3} respectively. Calculate the value of the equilibrium constant for the reaction $H_2(g) + I_2(g) \rightleftharpoons 2HI(g)$ at this temperature.

(b) Some iodine was introduced into the above system. After the equilibrium had been re-established it was found that the new concentrations of the hydrogen iodide and the hydrogen were 1.1 and 0.15 mol dm^{-3} respectively. Assuming that the measurements were all taken at the same temperature, what was the new concentration of the iodine?

Critical considerations

Do equilibrium constants have units?

If we consider a reaction where the total amounts (number of moles) of reactants and products in the stoichiometric equation are the same, then it is clear that the equilibrium constant will have no units. In other words, it is dimensionless. An example is the decomposition of gaseous hydrogen iodide into gaseous hydrogen and iodine:

$$2HI(g) \rightleftharpoons H_2(g) + I_2(g)$$

$$K_c = \frac{[H_2(g)] \times [I_2(g)]}{[HI(g)]^2}$$

Critical considerations (continued)

so that the units will be

$$\frac{\text{concentration}^2}{\text{concentration}^2}$$

However, if we now apply the same approach to the Haber process, where there are fewer moles of products than of reactants, the expression for the equilibrium constant is

$$K_c = \frac{[NH_3(g)]^2}{[H_2(g)]^3 \times [N_2(g)]}$$

When we simplify this it gives concentration^{-2}, and we would therefore expect the equilibrium constant to have the units mol^{-2} dm^6. Using the same approach you should be able to deduce that the units for the conversion of sulfur dioxide and oxygen to sulfur trioxide will be mol^{-1} dm^3.

In fact, all equilibrium constants are dimensionless and have no units!

The reason for this is that one of the assumptions we have made in arriving at the equilibrium expression is wrong. Sometimes it is necessary to simplify chemistry at this level. The square brackets do not truly represent the concentration of the species inside the brackets. They actually involve a quantity known as the **activity** of the species inside the bracket. Activity is a concept that essentially is beyond this level. In simple terms it is a measure of how different particles in a non-ideal gas or solution interact with each other. For dilute solutions the activity is very close to the molar concentration, so it is reasonable to assume that the brackets can represent concentration. The problem is that, unlike concentration, activity has no units. Unfortunately, simplifying the chemistry then leads to wrong chemistry. If we take the square brackets to mean concentration then logically some equilibrium constants will have to have units.

Until the current programme was introduced, some examination questions in the IB would ask for the units of a calculated equilibrium constant, and you would be penalized if you omitted them. Many books still include these units. Indeed, even in the next chapter in this book units are still given (wrongly?) for the equilibrium constant for the dissociation constant for water, K_w. You might like to consider whether wrong conclusions such as these, which have been reached as a result of necessary simplifications, are justified.

1 Logarithms to the base 10 can only be taken of numbers that have no units. Is it correct to state that pH $= -\log_{10}[H^+(aq)]$, or is it better at this level to state that pH $= -\log_{10}[H^+(aq)/\text{ mol dm}^{-3}]$, as some school textbooks do?

2 At standard level the electron arrangement of a sodium atom is 2.8.1. At higher level the electron arrangement (often called electron configuration) of a sodium atom is $1s^2 2s^2 2p^6 3s^1$. Which is right?

point, because we shall be considering mainly acids and bases in aqueous solution. When dealing with buffer solutions (solutions that help to maintain a constant hydrogen ion concentration) it can be helpful to use Brønsted–Lowry theory, and when dealing with the mechanisms of organic reactions Lewis theory is often used.

So what exactly is an acid? Whether a substance is an acid or not clearly depends to a large extent on how an acid is defined. Water to a non-chemist is not an acid, and yet under the Brønsted–Lowry definition water can show acidic properties, and under the Lewis definition it can donate a pair of electrons and is classified as a base. Most non-chemists would not even consider that water could be acidic, because the popular definition of an acid is that it is a corrosive substance. Chemically speaking, some acids are corrosive but by no means all. In everyday language "corrosive", "strong" and "concentrated", when applied to acids, all tend to imply a very reactive substance. Chemists distinguish very clearly between these three words:

- *Strong*: completely dissociated into ions in aqueous solution. The opposite of "strong" is "weak".
- *Concentrated*: a measure of the quantity of acid or hydrogen ions in a given volume of solution. Concentration is measured in moles per cubic decimeter, mol dm^{-3}. The opposite of "concentrated" is "dilute".
- *Corrosive*: very reactive, chemically. The opposite of "corrosive" is "unreactive" or "inert".

This can be illustrated by the following statements:

- A 0.01 mol dm^{-3} aqueous solution of hydrochloric acid is not very concentrated, and yet it is a solution of a strong acid that can be quite corrosive.
- A 0.01 mol dm^{-3} aqueous solution of hydrofluoric acid, HF(aq), is a dilute solution of a weak acid, and yet it is still corrosive as it can etch glass.
- A 10.0 mol dm^{-3} solution of ethanoic acid, CH_3COOH(aq), is a concentrated solution of a weak acid. It too is quite corrosive.

There are many words that have different and very precise meanings in chemistry compared with their meanings in everyday English. Examples include: *degenerate, spontaneous, reduction, element, oil, orbital, phase and volatile*. When chemists communicate with each other they need to use the precise word to convey their meaning clearly, but perhaps of even more importance is that when chemists and non-chemists communicate they need to ensure that they are using the same language.

1 Build on the list above by finding other words that have different meanings in chemistry from those in everyday English. Define both meanings for each word.
2 Consider the following reaction:
H_2S(g) + H_2O(l) $\rightleftharpoons$ HS$^-$(aq) + H_3O^+(aq)
   (a) Which theories of acids and bases would you choose to best describe this reaction?
   (b) Under your definition, which species is/are acting as acids?
   (c) Can water act as an acid or a base or both under the following theories:
       (i) Arrhenius; (ii) Brønsted–Lowry; (iii) Lewis?

Properties of acids and bases

Acids in dilute aqueous solution show several characteristic reactions. These include the following.

- *Neutralization reactions with bases.* Acids react with bases such as metal hydroxides to form a salt and water. For example:

$$HNO_3(aq) + NaOH(aq) \rightarrow NaNO_3(aq) + H_2O(l)$$

If, like nitric acid and sodium hydroxide, both the acid and the base are strong—that is, they are both completely dissociated in water—then the equation can be written

$$\mathbf{H^+(aq)} + NO_3^{-}(aq) + Na^+(aq) + \mathbf{OH^-(aq)}$$
$$\rightarrow Na^+(aq) + NO_3^{-}(aq) + \mathbf{H_2O(l)}$$

The nitrate ions and the sodium ions are "spectator" ions, and so the only reaction that is actually taking place is the neutralization of hydrogen ions by hydroxide ions to form water molecules:

$$H^+(aq) + OH^-(aq) \rightarrow H_2O(l) \quad \Delta H^{\ominus} = -57.3 \text{ kJ mol}^{-1}$$

The enthalpy change for the reaction, -57.3 kJ mol^{-1}, will therefore always be the same whenever a strong acid is neutralized by a strong base in aqueous solution.

A base can be defined as a substance that neutralizes an acid to form a salt and water, so other acid–base reactions include the reactions of acids with most metal oxides. For example:

$$CuO(s) + H_2SO_4(aq) \rightarrow CuSO_4(aq) + H_2O(l)$$

- *Reactions of acids with carbonates and hydrogen carbonates.* All carbonates and hydrogen carbonates will neutralize acids to produce carbon dioxide and water. For example:

$$CaCO_3(s) + 2HCl(aq) \rightarrow CaCl_2(aq) + H_2O(l) + CO_2(g)$$

$$2NaHCO_3(aq) + H_2SO_4(aq) \rightarrow Na_2SO_4(aq) + 2H_2O(l) + 2CO_2(g)$$

The carbonate ion (the conjugate base of the hydrogen carbonate ion, HCO_3^-) and the hydrogen carbonate ion (the conjugate base of carbonic acid, H_2CO_3) both behave as Brønsted–Lowry bases and accept protons from the acid:

$$CO_3^{2-}(s) + 2H^+(aq) \rightarrow H_2CO_3(aq) \rightarrow CO_2(g) + H_2O(l)$$

$$HCO_3^{-}(s) + H^+(aq) \rightarrow H_2CO_3(aq) \rightarrow CO_2(g) + H_2O(l)$$

- *Reaction with metals above hydrogen in the reactivity series.* Metals above hydrogen in the reactivity series react with acids to produce hydrogen gas and the metal salt. For example:

$$Mg(s) + 2HCl(aq) \rightarrow MgCl_2 + H_2(g)$$
$$Fe(s) + H_2SO_4(aq) \rightarrow FeSO_4(aq) + H_2(g)$$

or, ionically:

$$Mg(s) + 2H^+(aq) \rightarrow Mg^{2+}(aq) + H_2(g)$$
$$Fe(s) + 2H^+(aq) \rightarrow Fe^{2+}(aq) + H_2(g)$$

- *Reactions with indicators.* Indicators are substances that change colour when the concentration of hydrogen ions changes. Most

145

just indicate whether the solution is acidic or basic (see examples in Table 2).

Universal indicator is a mixture of different indicators, and produces a range of colours according to the hydrogen ion concentration.

Characteristic reactions of bases

The simplest definition of a base is that it is a substance that can neutralize an acid. A base that is soluble in water is known as an **alkali**. Typical reactions of bases in solution include the following.

- *Neutralization of acids.* For example:

$$2NaOH(aq) + H_2SO_4(aq) \rightarrow Na_2SO_4(aq) + 2H_2O(l)$$

$$MgO(s) + 2HCl(aq) \rightarrow MgCl_2(aq) + H_2O(l)$$

- *Displacement of ammonia from ammonium salts.* For example:

$$NH_4Cl(s) + NaOH(aq) \rightarrow NaCl(s) + H_2O(l) + NH_3(aq)/(g)$$

Ammonia is very soluble in water, so it may be formed in solution, or, if small amounts of sodium hydroxide are added to solid ammonium chloride, it can be collected as a gas. In this reaction the chloride ions and sodium ions are spectator ions, so the actual reaction taking place in solution is

$$NH_4^+(aq) + OH^-(aq) \rightarrow NH_3(g) + H_2O(l)$$

The equilibrium lies very much on the right, so a single arrow has been shown, but technically the reaction should be written

$$NH_4^+(aq) + OH^-(aq) \rightleftharpoons NH_3(g) + H_2O(l)$$

Under the Brønsted–Lowry definition, ammonium ions (NH_4^+) and water molecules (H_2O) are acting as acids, and hydroxide ions (OH^-) and ammonia molecules (NH_3) are acting as bases. As water is a weak acid, its conjugate base is strong, so the position of equilibrium lies very much to the right.

- *With indicators.* If a base is insoluble in water, for example copper(II) oxide (CuO), then there will be no reaction with indicators. Indicators actually measure the concentration of hydrogen ions in aqueous solution, $H^+(aq)$, so only soluble bases (alkalis) that produce hydroxide ions in water will change the colour of indicators.

$$H^+(aq) + OH^-(aq) \rightleftharpoons H_2O(l)$$

Increasing the concentration of hydroxide ions will lower the concentration of hydrogen ions.

Strong and weak acids and bases

Strong acids and bases are completely dissociated into their ions in aqueous solution. Examples include:

Strong acids	Strong bases
hydrochloric acid, HCl(aq)	Group 1 hydroxides, e.g. NaOH(aq) and KOH(aq)
nitric acid, HNO_3(aq)	barium hydroxide, $Ba(OH)_2$(aq)
sulfuric acid, H_2SO_4(aq)	

Table 2 Indicators

Indicator	Colour in acidic solution	Colour in basic solution
litmus	red	blue
phenolphthalein	colourless	pink/red
methyl orange	red	yellow

1 Antacids are used to combat indigestion by neutralizing excess acid in the stomach. Write chemical equations for the reactions of the following antacids with hydrochloric acid: aluminium hydroxide, $Al(OH)_3$; magnesium oxide, MgO; magnesium hydroxide, $Mg(OH)_2$; sodium hydrogen carbonate, $NaHCO_3$; and calcium carbonate, $CaCO_3$.

2 Ammonia, NH_3, is extremely soluble in water. Amines are compounds in which one of the hydrogen atoms in ammonia has been replaced by an organic group. An example is phenylamine, $C_6H_5NH_2$. The phenyl group, C_6H_5-, is non-polar, so phenylamine is almost insoluble in water. However, in the presence of hydrochloric acid phenylamine readily dissolves in aqueous solution. Suggest a reason why.

Hydrochloric and nitric acid are known as **monoprotic acids**, because 1 mol of acid dissociates to give 1 mol of hydrogen ions in solution. Sulfuric acid is a **diprotic acid**, because it dissociates to give 2 mol of hydrogen ions in solution.

Weak acids and bases are only slightly dissociated into their ions in aqueous solution. Examples include:

Weak acids	Weak bases
carboxylic acids, e.g. ethanoic acid, $CH_3COOH(aq)$	ammonia, $NH_3(aq)$
carbonic acid, (CO_2 in water), $H_2CO_3(aq)$	amines, e.g. ethylamine, $C_2H_5NH_2(aq)$

It is relatively easy to distinguish between equimolar solutions of strong and weak acids, because the amount of ions present in solution is significantly different. A strong acid will be a much better conductor of electricity, and this can be measured using a conductivity meter. Similarly a solution of a strong base will have a higher conductivity measurement than a solution of a weak base with the same molarity.

For equimolar solutions of acids the hydrogen ion concentration will be much greater for a strong acid than for a weak acid, and this can be measured using a pH meter. It is not, however, possible to distinguish between them easily by a simple acid–base titration. 25.0 cm³ of 0.100 mol dm⁻³ ethanoic acid will react with exactly the same volume of 0.100 mol dm⁻³ alkali as 25.0 cm³ of 0.100 mol dm⁻³ hydrochloric acid. Although there are fewer hydrogen ions in the ethanoic acid solution, once they react with hydroxide ions then, according to Le Châtelier's principle, more of the acid dissociates to replace them until eventually all of the acid has been neutralized.

There *is* a difference in the titrations, but it involves temperature rather than volume. Strong acids and bases in aqueous solution always release the same amount of heat per mole of hydrogen ions neutralized, because the only reaction taking place is the neutralization of hydrogen ions by hydroxide ions. Titrations involving weak acids or bases will release a different amount of heat, because heat energy is required to dissociate the molecules to form ions, and heat energy is released when the these ions become hydrated, in addition to the heat evolved during the neutralization process.

The pH scale

Pure water is only very slightly dissociated into its ions:

$$H_2O(l) \rightleftharpoons H^+(aq) + OH^-(aq)$$

At 25 °C the concentration of hydrogen ions is 1.0×10^{-7} mol dm⁻³. pH is defined as being equal to minus the logarithm to base 10 of the hydrogen ion concentration:

$$pH = -\log_{10}[H^+(aq)]$$

Hence the pH of pure water $= -\log_{10}(1.0 \times 10^{-7}) = 7$. Pure water is a neutral solution, and the pH of any neutral solution at 25 °C is 7.

If the solution is acidic the hydrogen ion concentration will be greater than 1.0×10^{-7} mol dm^{-3}, and the pH will be lower than 7. Similarly, if the hydroxide concentration is greater than 1.0×10^{-7} mol dm^{-3} the hydrogen ion concentration will be less than 1.0×10^{-7} mol dm^{-3}, and the pH will be greater than 7.

The pH scale runs from 0 to 14. Because the scale is logarithmic to base 10, a change in one unit of pH corresponds to a tenfold change in hydrogen ion concentration. Thus a 0.1 mol dm^{-3} aqueous solution of a strong monoprotic acid has a pH of 1, whereas a 0.01 mol dm^{-3} aqueous solution of the same acid will have a pH of 2. Similarly a 0.1 mol dm^{-3} aqueous solution of a strong base such as sodium hydroxide, NaOH(aq), will have a pH of 13, whereas a 0.01 mol dm^{-3} aqueous solution of the same base will have a pH of 12.

1 The pH of many carbonated drinks such as cola is approximately four. What is the hydrogen ion concentration of cola?
2 A solution of a strong acid has a pH of 2. If 10.0 cm^3 of this solution is placed in a flask, and 990 cm^3 of pure water is added with stirring, what will be the pH of the resulting solution?

The term "pH" was invented by the Danish chemist Søren Sørenson (1868–1939) while he was working as head of the laboratory at the Carlsberg brewery in Copenhagen. He realized that it was much easier to describe weakly acidic solutions in terms of pH rather than using tedious expressions such as "the concentration of hydrogen ions equals one-hundred-thousandth of a mole in one cubic decimetre of solution" when describing a solution with a pH of 5.

Figure 3 IB (who was born in 1985), one of the dray horses who pulls the lager carts at the Carlsberg brewery in Copenhagen.

The pH of a solution can be measured accurately using a pH meter (Figure 4) or, less accurately, by using universal indicator (Table 3).

Table 3 Use of universal indicator

pH	[H$^+$(aq)] / mol dm^{-3}	[OH$^-$(aq)] / mol dm^{-3}	Description	Colour of universal indicator
0	1	1×10^{-14}	very acidic	red
4	1×10^{-4}	1×10^{-10}	acidic	orange
7	1×10^{-7}	1×10^{-7}	neutral	green
10	1×10^{-10}	1×10^{-4}	basic	blue
14	1×10^{-14}	1	very basic	purple

Acid rain

Rain is naturally acidic, because it dissolves and reacts with carbon dioxide as it falls through the air. However, carbon dioxide forms only a weak acid in water, and even if water is completely saturated with carbon dioxide the pH will only be 5.6. Acid rain is therefore defined as precipitation (rain, snow, hail etc.) that has a pH lower than 5.6.

There are two main reasons why acid rain is formed. The first is the release of sulfur dioxide into the atmosphere from the combustion

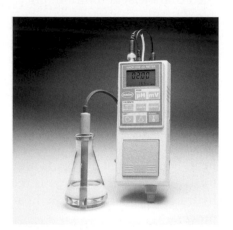

Figure 4 pH meter. Although collectively known as a "pH meter" a pH meter actually consists of a pH probe connected to a meter. The meter is a high resistance voltmeter that measures the potential difference between a reference electrode and a glass electrode. This potential difference changes as the hydrogen ion concentration changes.

of fossil fuels such as coal, which contain sulfur (Figure 5), or the smelting of sulfide ores.

$$S(s) + O_2(g) \rightarrow SO_2(g)$$

The sulfur dioxide is slowly oxidized in the atmosphere to form sulfur trioxide, which dissolves in the rain to form sulfuric acid.

$$2SO_2(g) + O_2(g) \rightarrow 2SO_3(g)$$
$$SO_3(g) + H_2O(l) \rightarrow H_2SO_4(aq)$$

The second major source of acid rain is the release of oxides of nitrogen, NO_x, into the atmosphere. Oxides of nitrogen are formed in jet engines and in internal combustion engines where the temperature reached is high enough for nitrogen to combine with the oxygen in the air to form nitrogen monoxide:

$$N_2(g) + O_2(g) \rightarrow 2NO(g)$$

Oxidation in the atmosphere then occurs to produce nitrogen dioxide, which then can dissolve and react with water to form nitric acid and nitrous acid:

$$2NO_2(g) + H_2O(l)_2 \rightarrow HNO_3(aq) + HNO_2(aq)$$

or be oxidized directly to nitric acid by oxygen in the presence of water:

$$4NO_2(g) + O_2(g) + 2H_2O(l) \rightarrow 4HNO_3(aq)$$

Acid rain is a worldwide problem, because it often precipitates many miles away from its source as a result of air currents. It causes damage to buildings by reacting with carbonates in building material (Figure 6); it affects vegetation by leaching out important minerals from the soil; it affects aquatic life by altering the pH of rivers and lakes; and it directly affects human health by increasing the risk of respiratory diseases.

Figure 5 A coal-burning power station in Australia. In addition to contributing to global warming by emitting carbon dioxide, coal-fired power stations also emit sulfur dioxide.

Figure 6 The caryatids—statues of women on the Erechtheum in Athens, Greece—which have been severely affected by acid rain.

Critical considerations

Read the following article on acid rain in China, and then answer the questions.

Acid rain polluting one-third of China
By Geoffrey Lean
Environment Editor

Acid rain caused by sulphur dioxide spewed from factories and power plants affected a third of China's vast land mass last year, posing a threat to food safety, according to a Chinese parliamentary report.

More than half the 696 cities and counties monitored had suffered acid rain, in some cases on a daily basis, the official Xinhua news agency said.

"Increased sulphur dioxide emissions meant that one third of China's territory was affected by acid rain, posing a major threat to soil and food safety," Xinhua cited standing committee vice chairman Sheng Huaren as saying.

Discharge of sulphur dioxide in booming China rose by 27 per cent between 2000 and 2005 to 25 million tonnes, making the country the world's top emitter of the pollutant.

> Sheng told lawmakers the sulphur dioxide emissions were double the acceptable limit.
>
> China has pledged to install desulphurisation facilities in coal-burning power plants and is planning pilot emissions trading schemes to help improve air quality. The capital, Beijing, has promised to replace its notorious smog with clear skies in time for the 2008 Olympics.
>
> Article from: *The Independent on Sunday*, 27 August 2006
>
> 1 Why do fossils fuels such as coal contain sulfur?
> 2 How does a "desulfurization facility" in a coal-burning power plant work?
> 3 What can a city such as Beijing do to reduce the amount of nitrogen oxide production from motorized vehicles?

HL

Calculations involving acids and bases

The ionic product of water

We have seen that pure water is only very slightly dissociated into its ions:

$$H_2O(l) \rightleftharpoons H^+(aq) + OH^-(aq)$$

The concentration of water can be regarded as constant, so the equilibrium expression is

$$K_w = [H^+(aq)] \times [OH^-(aq)] = 1.00 \times 10^{-14} \text{ mol}^2 \text{ dm}^{-6} \text{ at 298 K}$$

K_w is known as the **ionic product** of water. It has the value of 1.00×10^{-14} mol^2 dm^{-6} only at 298 K. As the dissociation of water is endothermic, increasing the temperature will cause the position of equilibrium to shift towards the products so that more ions are formed, and at higher temperatures the value of K_w increases (Figure 7).

If the value of K_w is known then by taking its square root the concentrations of the ions in a solution of pure water can be determined, as $[H^+(aq)] = [OH^-(aq)]$.

At 80 °C (378 K) the value of K_w is approximately 2.0×10^{-13} mol^2 dm^{-6}. Hence $[H^+(aq)] = (2.0 \times 10^{-13})^{1/2} = 4.47 \times 10^{-7}$ mol dm^{-3}.

pH and pOH calculations

Because strong acids are completely dissociated in aqueous solution, the concentration of hydrogen ions for an aqueous solution of a strong monoprotic acid will be the same as the concentration of the undissociated acid. For example:

$$HCl(aq) \quad \rightarrow \quad H^+(aq) \quad + \quad Cl^-(aq)$$
$$(0.10 \text{ mol dm}^{-3}) \quad (0.10 \text{ mol dm}^{-3}) \quad (0.10 \text{ mol dm}^{-3})$$

$$pH = -\log_{10}[H^+(aq)]$$

Therefore the pH of 0.10 mol dm^{-3} hydrochloric acid is equal to $-\log_{10}0.10 = 1.0$.

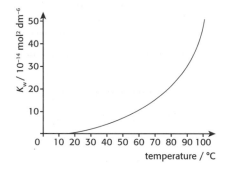

Figure 7 Variation of K_w with temperature.

It is sometimes stated that an aqueous solution is acidic if the hydrogen ion concentration is greater than 1.0×10^{-7} mol dm^{-3}. Do you think that pure water is therefore acidic at 80 °C?

For a strong diprotic acid the hydrogen ion concentration will be twice the concentration of the acid:

$$H_2SO_4(aq) \rightarrow 2H^+(aq) + SO_4^{2-}(aq)$$
$$(0.10 \text{ mol dm}^{-3}) \quad (0.20 \text{ mol dm}^{-3}) \quad (0.10 \text{ mol dm}^{-3})$$

Hence the pH of 0.10 mol dm^{-3} sulfuric acid is equal to $-\log_{10}0.20$ = 0.7.

(To simplify calculations the IB Diploma Programme chemistry syllabus assumes that both of the acidic hydrogen atoms in sulfuric acid are completely dissociated into ions. This is not strictly true, as hydrogen sulfate ions, HSO_4^-, are only partially dissociated in aqueous solution.)

For a strong base such as an aqueous solution of sodium hydroxide the hydroxide ion concentration will be the same as the concentration of the sodium hydroxide:

$$NaOH(aq) \rightarrow Na^+(aq) + OH^-(aq)$$
$$(0.10 \text{ mol dm}^{-3}) \quad (0.10 \text{ mol dm}^{-3}) \quad (0.10 \text{ mol dm}^{-3})$$

We know that $[H^+(aq)] \times [OH^-(aq)] = K_w = 1.0 \times 10^{-14} \text{ mol}^2 \text{ dm}^{-6}$ at 298 K. If $[OH^-(aq)] = 0.10 \text{ mol dm}^{-3}$ then $[H^+(aq)] = 1.0 \times 10^{-13} \text{ mol dm}^{-3}$. Hence the pH of a solution of 0.10 mol dm^{-3} sodium hydroxide will have a pH equal to $-\log_{10}(1.0 \times 10^{-13}) = 13$.

We can arrive directly at this answer by using the relationship between pOH, pH and pK_w:

$$pOH = -\log_{10}[OH^-(aq)]$$

and

$$pK_w = -\log_{10}K_w$$

Since

$$[H^+(aq)][OH^-(aq)] = K_w = 1.0 \times 10^{-14} \text{ at 298 K}$$

Then

$$pH + pOH = pK_w = 14$$

For 0.10 mol dm^{-3} sodium hydroxide solution the hydroxide concentration is 0.10 mol dm^{-3}, so that pOH = $-\log_{10}0.1 = 1$. The pH will therefore be equal to $14 - 1 = 13$.

Acid and base dissociation constants

A weak acid represented by HA will be in equilibrium with its ions in aqueous solution:

$$HA(aq) \rightleftharpoons H^+(aq) + A^-(aq)$$

The equilibrium expression will be given by

$$K_a = \frac{[H^+(aq)] \times [A^-(aq)]}{[HA(aq)]}$$

where K_a is known as the **acid dissociation constant** at a specified temperature.

We can use this expression to find the pH of a solution of a weak acid. For example, calculate the pH of 0.10 mol dm^{-3} ethanoic acid ($CH_3COOH(aq)$) given that the acid dissociation constant K_a for $CH_3COOH = 1.8 \times 10^{-5}$ mol dm^{-3} at 298 K.

1 What will be the pH of:
   (a) 10.0 cm^3 of an aqueous solution of 0.001 mol dm^{-3} nitric acid, $HNO_3(aq)$?
   (b) 100 cm^3 of an aqueous solution of 0.001 mol dm^{-3} nitric acid, $HNO_3(aq)$?
   (c) 0.02 mol dm^{-3} of potassium hydroxide solution, KOH(aq)?

2 A solution is made by dissolving 1.71 g of barium hydroxide, $Ba(OH)_2(s)$, in distilled water and then adding more distilled water until the total volume of the solution is one litre (1.00 dm^3). Calculate:
   (a) the molarity of the barium hydroxide solution
   (b) the hydroxide concentration in mol dm^{-3}
   (c) the pOH of the solution
   (d) the pH of the solution.

Assume that at equilibrium the concentration of hydrogen ions will be x mol dm^{-3}.

	CH$_3$COOH(aq) $\rightleftharpoons$ H$^+$(aq) + CH$_3$COO$^-$(aq)		
Initial concentration/ mol dm^{-3}	0.10	0	0
Equilibrium concentration/ mol dm^{-3}	0.10$-x$	x	x

$$K_a = \frac{[H^+(aq)] \times [CH_3COO^-(aq)]}{[CH_3COOH(aq)]}$$

$$= \frac{x^2}{0.10 - x}$$

$$= 1.8 \times 10^{-5} \text{ mol dm}^{-3}$$

This leads to the quadratic equation

$$x^2 + (1.8 \times 10^{-5})x - (1.8 \times 10^{-6}) = 0$$

By solving the quadratic equation, $x = 1.33 \times 10^{-3}$ mol dm^{-3}.

$$\text{pH} = -\log_{10}(1.33 \times 10^{-3}) = 2.88$$

It can be seen that, as the acid is only slightly dissociated, x is a very small value. In the example above the equilibrium concentration of the acid is $(0.10 - 0.00133)$ mol dm^{-3}, which is close to 0.10 mol dm^{-3}. When solving these types of question it is reasonable to assume that, for a weak acid, the concentration of the acid at equilibrium is the same as the initial concentration. If this assumption is made, then no quadratic expression is needed.

$$K_a = 1.8 \times 10^{-5} \text{ mol dm}^{-3}$$

$$= \frac{[H^+(aq)] \times [CH_3COO^-(aq)]}{[CH_3COOH(aq)]}$$

$$= \frac{x^2}{0.10 - x} \approx \frac{x^2}{0.10}$$

$x \approx (1.8 \times 10^{-6})^{1/2} \approx 1.34 \times 10^{-3}$ mol dm^{-3} and the pH $\approx$ 2.87.

This value of 2.87 is extremely close to the true value when no approximation was made.

For weak bases the base dissociation constant, K_b, is used. For example, ammonia reacts slightly with water to form ammonium ions and hydroxide ions:

$$NH_3(aq) + H_2O(l) \rightleftharpoons NH_4^+(aq) + OH^-(aq)$$

$$K_b = \frac{[NH_4^+(aq)] \times [OH^-(aq)]}{[NH_3(aq)]} = 1.8 \times 10^{-5} \text{ mol dm}^{-3} \text{ at 298 K}$$

Thus for a 0.10 mol dm^{-3} aqueous solution of ammonia we can calculate the pH. As ammonia is a weak base we can assume that the equilibrium concentration of ammonia will be the same as its initial concentration. Therefore:

$$K_b = 1.8 \times 10^{-5} = \frac{[NH_4^+(aq)] \times [OH^-(aq)]}{[NH_3(aq)]} \approx \frac{[OH^-(aq)]^2}{0.10}$$

Hence

$$[OH^-(aq)] = (1.8 \times 10^{-6})^{1/2} = 1.34 \times 10^{-3} \text{ mol dm}^{-3}$$

Therefore pOH = 2.87 and pH = (14.00 − 2.87) = 11.13.

Ammonia is the conjugate base of the ammonium ion. If we write the reaction of ammonium ions dissociating to give ammonia and hydrogen ions in aqueous solution, then we can write the expression for K_a:

$$K_a = \frac{[NH_3(aq)] \times [H^+(aq)]}{[NH_4^+(aq)]}$$

Then

$$K_a \times K_b = \frac{[NH_3(aq)] \times [H^+(aq)]}{[NH_4^+(aq)]} \times \frac{[NH_4^+(aq)] \times [OH^-(aq)]}{[NH_3(aq)]}$$
$$= [H^+(aq)] \times [OH^-(aq)]$$
$$= K_w$$

If we take logarithms to base 10 and rearrange, then

$$-\log_{10} K_a - \log_{10} K_b = -\log_{10} K_w$$

This is best expressed as

$$pK_a + pK_b = 14$$

We can use this expression to calculate the pH of any base if we are given its pK_a value.

For example, calculate the pH of a 1.00×10^{-3} mol dm^{-3} aqueous solution of ethylamine, $C_2H_5NH_2$, given that the pK_a value for ethylamine is 10.73 at 298 K.

$$C_2H_5NH_2(aq) + H_2O(l) \rightleftharpoons C_2H_5NH_3^+(aq) + OH^-(aq)$$
$$K_b \approx \frac{[OH^-(aq)]^2}{1.00 \times 10^{-3}}$$
$$\log_{10} K_b \approx 2 \times \log_{10}[OH^-(aq)] - \log_{10}(1.00 \times 10^{-3})$$

or

$$-\log_{10} K_b \approx -2 \times \log_{10}[OH^-(aq)] + \log_{10}(1.00 \times 10^{-3})$$

which becomes $pK_b \approx 2pOH - 3.00$

As the pK_a for ethylamine is 10.73, the pK_b for ethylamine = 14.00 − 10.73 = 3.27.

Therefore the pOH = (3.27 + 3.00)/2 = 3.14, and the pH = 14.00 − 3.14 = 11.90.

❓

1 The pK_a of propanoic acid at 298 K is 4.87.
   (a) Write the equation for the dissociation of propanoic acid, C_2H_5COOH, in water.
   (b) What is the value of K_a for propanoic acid at 298 K?
   (c) Write the equilibrium expression for the dissociation of propanoic acid in water.
   (d) Calculate the pH of a 0.05 mol dm^{-3} aqueous solution of propanoic acid, stating any assumptions you make in arriving at your answer.

2 The pK_b values for amines are given in the IB Data Booklet. What will be the pH of a 3.00×10^{-3} mol dm^{-3} aqueous solution of methylamine? If it is compared against 3.00×10^{-3} mol dm^{-3} aqueous solutions of ammonia and ethylamine, which of the three solutions would have the lowest pH?

> **Critical considerations**
>
> **The pH of 2.0 mol dm^{-3} hydrochloric acid**
> The typical concentration for a laboratory bottle of "dilute" hydrochloric acid is in the region of 2.0 mol dm^{-3}. If we apply the definition for pH, then the pH of 2.0 mol dm^{-3} hydrochloric acid equals $-\log_{10} 2.0$, which has the value −0.30. However, it has previously been stated that the pH scale runs from 0 to 14, so one of these statements must be untrue.
>
> The problem lies in the assumptions that have been made. One assumption is that the hydrogen ion concentration of 2.0 mol dm^{-3} hydrochloric acid is also 2.0 mol dm^{-3}. More concentrated solutions of strong acids are not in fact completely dissociated, and solutions of all

weak electrolytes become more dissociated (that is, stronger) in aqueous solution as the solution becomes more dilute. This is known as **Ostwald's dilution law**. Consider a weak acid HA that dissociates to form hydrogen ions and A^- ions, and that for 1 mol of acid in a total volume of V dm^3 we shall assume that x mol of acid have dissociated.

$$HA(aq) \rightleftharpoons H^+(aq) + A^-(aq)$$

	HA(aq)	H$^+$(aq)	A$^-$(aq)
Initial amount/ mol	1	0	0
Equilibrium amount/ mol	$1 - x$	x	x
Equilibrium concentration/ mol dm^{-3}	$\dfrac{1 - x}{V}$	$\dfrac{x}{V}$	$\dfrac{x}{V}$

$$K_a = \frac{[H^+(aq)] \times [OH^-(aq)]}{[HA(aq)]}$$

$$= \frac{(x/V)^2}{(1 - x)/V}$$

$$= \frac{x^2}{(1 - x)/V}$$

It can be seen from this expression that if the dilution increases (that is, V is increased), then in order for the acid dissociation constant K_a to remain constant x must also increase. In fact at infinite dilution even a very weak acid will become completely dissociated. It might be that a 2.0 mol dm^{-3} solution of hydrochloric acid is not in fact completely dissociated, and so its pH value is not less than zero. This questioning of the actual concentration of the hydrogen ion concentration also ties in with what we discussed in the previous chapter, where we saw that the square brackets actually represent activity not concentration: the more concentrated the solution, the more the value of the activity deviates from the concentration. Of course, another assumption is that the pH scale cannot go below zero. Which assumption do you think is most likely to be wrong, and how could you test it?

Salt hydrolysis

Salts are ionic, and are therefore already completely dissociated. When they dissolve in water they are strong electrolytes. For example:

$$Na^+Cl^- \xrightarrow{H_2O\,(l)} Na^+(aq) + Cl^-(aq)$$

Salts like sodium chloride, which are derived from a strong acid (hydrochloric acid) and a strong base (sodium hydroxide), form neutral solutions when they dissolve in water. However, salts that are derived from a weak acid and a strong base produce alkaline solutions when they are dissolved in water. An example is sodium ethanoate, which can be formed from the reaction between ethanoic acid and sodium hydroxide. When the salt dissolves in water the ethanoate ions combine with hydrogen ions from the water to form mainly undissociated ethanoic acid. This leaves an excess of hydroxide ions in the solution, which do not combine with the sodium ions, because sodium hydroxide is a strong base. See Figure 8.

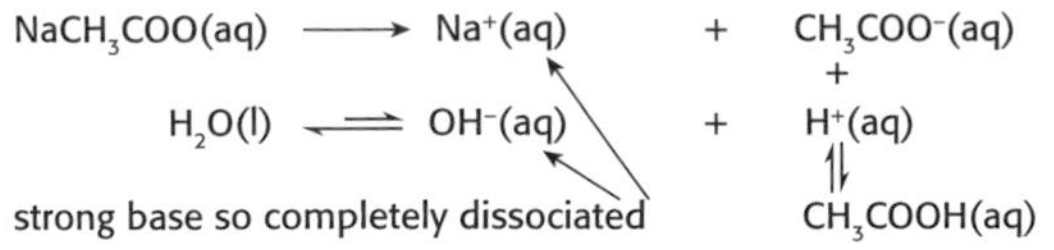

$$NaCH_3COO(aq) \longrightarrow Na^+(aq) \quad + \quad CH_3COO^-(aq)$$
$$+$$
$$H_2O(l) \rightleftharpoons OH^-(aq) \quad + \quad H^+(aq)$$
$$\Updownarrow$$
strong base so completely dissociated $\qquad CH_3COOH(aq)$

Figure 8 The hydrolysis of sodium ethanoate

Similarly salts such as ammonium chloride, which are derived from a strong acid and a weak base, will be acidic in solution (Figure 11).

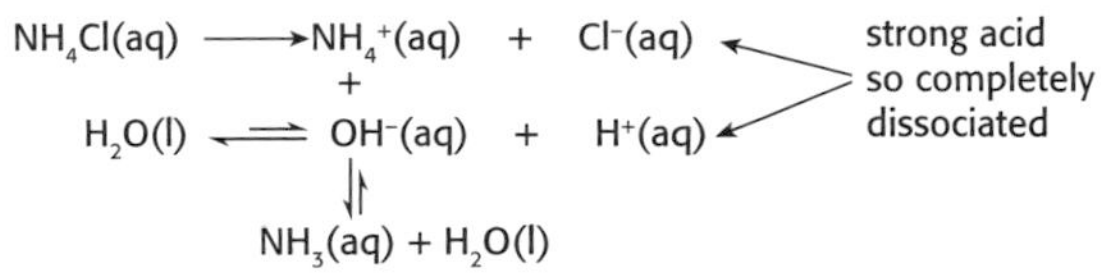

$$NH_4Cl(aq) \longrightarrow NH_4^+(aq) \quad + \quad Cl^-(aq) \quad \longleftarrow \text{ strong acid}$$
$$+ \qquad\qquad\qquad\qquad\qquad \text{so completely}$$
$$H_2O(l) \rightleftharpoons OH^-(aq) \quad + \quad H^+(aq) \longleftarrow \text{ dissociated}$$
$$\Updownarrow$$
$$NH_3(aq) + H_2O(l)$$

Figure 9 The hydrolysis of ammonium chloride

The acidity of salts also depends on the size and charge of the cation. Aluminium chloride reacts vigorously with water to give a strongly acidic solution:

$$AlCl_3(s) + 3H_2O(l) \rightarrow Al(OH)_3(s) + 3HCl(aq)$$

The tripositive (+3) charge is spread over a very small ion, which gives the aluminium ion, Al^{3+}, a very high charge density. This makes it an excellent Lewis acid, and it will attract a non-bonding pair of electrons on six water molecules to form the hexahydrated ion (Figure 10). The non-bonding pair of one of the six water molecules surrounding the ion is strongly attracted to the ion and loses a hydrogen ion in the process. This process will continue until three hydrogen ions have been released and aluminium hydroxide is formed. The equilibrium can be moved further to the right by adding hydroxide ions, $OH^-(aq)$, or back to the left by adding hydrogen ions, $H^+(aq)$, which explains the amphoteric nature of aluminium hydroxide.

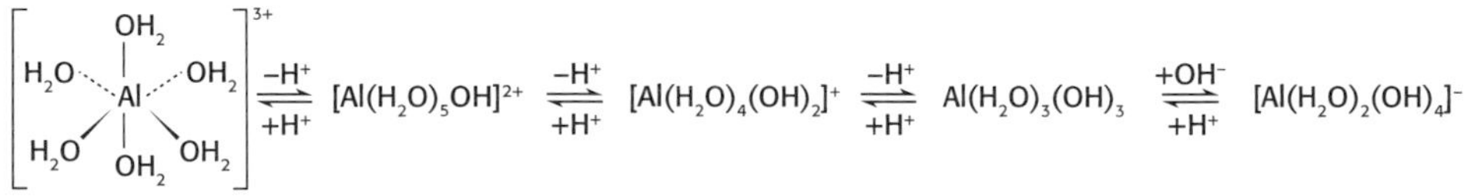

$$\left[\begin{array}{c} OH_2 \\ H_2O \cdots \overset{|}{\underset{|}{Al}} \diagup OH_2 \\ H_2O \diagup \overset{}{OH_2} \diagdown OH_2 \end{array}\right]^{3+} \overset{-H^+}{\underset{+H^+}{\rightleftharpoons}} [Al(H_2O)_5OH]^{2+} \overset{-H^+}{\underset{+H^+}{\rightleftharpoons}} [Al(H_2O)_4(OH)_2]^+ \overset{-H^+}{\underset{+H^+}{\rightleftharpoons}} Al(H_2O)_3(OH)_3 \overset{+OH^-}{\underset{+H^+}{\rightleftharpoons}} [Al(H_2O)_2(OH)_4]^-$$

Figure 10 The hydrolysis of the hexahydrated aluminium(III) ion.

Similar reactions occur with other small highly charged ions such as the iron(III) ion, Fe^{3+}. Even magnesium chloride, $MgCl_2$, is slightly acidic in aqueous solution for the same reason.

	Charge	Ionic radius /nm	Aqueous solution
Na^+	+1	0.098	neutral
Mg^{2+}	+2	0.065	acidic
Al^{3+}	+3	0.045	acidic

1 State whether aqueous solutions of the following salts will be neutral, acidic or alkaline: potassium nitrate, sodium carbonate, ammonium sulfate and iron(III) sulfate.

2 It is quite common to warm solutions in order to help the solute dissolve. Suggest a reason why it is potentially dangerous to add hot water to crystals of potassium cyanide, $KCN(s)$.

at 298 K (given that K_a for ethanoic acid = 1.8×10^{-5} mol dm^{-3} at 298 K).

$$M_r(NaCH_3COO) = [22.99 + 12.01 + (3 \times 1.01) + 12.01 + (2 \times 16.00)] = 82.04$$

Amount of sodium ethanoate in 100 cm^3 of solution

$$= \frac{3.00}{82.04} = 3.66 \times 10^{-2} \text{ mol}$$

Concentration of sodium ethanoate = 3.66×10^{-1} mol dm^{-3}

Assuming all the ethanoate ions originate from the salt, and the concentration of the acid at equilibrium is the same as the undissociated acid, then

$[CH_3COO^-(aq)] = 3.66 \times 10^{-1}$ mol dm^{-3} and
$[CH_3COOH(aq)] = 0.05$ mol dm^{-3}

$$K_a = \frac{[H^+(aq)] \times 0.366}{0.05} = 1.8 \times 10^{-5} \text{ mol dm}^{-3}$$

$[H^+(aq)] = 2.46 \times 10^{-6}$ mol dm^{-3}

pH = 5.61

2 What mass of ammonium chloride must be dissolved in 1.00 dm^3 of 0.100 mol dm^{-3} ammonia solution to produce a buffer solution with a pH equal to 9.00?

From the IB Data Booklet:

pK_b for ammonia = 4.75: hence $K_b = 10^{-4.75} = 1.78 \times 10^{-5}$ mol dm^{-3}

$$K_b = \frac{[NH_4^+(aq)] \times [OH^-(aq)]}{[NH_3(aq)]} = 1.78 \times 10^{-5} \text{ mol dm}^{-3}$$

pH = 9.00: hence pOH = (14.00 − 9.00) = 5.00 and $[OH^-(aq)]$ = 1.00×10^{-5} mol dm^{-3}

Assuming that the concentration of the ammonia at equilibrium is the same as the original concentration, then

$$[NH_4^+(aq)] = \frac{(1.78 \times 10^{-5}) \times 0.100}{1.00 \times 10^{-5}}$$

$$= 1.78 \times 10^{-1} \text{ mol dm}^{-3}$$

Assuming all the ammonium ions originate from ammonium chloride, then the amount of ammonium chloride required in 1.00 dm^3 of solution = 1.78×10^{-1} mol

$$M_r(NH_4Cl) = [14.01 + (4 \times 1.01) + 35.45] = 53.5$$

Mass of ammonium chloride = $53.5 \times 1.78 \times 10^{-1}$ = 9.53 g

Theory of acid–base titrations

One of the most common laboratory techniques is a simple acid–base titration in which an acid of unknown strength is titrated against a standard solution of an alkali, or vice versa, in order to find the unknown concentration. The method of calculating the unknown concentration has been discussed with an example in Chapter 1, but

1 Calculate the pH of the buffer solution formed when 25.0 cm^3 of 0.100 mol dm^{-3} sodium hydroxide solution, NaOH(aq), is added to 50.0 cm^3 of 0.100 mol dm^{-3} propanoic acid, $C_2H_5COOH(aq)$.

2 Calculate the pH of the buffer solution formed when 13.0 g of solid ammonium chloride, $NH_4Cl(s)$, is dissolved in 500 cm^3 of 0.05 mol dm^{-3} ammonia solution, $NH_3(aq)$. (Assume that there is no volume change when the salt dissolves.)

3 Explain how a system containing dissolved carbon dioxide ($CO_2(aq)$), hydrogencarbonate ions ($HCO_3^-(aq)$) and carbonic acid ($H_2CO_3(aq)$) can act as a buffer when small amounts of acid or alkali are added.

how do we decide which standard solution of acid or base to choose, and which indicator should be used?

Acid–base indicators

An indicator is a solution of a weak acid in which the conjugate base has a different colour from that of the undissociated acid.

If we represent an aqueous solution of an indicator by HIn(aq), then the equation for its dissociation in water is

HIn(aq) $\rightleftharpoons$ H$^+$(aq) + In$^-$(aq)
colour A   colour B
(colour in acid solution)   (colour in alkali solution)

and the equilibrium expression is given by

$$K_{in} = \frac{[H^+(aq)] \times [In^-(aq)]}{[HIn(aq)]}$$

where K_{in} represents the acid dissociation constant of the indicator.

Assuming the colour changes when the concentration of the acid is approximately equal to the concentration of the conjugate base ([HIn(aq)] ≈ [In$^-$(aq)]), then the end point of the indicator will be when [H$^+$(aq)] ≈ K_{in}: that is, when pH ≈ pK_{in}.

Several indicators are listed in the IB Data Booklet, and it can be seen that the various indicators have different K_{in} values, and so change colour within different pH ranges. Two common indicators are methyl orange and phenolphthalein:

Indicator	pK_{in}	pH range	Colour change	
			Acid	Alkali
Methyl orange	3.7	3.1–4.4	red	yellow
Phenolphthalein	9.6	8.3–10.0	colourless	pink

Extension material
Why do indicators change colour at a certain pH?

A good chemist will ask why the conjugate base has a different colour from that of the acid it is derived from. Many organic compounds absorb light in the ultraviolet region of the electromagnetic spectrum as electrons in a chemical bond are excited to a higher empty orbital, usually an antibonding orbital. If there is extensive delocalization of the pi electrons then the absorption is of lower energy and occurs in the visible region. Delocalization of pi electrons occurs when there are alternate double and single carbon-to-carbon bonds. When these are linked together it is known as **conjugation**. In its undissociated form phenolphthalein contains some delocalized pi electrons, but they are restricted to the three separate aromatic rings, and hence phenolphthalein in its undissociated form in acid solution is colourless. Once it loses the hydrogen ion the conjugate base shows more extensive conjugation, and it absorbs light in the visible region, so the anion is coloured (Figure 11).

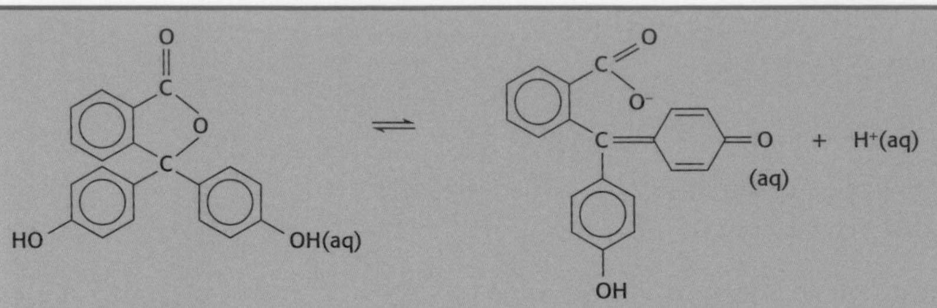

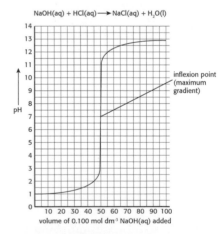

Figure 11 The action of phenolphthalein as an indicator. In acidic solution it is colourless. In alkaline solution the anion (conjugate base) contains more conjugation and appears pink.

HL

Strong acid–strong base titrations

The change in pH during an acid–base titration can be followed using a pH meter. Consider starting with 50.0 cm^3 of 0.100 mol dm^{-3} hydrochloric acid and then titrating with 0.100 mol dm^{-3} sodium hydroxide solution.

The initial pH will be 1.00, because $[H^+(aq)] = 0.100$ mol dm^{-3}.

After 49.0 cm^3 of 0.100 mol dm^{-3} sodium hydroxide has been added there will be 1.0 cm^3 of the original 0.100 mol dm^{-3} hydrochloric acid remaining in 99.0 cm^3 of solution. At this point $[H^+(aq)] \approx 1.0 \times 10^{-3}$ mol dm^{-3}, so the pH ≈ 3.00.

When 49.1 cm^3 of the 0.100 mol dm^{-3} sodium hydroxide has been added there will be 0.9 cm^3 of the original 0.100 mol dm^{-3} hydrochloric acid remaining in 99.1 cm^3 of solution. Now $[H^+(aq)] \approx 0.9 \times 10^{-3}$ mol dm^{-3}, so the pH ≈ 3.05.

You may now like to work out the pH after successive further additions of 0.1 cm^3 of 0.100 mol dm^{-3} sodium hydroxide solution.

After 50.0 cm^3 of sodium hydroxide solution has been added the equivalence point has been reached and the solution is neutral, so the pH is equal to 7.00. After a further 1.0 cm^3 of 0.100 mol dm^{-3} sodium hydroxide solution has been added, bringing the total amount added to 51.0 cm^3, you should be able to calculate that the pOH will be 3.00, and therefore the pH will be 11.0.

A graph of pH against volume of sodium hydroxide added can be plotted, and it can be seen that there is a very rapid change in pH at the equivalence point (Figure 12). The **end-point** is when the indicator changes colour accurate to the addition of one drop. It can be seen that almost any common acid–base indicator will work for this type of titration, because they all change colour within the pH range.

Weak acid–strong base titrations

The same procedure as for the titration of a strong acid with a strong base can be followed, except that in this case 0.100 mol dm^{-3} ethanoic acid is substituted for the hydrochloric acid.

$$K_a(CH_3COOH) = 1.8 \times 10^{-5} \text{ mol dm}^{-3} \text{ at 298 K.}$$

Making the usual assumptions, the initial hydrogen ion concentration, $[H^+(aq)]$, is given by

$$[H^+(aq)] = (K_a \times [CH_3COOH(aq)])^{1/2}, \text{ giving the pH} = 2.87.$$

Figure 12 Titration curve for a strong acid with a strong base.

When 25.0 cm³ of the sodium hydroxide solution has been added, we shall have the ideal buffer solution, because the half-equivalence point has been reached. At this point the pH is equal to the pK_a of the acid, and so has a value of 4.76.

When 49.0 cm³ of the 0.100 mol dm⁻³ sodium hydroxide has been added, $[CH_3COO^-(aq)] \approx 0.05$ mol dm⁻³ and $[CH_3COOH(aq)]$ $\approx 1.0 \times 10^{-3}$ mol dm⁻³.

$$[H^+(aq)] = \frac{[K_a \times [CH_3COOH(aq)]}{[CH_3COO^-(aq)]}$$

$$\approx \frac{(1.8 \times 10^{-5}) \times (1 \times 10^{-3})}{0.05}$$

$$\approx 3.6 \times 10^{-7} \text{ mol dm}^{-3}$$

This gives a pH of 6.44.

This time the equivalence point will not have a pH equal to 7, owing to salt hydrolysis. After the equivalence point the graph will follow the same pattern as the strong acid–strong base curve, as more sodium hydroxide is simply being added to the solution. See Figure 13. The end-point occurs in a much more narrow pH range, and a suitable indicator to use is phenolphthalein, which changes colour in the pH range 8.3–10.0.

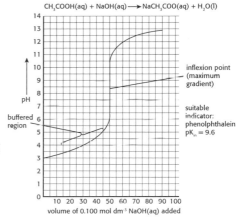

Figure 13 Titration curve for a weak acid with a strong base.

Strong acid–weak base titrations

Now 0.100 mol dm⁻³ ammonia solution is added to 50.0 cm³ of 0.100 mol dm⁻³ hydrochloric acid. Initially the graph will follow the curve for a strong acid. The equivalence point will have a pH less than 7 owing to salt hydrolysis of the salt derived from a strong acid and a weak base. After 50.1 cm³ of the solution of the weak base has been added:

$$[NH_4^+(aq)] \approx 0.05 \text{ mol dm}^{-3} \text{ and } [NH_3(aq)] \approx 1.00 \times 10^{-3} \text{ mol dm}^{-3}$$

Using the value for the pK_b of ammonia given in the IB Data Booklet you should be able to calculate the pOH value and then obtain the value of 7.56 for the pH.

The end-point will occur in the pH range 4–6, so methyl orange or another indicator that changes colour within this range should be used for this type of titration. See Figure 14.

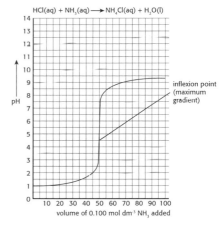

Figure 14 Titration curve for a strong acid with a weak base.

Weak acid–weak base titrations

It should be obvious from the arguments above that there will hardly be a pH change around the equivalence point. When 49.9 cm³ of the 0.100 mol dm⁻³ ammonia solution has been added to 50.0 cm³ of 0.100 mol dm⁻³ ethanoic acid solution the pH is 6.44, and when 50.1 cm³ has been added the pH has only increased to 7.56. This is a change of only one unit of pH during the addition of 0.2 cm³ of alkali. No indicator will detect the end-point accurate to one drop. However, this is not a problem, as it is never necessary to perform this type of titration because, if the unknown is weak, the standard solution can always be chosen to be strong.

such a table is known as the **electrochemical series**, and that the standard hydrogen half-cell (not the copper half-cell) is used as the reference half-cell.

Practical work with voltaic cells

Voltaic cells offer much scope for laboratory investigations, because they use relatively simple materials. For example, as a design experiment that will count as part of your internally assessed coursework, your teacher may simply say to you, "Investigate an aspect of voltaic cells." From now on you are on your own, and a good investigation should mean that each person in the class can come up with a different problem to investigate.

To focus on a particular research question you need to be able to identify which variables you can manipulate and which you can keep constant. You need to think what can be easily measured, and it may also be helpful to formulate a sensible hypothesis. If you do propose a hypothesis it does not have to be correct, but you should be able to justify it using a sound chemical argument. If you read carefully what has been written about voltaic cells you will see that potential variables include the metals used in the half-cells, the concentration of the solutions of metal ions, the temperature of the reaction, the surface area of the pieces of metal, the mass of the pieces of metal, the composition of the salt bridge, and the surrounding pressure. Some of these you may not easily be able to control but others can easily be manipulated.

Once you have identified your own problem or research question to investigate, you need to plan how to carry this out experimentally. You will need to decide and record exactly what equipment you will need, together with the uncertainties associated with each piece of apparatus. You will also need to devise and record a method that enables you to alter the independent variable and measure the effect on the dependent variable while keeping control of all the other variables. Don't forget that you will also need to ensure that you have sufficient measurements for your result to be scientifically valid. Because each student must define their own problem or research question it is difficult to be any more prescriptive, but it might help you to design similar experiments if you can see what some students in the past have investigated when faced with this problem.

Figure 7 Two students from Atlantic College (Sarah Mohammed Dawam from Malaysia and Richard Onekayon from the Sudan) working on their design experiment to see the effect that having one or many salt bridges would have on the e.m.f. of a cell (photo by Geoff Neuss).

1 What is the effect of a change in temperature of a zinc/copper cell on the voltage when one half-cell stays constant and the other varies in temperature?
2 How can the change in concentration of $Pb^{2+}(aq)$ affect the direction of flow of electrons when a tin half-cell is connected to a lead half-cell?
3 Does brass, when used in electrochemical cells, tend to behave as copper or as zinc?
4 What is the relationship between the surface area of the electrodes in the solution and the voltage produced by an electrochemical cell?
5 What is the effect of the length of one or many salt bridges on the e.m.f. of a cell?
6 How does changing the concentration of $Fe^{2+}(aq)$ while keeping the concentration of $Mg^{2+}(aq)$ constant affect the overall voltage when a magnesium half-cell is connected to an iron half-cell?

Electrolytic cells

In a voltaic cell, electricity is produced from a spontaneous chemical reaction, whereas in an electrolytic cell electricity is supplied from an external source and is used to make a non-spontaneous reaction take place. The substance that conducts electricity in an electrolytic cell is known as the **electrolyte**. Electrolytes are substances that contain ions. They do not conduct electricity in the solid state, because the ions are in fixed positions and have no translational velocity, and so cannot move towards oppositely charged electrodes; and unlike metallic conductors there are no delocalized electrons. However, when the substance become molten or is dissolved in aqueous solution the ions are free to move. The electrolyte now conducts electricity by the movement of ions within it, and chemical reactions take place at each electrode so that the electrolyte is decomposed in the process. This can be illustrated by using molten sodium chloride as the electrolyte (Figure 8).

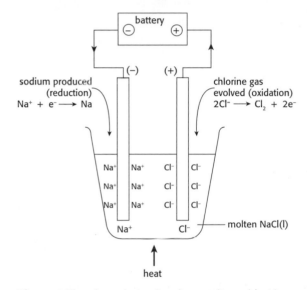

Figure 8 The electrolysis of molten sodium chloride

Oxidation occurs at the positive electrode. The positive electrode is sometimes known as the **anode**, because negative ions (anions) are attracted to it. In this example chloride ions are attracted to the positive electrode, and for every two chloride ions that release electrons one molecule of chlorine gas will be produced.

Positive electrode (anode): $2Cl^-(l) \rightarrow Cl_2(g) + 2e^-$ Oxidation

Reduction occurs at the negative electrode. This electrode is sometimes known as the **cathode**, because positive ions (cations) are attracted to it. The sodium ions each gain one electron, and sodium metal is produced.

Negative electrode (cathode): $Na^+(l) + e^- \rightarrow Na(l)$ Reduction

For every 2 mol of electrons that flow through the external circuit 1 mol of chlorine gas and 2 mol of sodium metal will be produced.

1 Explain why solid lead(II) bromide does not conduct electricity. What will be observed at the positive electrode and the negative electrode when an electric current is passed through molten lead(II) bromide? Give the half-equations for the reactions taking place at each electrode, state whether they are oxidation or reduction processes, and deduce the molar ratios of the two products formed.

2 Explain why the anode in a voltaic cell is the negative electrode whereas in an electrolytic cell it is the positive electrode.

Extension material

Electrolysis and the isolation of elements

There is a very close correlation between the date of discovery of metallic elements and their position in the reactivity series. Some elements, such as copper, gold, silver, lead, iron and tin, have been known in their pure form for several thousand years. These are all elements that are low in the reactivity series. Those that are very low, such as silver and gold, occur in their elemental state in nature: that is, they are not combined with any other element. Copper, which is slightly higher in the series, is sometimes found uncombined and sometimes in compounds such as its carbonate or sulfide. However, these compounds need only to be heated and they decompose to elemental copper. Metals such as iron, lead and tin are more reactive, and are not found uncombined in nature. Their ores, such as their oxides or sulfides, can be reduced to the metal by chemical reducing agents. The oxides of lead can be reduced using carbon, and iron oxides are reduced to iron using a mixture of carbon monoxide and hydrogen in a modern blast furnace.

As the reactivity of the metal increases, it gets progressively harder to reduce the ore of a metal by chemical means. Aluminium provides a good example of this. The existence of aluminium was first proposed by Sir Humphry Davy (1778–1829) in 1808 and it was he who suggested the name "aluminium". However, the metal was not actually isolated from its ore until 1825. The Danish physicist and chemist Hans Christian Oersted (1777–1851) produced minute quantities of aluminium metal by using dilute potassium amalgam (an alloy of potassium with mercury) to react with anhydrous aluminium chloride, and then distilling the resulting mercury away to leave a residue of slightly impure aluminium. The properties of aluminium soon became known but, because of the difficulty and expense of producing it, it was more expensive to buy than gold. Two famous authors at the time—Charles Dickens (1812–1870) and Jules Verne (1828–1905)—wrote about its properties. Jules Verne even went so far as to describe it as the ideal metal with which to build a rocket to go to the moon (Figure 9).

However, there is one better reducing agent than any chemical substance, and that is electricity. The discovery of electricity revolutionized chemistry—and indeed society itself. In 1886 the American Charles Hall (1863–1914) and independently the Frenchman Paul Héroult (1863–1914) first produced aluminium electrolytically by passing an electric current through a solution of aluminium oxide dissolved in molten cryolite, Na_3AlF_6.

By the use of this process aluminium became more readily available and thus much cheaper. Unlike iron and steel it does not corrode, and so is rapidly replacing iron for many purposes, although it is still more expensive than steel to produce. Its uses are changing. Whereas once its main use was to build aeroplanes, now it is used extensively in packaging and in architecture, and for luxury goods such as kitchens and cars (Figure 10).

The ready availability of metals such as aluminium through the use of electrolysis has enabled engineering and technology to develop enormously during the past century, with the consequent increase in the quality of life. However, as countries develop their economies the demand for the raw materials to make these metals increases, and the price of the ores becomes more expensive.

"This valuable metal possesses the whiteness of silver, the indestructibility of gold, the tenacity of iron, the fusibility of copper, the lightness of glass. It is easily wrought, is very widely distributed, forming the base of most of the rocks, is three times lighter than iron, and seems to have been created for the express purpose of furnishing us with the material for our projectile."

"De la terre á la lune"

(From the Earth to the Moon)

Figure 9 The author Jules Verne writing about aluminium in 1865

Figure 10 A new kitchen or a Mercedes?

There is another price to pay too. The electrodes used in the production of aluminium are made of carbon and break down to form carbon dioxide. Coupled with the fact that the electricity required for the electrolysis is often produced from the combustion of fossil fuels, it can be seen that the production of aluminium contributes hugely to global warming. As the energy required to recycle aluminium is only about 5% of the energy required to electrolyse it from its ore, it makes considerable sense to recycle as much aluminium as possible. Currently 20% of aluminium is being recycled worldwide (Figure 11). The aim of the International Aluminium Institute (IAI) is to increase this to 50% by the year 2020.

Figure 11 Recycling containers outside Selvatura Park, Monteverde, Costa Rica (photo by Geoff Neuss).

1 The first voltaic cell was developed in 1800 by Alessandro Volta (1745–1827), an Italian professor of physics. Shortly afterwards, in 1807, Sir Humphry Davy first isolated sodium by passing electricity through molten sodium hydroxide. However, despite many attempts, Davy and others were unable to isolate fluorine from calcium fluoride or other ionic salts containing fluorine. Fluorine was not in fact isolated until 1886 by the Frenchman Henri Moissan (1852–1907). Explain why it was so difficult to isolate fluorine, and find out how Moissan actually achieved it.

2 Why do manufacturers still continue to make most cars out of steel rather than aluminium, even though aluminium cars do not rust?

HL

Standard electrode potentials

The idea of an electrochemical series was introduced in the last section. It is impossible to measure the actual electrode potential of a half-cell, as any attempt to do so alters the half-cell, so the relative electrode potential must be measured by comparing one half-cell with another. By convention all half-cells are compared under standard conditions with the **standard hydrogen electrode**.

A standard hydrogen electrode consists of a platinum electrode surrounded by hydrogen gas at one atmosphere (1.01×10^5 Pa) pressure. The electrode is immersed in an aqueous solution of acid in which the concentration of hydrogen ions, H+(aq), is exactly 1.00 mol dm^{-3}. The temperature is maintained at 298 K (25 °C).

The diagram shown of the standard hydrogen electrode shown in Figure 12 is similar to many others that can be found in typical textbooks. Look at it carefully and critically. Can you see anything that is not correct science? (Check your findings with the comments made at the bottom of the next page.)

The standard hydrogen electrode is assigned a standard electrode potential of 0.00 V. The standard electrode potential of another half-cell is determined by connecting the half-cell to the standard hydrogen electrode using a salt bridge and a connecting wire with a high-resistance voltmeter, and measuring the cell e.m.f. under standard conditions. If the electrons flow in the external wire *from* the half-cell *to* the hydrogen electrode the standard electrode potential of the half-cell is assigned a negative value. An example of this is a zinc half-cell connected to the hydrogen half-cell (Figure 13). Zinc is a better reducing agent than hydrogen, so the zinc metal forms the negative electrode and the platinum surrounded by hydrogen molecules and ions will be the positive

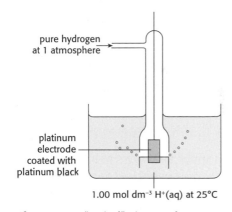

pure hydrogen at 1 atmosphere

platinum electrode coated with platinum black

1.00 mol dm^{-3} H+(aq) at 25°C

Figure 12 A "typical" picture of a standard hydrogen electrode, which can be found in many textbooks.

electrode. The measured e.m.f. from this cell is 0.76 V, so the standard electrode potential of the zinc half-cell is assigned the value of −0.76 V.

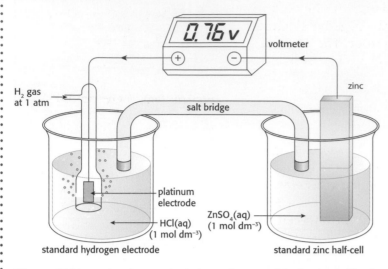

Figure 13 Measuring the standard electrode potential of a zinc half-cell

Chemists have developed a shorthand way of showing this using a cell-diagram. A slash, /, represents the boundary between the reduced and oxidized parts of a half-cell and the two vertical lines represent the salt bridge between the two half-cells:

$$Zn(s)/Zn^{2+}(aq) \parallel H^+(aq)/\frac{1}{2} H_2(g)(Pt)$$

By convention the half-cell undergoing oxidation is placed on the left in the cell diagram, but—as we shall see—the convention can be ignored, as it is really not important which way round they are written. As the vertical lines represent the salt bridge it is usual to put the aqueous solutions of the ions on either side of them, but this is not really important either.

Copper is below hydrogen in the electrochemical series, and so is a worse reducing agent than hydrogen. When a copper half-cell is connected to a hydrogen half-cell the electrons will flow from the hydrogen half-cell to the copper half-cell. The e.m.f. is found to be 0.34 V under standard conditions. The standard electrode potential of the copper half-cell is thus assigned the value of +0.34 V.

We have already seen that the e.m.f. produced by a copper half-cell connected to a zinc half-cell is 1.10 V. How can we use standard electrode potentials to arrive at this value? First, draw out the cell diagram. Then look up the values for the standard electrode

Comment on the diagram of the standard hydrogen electrode on the previous page

In the apparatus as drawn, the hydrogen gas will not surround the electrode and bubble out. The pressure on the surface of the acid solution is one atmosphere and the hydrogen gas is at one atmosphere pressure. For the hydrogen to go below the surface of the acid solution and bubble out as drawn in this diagram it must be at a slightly higher pressure than one atmosphere.

potentials. For the spontaneous reaction *the electrons in the external circuit will always flow from the more negative half-cell to the more positive half-cell*. This is the most important rule to remember, and because of this it does not matter which cell you have drawn on the right or the left. The more negative half-cell will form the negative electrode and the more positive half-cell the positive electrode. The e.m.f. of the cell, $E^{\ominus}_{cell}$, is simply the arithmetical difference between the two values. See Figure 14.

Predicting the spontaneity of a chemical reaction using standard electrode potentials

Simply by looking at the table of standard electrode potentials given in the IB Data Booklet (Table 3) it is easy to determine whether or not a redox reaction will be spontaneous. The spontaneous reaction will be the reaction in which the half-cell with the more negative standard electrode potential gives up electrons and the half-cell with the more positive standard electrode potential receives electrons.

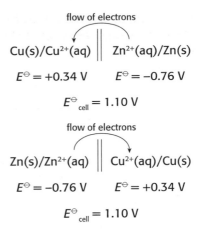

Figure 14 Cell diagrams for the copper/zinc cell. Whichever way it is drawn, the electrons flow from the half-cell with the more negative $E^{\ominus}$ value, and the total e.m.f. will always be 1.10 V.

Table 3 Standard electrode potentials

Oxidized species	Reduced species	$E^{\ominus}$ / V
$Li^+(aq) + e^-$	$\rightleftharpoons$ $Li(s)$	−3.03
$K^+(aq) + e^-$	$\rightleftharpoons$ $K(s)$	−2.92
$Ca^{2+}(aq) + 2e^-$	$\rightleftharpoons$ $Ca(s)$	−2.87
$Na^+(aq) + e^-$	$\rightleftharpoons$ $Na(s)$	−2.71
$Mg^{2+}(aq) + 2e^-$	$\rightleftharpoons$ $Mg(s)$	−2.36
$Al^{3+}(aq) + 3e^-$	$\rightleftharpoons$ $Al(s)$	−1.66
$Mn^{2+}(aq) + 2e^-$	$\rightleftharpoons$ $Mn(s)$	−1.18
$H_2O(l) + e$	$\rightleftharpoons$ $\frac{1}{2}H_2(g) + OH^-(aq)$	−0.83
$Zn^{2+}(aq) + 2e^-$	$\rightleftharpoons$ $Zn(s)$	−0.76
$Fe^{2+} + 2e^-$	$\rightleftharpoons$ $Fe(s)$	−0.44
$Ni^{2+}(aq) + 2e^-$	$\rightleftharpoons$ $Ni(s)$	−0.23
$Sn^{2+}(aq) + 2e^-$	$\rightleftharpoons$ $Sn(s)$	−0.14
$Pb^{2+}(aq) + 2e^-$	$\rightleftharpoons$ $Pb(s)$	−0.13
$H^+(aq) + e^-$	$\rightleftharpoons$ $\frac{1}{2}H_2(g)$	0.00
$Cu^{2+}(aq) + e^-$	$\rightleftharpoons$ $Cu^+(s)$	+0.15
$SO_4^{2-}(aq) + 4H^+(aq) + 2e^-$	$\rightleftharpoons$ $H_2SO_3(aq) + H_2O(l)$	+0.17
$Cu^{2+}(aq) + 2e^-$	$\rightleftharpoons$ $Cu(s)$	+0.34
$\frac{1}{2}O_2(g) + H_2O(l) + 2e^-$	$\rightleftharpoons$ $2OH^-(aq)$	+0.40
$Cu^+(aq) + e^-$	$\rightleftharpoons$ $Cu(s)$	+0.52
$\frac{1}{2}I_2(s) + e^-$	$\rightleftharpoons$ $I^-(aq)$	+0.54
$Fe^{3+}(aq) + e^-$	$\rightleftharpoons$ $Fe^{2+}(aq)$	+0.77
$Ag^+ + e^-$	$\rightleftharpoons$ $Ag(s)$	+0.80
$\frac{1}{2}Br_2(l) + e^-$	$\rightleftharpoons$ $Br^-(aq)$	+1.09
$\frac{1}{2}O_2(g) + 2H^+(aq) + 2e^-$	$\rightleftharpoons$ $H_2O(l)$	+1.23
$Cr_2O_7^{2-}(aq) + 14H^+(aq) + 6e^-$	$\rightleftharpoons$ $2Cr^{3+}(aq) + 7H_2O(l)$	+1.33
$\frac{1}{2}Cl_2(g) + e^-$	$\rightleftharpoons$ $Cl^-(aq)$	+1.36
$MnO_4^-(aq) + 8H^+(aq) + 5e^-$	$\rightleftharpoons$ $Mn^{2+}(aq) + 4H_2O(l)$	+1.51

Can an acidified aqueous solution of potassium dichromate(VI)
spontaneously oxidize a solution of bromide ions to bromine?

The relevant half-equations are

$$Cr_2O_7^{2-}(aq) + 14H^+(aq) + 6e^- \rightleftharpoons 2Cr^{3+}(aq) + 7H_2O(l) \quad E^\ominus = +1.33 \text{ V}$$
$$Br_2(aq) + 2e^- \rightleftharpoons 2Br^-(aq) \qquad\qquad E^\ominus = +1.07 \text{ V}$$

Both of the standard electrode potentials have positive values, but
the one with the more negative value (i.e. the less positive) is the
second one, involving bromine. The second reaction will therefore
provide the electrons, and bromide ions will be converted into
bromine. The dichromate(VI) ions will be gaining electrons. When
the number of electrons has been balanced, the overall spontaneous
reaction will be

$$Cr_2O_7^{2-}(aq) + 14H^+(aq) + 6Br^-(aq) \rightarrow 2Cr^{3+}(aq) + 3Br_2(aq) + 7H_2O(l)$$
$$E^\ominus_{cell} = 0.26 \text{ V}$$

Thus an acidified solution of dichromate(VI) ions can spontaneously
oxidize a solution of bromide ions, and the total e.m.f. for the
reaction under standard conditions is 0.26 V. Using standard
electrode potentials in this way enables the spontaneity of a reaction
to be determined, but remember that it gives no information about
the rate of the reaction, so even though it is spontaneous the
reaction may proceed very slowly.

Can a solution of tin(II) ions reduce a solution of iron(III) ions? If so,
are the iron(III) ions reduced to iron(II) ions or to iron metal?

The two half-equations for the first part of the question are

$$Sn^{4+}(aq) + 2e^- \rightleftharpoons Sn^{2+}(aq) \qquad E^\ominus = +0.15 \text{ V}$$
$$Fe^{3+}(aq) + e^- \rightleftharpoons Fe^{2+}(aq) \qquad E^\ominus = +0.77 \text{ V}$$

The first equation has the more negative standard electrode
potential, so tin(II) ions will give up electrons, which will be gained
by iron(III) ions.

The spontaneous reaction will be

$$Sn^{2+}(aq) + 2Fe^{3+}(aq) \rightarrow Sn^{4+}(aq) + 2Fe^{2+}(aq) \qquad E^\ominus_{cell} = 0.62 \text{ V}$$

Therefore tin(II) ions can spontaneously reduce iron(III) ions.

To answer the second part of the question we need the following
two half-equations:

$$Sn^{4+}(aq) + 2e^- \rightleftharpoons Sn^{2+} \qquad E^\ominus = +0.15 \text{ V}$$
$$Fe^{2+}(aq) + 2e^- \rightleftharpoons Fe(s) \qquad E^\ominus = -0.44 \text{ V}$$

Now it can be seen that the electrons must come from the iron
metal, and they will be gained by the tin(II) ions. So tin(II) ions
cannot spontaneously reduce iron(III) ions to iron metal. They can
only spontaneously reduce them to iron(II) ions. In fact, what we
have shown is that iron metal can spontaneously reduce tin(IV) ions
to tin(II) ions:

$$Fe(s) + Sn^{4+}(aq) \rightarrow Fe^{2+}(aq) + Sn^{2+}(aq) \qquad E^\ominus = +0.59 \text{ V}$$

What will happen when solid copper(I) sulfate, $Cu_2SO_4(s)$, dissolves in water?

It is not obvious at first how to proceed with this question. We could guess that, apart from dissolving, nothing much will happen, or possibly some sort of redox reaction might take place. The two half-equations involving copper(I) ions are

$$Cu^+(aq) + e^- \rightleftharpoons Cu(s) \qquad\qquad E^\ominus = +0.52 \text{ V}$$
$$Cu^{2+}(aq) + e^- \rightleftharpoons Cu^+(aq) \qquad E^\ominus = +0.15 \text{ V}$$

From the relative values of their standard electrode potentials, copper(I) ions in the second equation (more negative $E^\ominus$ value) are donating electrons, and copper(I) ions in the first equation (more positive $E^\ominus$ value) are gaining electrons.

The equation of the spontaneous reaction will be

$$2Cu^+(aq) \rightarrow Cu^{2+}(aq) + Cu(s)$$
$$(+1) \qquad\quad (+2) \qquad (0)$$

From this we can predict that the copper(I) ions will be both oxidized and reduced at the same time: that is, copper(I) ions will disproportionate. We would expect to see the white solid react when it comes into contact with the water to form a blue solution of copper(II) sulfate, $CuSO_4(aq)$, and a reddish-brown precipitate of copper metal.

Electrolysis in aqueous solution

An electrolyte will conduct electricity when molten or in aqueous solution and be chemically decomposed in the process. For molten electrolytes it is relatively easy to determine the products of electrolysis, because there will be only one type of positive ion and one type of negative ion; however, there will be more types of ion present in aqueous solution. Water is a poor conductor of electricity. Nevertheless, it is dissociated slightly into its ions:

$$H_2O(l) \rightleftharpoons H^+(aq) + OH^-(aq)$$

Thus in any aqueous solution of an electrolyte there will also be hydrogen ions and hydroxide ions present in low concentration, as well as the ions from the electrolyte itself.

Factors affecting the discharge of ions during the electrolysis of aqueous solutions

- *Position in the electrochemical series.* Positive ions lower in the series will gain electrons at the negative electrode (cathode) in preference to those higher in the series. Hydroxide ions normally release electrons to form oxygen and water in preference to other anions at the positive electrode (anode).
- *Concentration of the electrolyte.* Normally this has little effect, but in some cases (e.g. sodium chloride solution) changing the concentration of the electrolyte can alter the relative proportions of products formed.

1 Calculate the total e.m.f., $E^\ominus_{cell}$, when a cobalt half-cell, $Co(s) / Co^{2+}(aq)$ ($E^\ominus = -0.28$ V), is connected, using an external wire and a salt bridge, to a manganese half-cell, $Mn(s) / Mn^{2+}(aq)$ ($E^\ominus = -1.18$ V) under standard conditions. What will be the spontaneous reaction that occurs?

2 Show that under standard conditions an acidified aqueous solution of potassium dichromate(VI) cannot oxidize chloride ions to chlorine. Suggest a reason why in the laboratory it is preferable to acidify the solution of dichromate(VI) ions with sulfuric acid rather than hydrochloric acid when performing a redox titration.

3 Predict and explain what will be observed when fluorine is dissolved in water.

● *The nature of the electrode.* Usually the electrodes, which are often made of platinum or graphite, are chemically inert and do not react. However, in certain cases (e.g. the use of copper electrodes in the electrolysis of copper(II) sulfate solution) the electrode itself reacts.

The first of these factors is by far the most important, because it always needs to be applied to explain the products of electrolysis in aqueous solution. The other two are rather specialized, and are important only in certain reactions. These factors can be illustrated by the following examples.

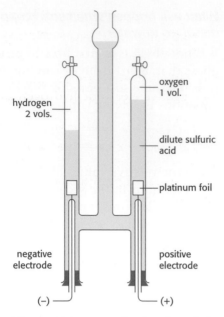

Figure 15 Apparatus for the electrolysis of "water".

Electrolysis of "water"

Pure water itself is an extremely poor electrolyte, and essentially does not conduct electricity at low voltage. However, once a small quantity of ions are dissolved in it the conductivity increases considerably. The electrolysis of "water" is usually carried out on a dilute solution of sulfuric acid (Figure 15). It can also be done with a dilute solution of sodium hydroxide. When trying to predict what will happen, it is useful to list all the cations and all the anions in solution.

For dilute sulfuric acid:

$$H_2SO_4(aq) \rightarrow 2H^+(aq) + SO_4^{2-}(aq)$$
$$H_2O(l) \rightleftharpoons H^+(aq) + OH^-(aq)$$

The only cations present are hydrogen ions, so hydrogen will be given off at the negative electrode:

$$(-) \text{ electrode: } 2H^+(aq) + 2e^- \rightarrow H_2(g)$$

There are two anions present: sulfate ions and hydroxide ions. Hydroxide ions are discharged in preference to sulfate ions, so oxygen gas will be evolved at the positive electrode:

$$(+) \text{ electrode: } 4OH^-(aq) \rightarrow 2H_2O(l) + O_2(g) + 4e^-$$

As the concentration of hydroxide ions is extremely low (although it is maintained via the equilibrium process), some people prefer to write the equation for the breakdown of water itself:

$$2H_2O(l) \rightarrow 4H^+(aq) + O_2(g) + 4e^-$$

Both equations are acceptable, and what is important is that they both involve four electrons. From this it can be seen that the electrolysis of dilute sulfuric acid produces two moles of hydrogen for every mole of oxygen produced. This is why it is known as the electrolysis of "water", H_2O.

If we use dilute sodium hydroxide instead, the only anions present will be hydroxide ions, so the same reaction will take place at the positive electrode. There are now two types of positive ion (cation) present: hydrogen ions and sodium ions. Hydrogen is much lower than sodium in the electrochemical series, so hydrogen ions will preferentially gain electrons, and again hydrogen gas is produced at the negative electrode.

Electrolysis of sodium chloride solution

$$NaCl(aq) \rightarrow Na^+(aq) + Cl^-(aq)$$
$$H_2O(l) \rightleftharpoons H^+(aq) + OH^-(aq)$$

The cations present are sodium ions and hydrogen ions. Therefore, at the negative electrode hydrogen gas will be evolved:

$$(-) \text{ electrode: } 2H^+(aq) + 2e^- \rightarrow H_2(aq)$$

Both the hydroxide ions and the chloride ions will be attracted towards the positive electrode. When the solution is dilute, oxygen is the principal product, because hydroxide ions are preferentially discharged:

$$(+) \text{ electrode: } 4OH^-(aq) \rightarrow 2H_2O(l) + O_2(g) + 4e^-$$

When the solution becomes more concentrated, then chlorine rather than oxygen is the principal product:

$$(+ \text{ electrode}): 2Cl^-(aq) \rightarrow Cl_2(g) + 2e^-$$

In the past, the electrolysis of sodium chloride solution also provided a good example of different products being produced if the electrodes were changed. The products above assume the electrodes are inert. To manufacture sodium hydroxide, electricity used to be passed through sodium chloride solution using mercury as the negative electrode. At this electrode, instead of hydrogen being evolved the sodium ions reacted to form an alloy of sodium and mercury called an **amalgam**. The mercury flowed through the cell in a continuous process, and in a separate container the sodium amalgam was made to react with water, producing hydrogen gas and a solution of sodium hydroxide.

$$2Na/Hg(l) + 2H_2O(l) \rightarrow H_2(g) + NaOH(aq) + Hg(l)$$

In theory, none of the mercury was lost, as it was simply recycled back into the electrolytic cell. However, in Minamata in Japan in the late 1960s and early 1970s many people suffered from mercury poisoning as it escaped into the sea and was concentrated in the food chain. Because of the danger of something similar happening again this process is now almost obsolete.

Electrolysis of copper(II) sulfate solution

The classic example of the nature of the electrodes affecting the products is the electrolysis of copper(II) sulfate solution:

$$CuSO_4(aq) \rightarrow Cu^{2+}(aq) + SO_4^{2-}(aq)$$
$$H_2O(l) \rightleftharpoons H^+(aq) + OH^-(aq)$$

If inert platinum or graphite electrodes are used, the products are copper and oxygen. Copper is formed at the negative electrode as copper(II) ions gain electrons in preference to hydrogen ions, because copper is below hydrogen in the reactivity series. At the positive electrode hydroxide ions donate electrons in preference to sulfate ions.

$$(-) \text{ electrode: } Cu^{2+}(aq) + 2e^- \rightarrow Cu(s)$$
$$(+) \text{ electrode: } 4OH^-(aq) \rightarrow 2H_2O(l) + O_2(g) + 4e^-$$

What will be observed in this case is a reddish-brown deposit of copper on the negative electrode (cathode) and bubbles of colourless oxygen gas at the positive electrode (anode), and, as the concentration of the copper(II) ions decreases, the solution loses its blue colour. Although it will not be observed, the solution will also become more acidic as the concentration of hydrogen ions increases.

If copper electrodes are used, however, it is observed that the blue colour of the solution remains constant, and the solution does not become acidic. No gases are evolved, and, although the colour of the electrodes remains the same, the mass of the positive electrode decreases by a certain amount and the mass of the negative electrode increases by the same amount. The explanation for this is that the positive copper electrode is itself reacting to give up electrons and produce copper(II) ions:

(−) electrode: $Cu^{2+}(aq) + 2e^- \rightarrow Cu(s)$
(+) electrode: $Cu(s) \rightarrow Cu^{2+}(aq) + 2e^-$

The net effect is that the concentration of copper(II) ions in the solution does not change. Use is made of this in the refining of copper. One of the main uses of copper is for electrical wiring, where it needs to be very pure, otherwise the resistance of the wire increases. To purify the copper the positive electrode is made of impure copper and the negative electrode is a small piece of pure copper.

This is the basis of **electroplating**. The object to be electroplated is made the negative electrode, and it is placed in a solution of the ions of the metal used to plate it. For example, to silver-plate an object, silver(I) ions, $Ag^+(aq)$, would be used.

Factors affecting the relative amounts of products formed during electrolysis

Provided sufficient ions are present, either in the molten state or in aqueous solution, then the quantity of products formed at the electrodes depends on the charge carried by the ions themselves and on the quantity of electrons flowing through the external circuit.

Charge on the ions

From the relevant balanced half-equation it is easy to determine how many moles of electrons must be involved to produce one mole of product:

$Na^+(l) + e^- \rightarrow Na(l)$
1 mol of electrons required to produce 1 mol of sodium atoms

$Cu^{2+}(aq) + 2e^- \rightarrow Cu(s)$
2 mol of electrons required to produce 1 mol of copper atoms

$Al^{3+}(aq) + 3e^- \rightarrow Al(s)$
3 mol of electrons required to produce 1 mol of aluminium atoms

$4OH^-(aq) \rightarrow 2H_2O(l) + O_2(g) + 4e^-$
4 mol of electrons required to produce 1 mol of oxygen molecules

Quantity of electrons

The quantity of electrons depends on the current (how fast they are flowing) and the time (how long they are flowing for).

If two electrolytic cells are connected in series, then the same current will flow through both and for the same length of time (Figure 16). The relative amounts can then easily be determined.

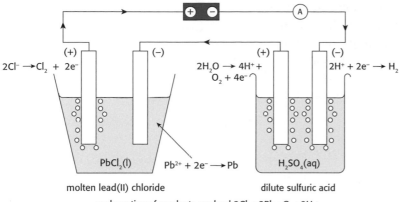

molar ratios of products evolved $2Cl_2 : 2Pb : O_2 : 2H_2$

Figure 16 The electrolysis of two different electrolytes in series

1 During the electrolysis of molten zinc chloride, $ZnCl_2(l)$, 112 cm³ of chlorine gas measured at 273 K and 1.01×10^5 Pa was evolved at the positive electrode. What mass of zinc would have been formed at the negative electrode?

2 Each electron carries a charge of 1.602189×10^{-19} coulombs. One coulomb is the charge produced when a current of one ampere flows for one second. For how long would a steady current of 1.0 A need to be passed through an aqueous solution of copper(II) sulfate to deposit 6.4 g of copper on the negative electrode? If both electrodes were made of copper, what would happen to the positive electrode?

Critical considerations

Why is potassium above lithium in the reactivity series but below lithium in the electrochemical series?

All the alkali metals react with water to give hydrogen and the metal hydroxide, but lithium is clearly the least reactive. When lithium is placed in water, hydrogen is evolved relatively slowly and no flame is seen. The reaction is more vigorous with sodium. With potassium the reaction is even more vigorous, and the heat produced ignites the hydrogen, which burns with a lilac flame owing to the presence of the potassium ions in the gaseous phase. For this reason potassium is placed above sodium and lithium in the reactivity series.

$$M(s) + H_2O(l) \rightarrow M^+(aq) + \tfrac{1}{2} H_2(g) + OH^-(aq)$$

where M represents Li, Na or K.

The enthalpy change for the reaction of water producing hydrogen and hydroxide ions will be the same whatever the metal, so we can attempt to arrive at an explanation by just considering what happens when different alkali metals form their metal ions.

A simple explanation is that, upon descending the group, the outer electron gets progressively further away from the nucleus, so the first ionization energy decreases. As it becomes easier to remove the outer electron so the reactivity increases, because it requires less energy to form the metal ion.

However, is it this simple? If this explanation is true, then why is lithium a better reducing agent than potassium? That is, why is lithium above potassium in the electrochemical series?

$$Li^+(aq) + e^- \rightleftharpoons Li(s) \qquad E^\ominus = -3.03 \text{ V}$$
$$K^+(aq) + e^- \rightleftharpoons K(s) \qquad E^\ominus = -2.92 \text{ V}$$

One answer that is sometimes given is that ionization energies refer only to the gaseous state. In the electrochemical series the solid metal is losing an electron to form the aqueous ion, so two other factors need to be taken into account: the atomization enthalpy change for the metal, and the enthalpy change of hydration of the metal ion. This can be shown in an energy cycle (Figure 16).

From the values it can be seen that the hydration energy of lithium ions has a more negative value (that is, the reaction is more exothermic) than for potassium ions. This is because lithium ions are smaller than sodium ions, so they have a greater charge density Overall, the enthalpy change for the reaction $M(s) \rightarrow M^+(aq) + e^-$ is less endothermic for lithium than for potassium: hence lithium will be above potassium in the electrochemical series. This explanation seems plausible at first sight, but it is worth looking at the values for all the alkali metals:

$\Delta H(Li) = +161 + 519 - 519 = +161$ kJ mol^{-1}
$\Delta H(Na) = +109 + 494 - 406 = +197$ kJ mol^{-1}
$\Delta H(K) = +90.0 + 418 - 322 = +186$ kJ mol^{-1}
$\Delta H(Rb) = +85.8 + 402 - 301 = +187$ kJ mol^{-1}
$\Delta H(Cs) = +78.7 + 376 - 276 = +177$ kJ mol^{-1}

In fact all the values are very similar, and we might expect from this that the alkali metals would all react in a very similar way—but clearly they do not.

When caesium is placed in water the reaction is so explosive that the glass container can shatter. (Video clips of this reaction can be seen by typing "reaction of caesium with water" in an Internet search engine.) It seems likely that the explanation has more to do with the rate of the reaction than with enthalpy changes. Probably they all give out a similar amount of heat when they react with water, but caesium is potentially explosive because the reaction is so much faster. One possibility is that the activation energy for the reaction decreases down the group. Certainly the amount of energy that has to be put in to atomize and then ionize the metals before they can be hydrated decreases down the group, but it is difficult to see that this should have such an effect on their reactivity.

Another explanation for the faster rate is to consider the available surface area. Only the metal atoms on the surface come into contact with the water and are able to react. Lithium has the highest melting point (180 °C), whereas caesium with a melting point of just 28.7 °C is almost a liquid at room temperature, so caesium atoms are much more available to react.

Which explanation do you think is most likely? Can you think of an even better one?

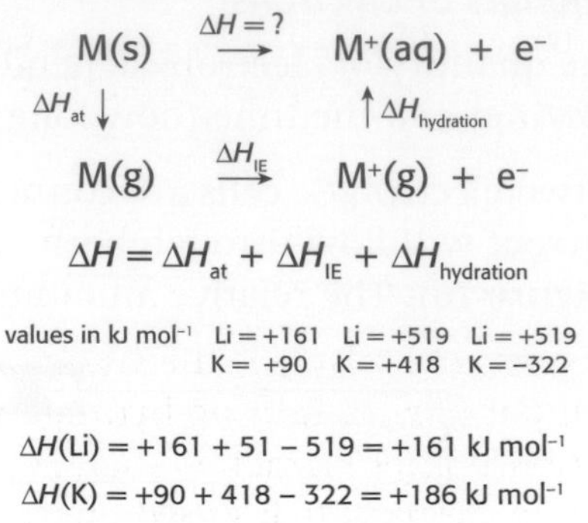

$$M(s) \xrightarrow{\Delta H = ?} M^+(aq) + e^-$$
$$\Delta H_{at} \downarrow \qquad \uparrow \Delta H_{hydration}$$
$$M(g) \xrightarrow{\Delta H_{IE}} M^+(g) + e^-$$

$$\Delta H = \Delta H_{at} + \Delta H_{IE} + \Delta H_{hydration}$$

values in kJ mol^{-1} Li = +161 Li = +519 Li = +519
K = +90   K = +418   K = −322

$\Delta H(Li) = +161 + 51 - 519 = +161$ kJ mol^{-1}
$\Delta H(K) = +90 + 418 - 322 = +186$ kJ mol^{-1}

Figure 16 Energy cycle for lithium or potassium metal forming aqueous ions in water

10 Organic chemistry

Introduction

Organic chemistry is the study of the chemistry of carbon compounds containing hydrogen. The compounds may also contain other elements, such as oxygen, nitrogen, a halogen, or sulfur. Originally it was thought that these compounds were only produced by some vital force in living organisms, but in 1828 a German chemist, Friedrich Wöhler (1800–1882), accidentally synthesized urea from inorganic chemicals. Urea occurs in urine and has the formula $(NH_2)_2CO$. Wöhler was actually attempting to synthesise ammonium cyanate, NH_4CNO, and as well as destroying the vitalism theory he also discovered one of the earliest examples of isomerism, as both urea and ammonium cyanate have the same molecular formula, CH_4N_2O.

Carbon is unique, in that it forms more compounds than are known of all the other elements in the Periodic Table put together. Many millions of organic compounds have been identified or synthesized, and it is worth considering why carbon is so unique compared with all the other elements.

- *Carbon can form multiple bonds to itself and with atoms of other elements*
  Carbon can form single bonds (C–C), double bonds (C=C), and triple bonds (C≡C) to itself. As well as forming single bonds it can form double and triple bonds with other elements such as oxygen in ethanoic acid, CH_3COOH, and nitrogen in hydrogen cyanide, HCN.
- *The strength of the C–C single bond and the C–H bond*
  The strength of the C–C bond means that it is thermally quite stable, particularly when compared with the single bond formed between two atoms of other elements in group 4 of the Periodic Table (see Table 1).
  The single covalent bond that carbon forms with hydrogen is also very strong, 412 kJ mol^{-1}, when compared with the silicon to hydrogen bond, 318 kJ mol^{-1}. This also contributes to the thermal stability of carbon compounds.
- *Carbon can form chains and rings*
  The ability of carbon to form strong bonds to itself means that it is able to form long chains of carbon atoms. In some polymers there can be well over 1000 carbon atoms in a single chain. The chains can also link to themselves to form cyclic or ring compounds, such as cyclohexane, C_6H_{12}, and benzene, C_6H_6. See Figure 1. This ability to form chains and rings is known as catenation.
- *Carbon cannot expand its octet of valence electrons*
  When it is bonded, a carbon atom has a share in a full octet of electrons, and the outer shell cannot expand to include any more electrons, so it cannot form any more bonds. Silicon,

Table 1

Bond	Average bond enthalpy / kJ mol^{-1}
C–C	348
Si–Si	226
Ge–Ge	188
Sn–Sn	151

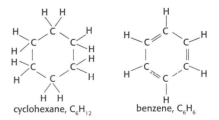

cyclohexane, C_6H_{12} benzene, C_6H_6

Figure 1 The structures of cyclohexane and benzene.

The more spherical the isomer, then the lower the boiling point will be, because there will be less surface area to form intermolecular forces of attraction.

Because the alkanes are non-polar, or have extremely low polarity, they do not dissolve in polar solvents such as water. If they are liquids, then when mixed with water they simply form two immiscible layers. However, they can be good solvents in their own right, because they dissolve other non-polar compounds such as fats and oils.

Chemically, alkanes are very unreactive. Their old name was **paraffins**, which literally means "little activity" (from the Latin *parum affinis*). The reasons for this are mainly the relatively strong carbon-to-carbon and carbon-to-hydrogen bonds within them, and their inability to expand their octet, as explained at the beginning of the chapter. At first sight their relative inertness seems surprising, because they burn extremely well, and their principal use is as fuels. Because of the relatively strong bonds within the alkanes they are stable in air until sufficient activation energy is provided. As explained in Chapter 5, the combustion reactions are then strongly exothermic, owing to the formation of the even stronger carbon-to-oxygen double bonds in carbon dioxide, and the strong oxygen-to-hydrogen bonds in water. Figure 10 shows the enthalpy diagram for the complete combustion of the simplest alkane, methane.

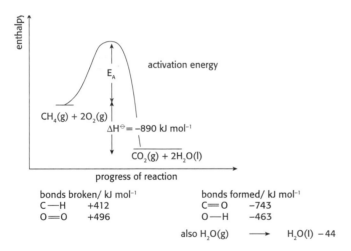

Figure 10 Enthalpy diagram for the complete combustion of methane.

If there is insufficient oxygen for complete combustion, alkanes will still burn, but undergo **incomplete combustion**. The products of incomplete combustion are carbon monoxide and carbon itself in addition to water and some carbon dioxide. For example, the equation for the complete combustion of octane is

$$2C_8H_{18}(l) + 25O_2(g) \rightarrow 16CO_2 + 18H_2O(l)$$

A possible equation for a reaction involving the incomplete combustion of octane is

$$C_8H_{18}(l) + 9O_2(g) \rightarrow C(s) + 5CO(g) + 2CO_2(g) + 9H_2O(l)$$

The burning of gasoline and diesel in internal combustion engines causes considerable pollution. This is particularly so when the engine is cold, because the ratio of petrol to air in the mixture needs to be greater, and both carbon monoxide and carbon (in the form of particulates) are emitted. Because of this, using a car for a short journey is relatively much more polluting than using it for a longer journey. Unfortunately, even when the engine has reached its normal operating temperature some incomplete combustion still takes place, and now another pollutant is added, because nitrogen and oxygen in the air can react at the high temperature reached inside the internal combustion to produce oxides of nitrogen.

Apart from combustion, the only other type of reaction that alkanes readily undergo is with the halogens. In the dark there is no reaction, but in ultraviolet light a rapid substitution reaction takes place. For example, methane will react with chlorine in ultraviolet light to produce chloromethane and hydrogen chloride:

$$CH_4(g) + Cl_2(g) \xrightarrow{uv} CH_3Cl(g) + HCl(g)$$

This reaction takes places in a series of separate steps. The **reaction mechanism** of an organic reaction describes and explains these steps. Overall, this particular reaction mechanism is known as **free radical substitution**. The first step, known as the **initiation step**, involves the formation of free radicals. It is for this reason that the reaction occurs only in the presence of ultraviolet light, because this provides the energy to break the chlorine-to-chlorine bond **homolytically** (Figure 11).

This produces chlorine atoms in a form known as **free radicals**. Free radicals contain an unpaired electron in one of their orbitals, and are a highly energetic and reactive species. When a chlorine free radical collides with a methane molecule it reacts to form hydrogen chloride and another free radical. This methyl radical can react with a molecule of chlorine, and the product, chloromethane, is formed, and at the same time another chlorine free radical has been generated. These two steps are known as **propagation steps**, because in each step radicals react to generate (propagate) new radicals. Once the ultraviolet light has initiated the formation of free radicals, then one radical can effectively go on to produce many molecules of products. However, there will also be **termination steps**. The radicals may collide with the walls of the vessel, or escape completely, or they may react with another free radical to produce a non-radical product. In fact, a trace of ethane can be found in the products, which provides good evidence for the free radical mechanism. See Figure 12.

A chemical bond between two atoms X and Y involves a shared pair of electrons. when it breaks there are two ways it can break heterolytically so that one atom has both electrons:

$$X\text{-}Y \longrightarrow X^+ + Y^-$$
$$X\text{-}Y \longrightarrow X^- + Y^+$$
heterolytic fission

It can also break homolytically with the result that each atom contains one unpaired electron:

$$X\text{-}Y \longrightarrow X^{\bullet} + Y^{\bullet} \text{ homolytic fission}$$

X˙ and Y˙ are known as free radicals.

Figure 11 Heterolytic and homolytic fission of bonds.

initiation $Cl\text{—}Cl(g) \xrightarrow{uv} 2Cl^{\bullet}(g)$

propagation $\begin{cases} Cl^{\bullet}(g) + CH_4(g) \longrightarrow CH_3^{\bullet}(g) + HCl(g) \\ CH_3^{\bullet}(g) + Cl_2(g) \longrightarrow CH_3Cl(g) + Cl^{\bullet}(g) \end{cases}$

possible termination steps $\begin{cases} Cl^{\bullet}(g) + Cl^{\bullet}(g) \longrightarrow Cl_2(g) \\ Cl^{\bullet}(g) + CH_3^{\bullet}(g) \longrightarrow CH_3Cl(g) \\ CH_3^{\bullet}(g) + CH_3^{\bullet}(g) \longrightarrow C_2H_6(g) \end{cases}$

overall reaction $Cl_2(g) + CH_4(g) \xrightarrow{uv} CH_3Cl(g) + HCl(g)$

Figure 12 Mechanism for the free radical substitution reaction between methane and chlorine in ultraviolet light.

If there is sufficient chlorine in the reaction mixture, then the reaction does not stop at just monosubstitution to produce chloromethane, but successive substitution can occur to produce—ultimately—tetrachloromethane:

$$Cl_2(g) + CH_4(g) \xrightarrow{UV} CH_3Cl(g) + HCl(g)$$
$$\text{chloromethane}$$

$$CH_3Cl(g) + Cl_2(g) \xrightarrow{UV} CH_2Cl_2(l) + HCl(g)$$
$$\text{dichloromethane}$$

$$CH_2Cl_2(g) + Cl_2(g) \xrightarrow{UV} CHCl_3(l) + HCl(g)$$
$$\text{trichloromethane}$$

$$CHCl_3(l) + Cl_2(g) \xrightarrow{UV} CCl_4(l) + HCl(g)$$
$$\text{tetrachloromethane}$$

Extension material

Free radical substitution and the destruction of the ozone layer

In the second half of the twentieth century much use was made of chlorofluorocarbons (CFCs). These were compounds containing carbon atoms bonded to both chlorine and fluorine atoms. One of the simplest was dichlorodifluoromethane, CF_2Cl_2 (Figure 13).

Figure 13 Dichlorodifluoromethane.

They were useful because they were chemically inert, so they were used as refrigerator gases and as carrier gases in pressurized containers such as spray-on deodorants.

In the early 1970s two chemists working at the University of California in the USA, Frank Rowland (1927–) and Mario Molina (1943–), put forward the hypothesis that CFCs would stay in the atmosphere for a long time, because they were so unreactive, but as they moved up to the ozone layer in the stratosphere (12–50 km above the Earth's surface) the weaker carbon-to-chlorine bond would be broken homolytically by the ultraviolet light:

$$CF_2Cl_2(g) \rightarrow CF_2Cl^\bullet(g) + Cl^\bullet(g)$$

The chlorine free radicals formed would then have the capacity to affect the ozone layer by breaking down the ozone by a free radical mechanism:

$$Cl^\bullet(g) + O_3(g) \rightarrow ClO^\bullet(g) + O_2(g)$$
$$ClO^\bullet(g) + O^\bullet(g) \rightarrow Cl^\bullet(g) + O_2(g)$$

Overall:

$$O_3(g) + O^\bullet(g) \rightarrow 2O_2(g)$$

This would be serious, because the ozone in the stratosphere absorbs much of the dangerous ultraviolet light from the Sun. It has been estimated that one chlorine free radical could cause the destruction of up to 100 000 ozone molecules before the radical reaction is terminated. At the time these ideas were not taken seriously, and were heavily criticized, particularly by the manufacturers of chlorofluorocarbons. However, in 1985 satellite evidence showed that the decrease in the concentration of ozone above the Antarctic was actually far larger than anyone had predicted. This much lower region of ozone concentration became

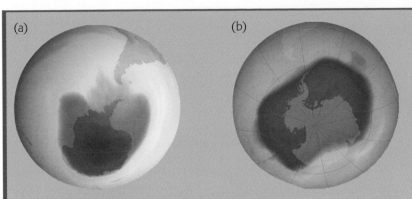

(a)   (b)

Figure 14 (a) An image from a NASA satellite, taken in 1999, showing the "ozone hole" over Antarctica. (b) An image taken in September 2006 of the largest Antarctic "ozone hole" so far recorded.

known as the **ozone hole** (Figure 14a). Indirect evidence that the ozone layer was being affected also came from the increase in the incidence of skin cancer caused by the increasing amount of ultraviolet light reaching ground level.

In 1987, at the headquarters of the International Civil Aviation Organization in Montreal, governments from around the world signed up to the **Montreal Protocol** on Substances that Deplete the Ozone Layer, thereby agreeing to phase out the use of CFCs. Because CFCs can remain in the atmosphere for up to 80 years, it is too early to establish whether the rate of decrease in ozone concentrations has been slowed, halted or reversed. In fact, in September 2006 the largest hole ever was reported over the Antarctic (Figure 14b).

Frank Rowland and Mario Molina were awarded the Nobel Prize in Chemistry in 1995 for their work on stratospheric ozone. The third recipient of the Nobel Prize in Chemistry that year was Paul Crutzen (1933–), a Dutch chemist, who had shown that oxides of nitrogen released from high-flying jet airplanes would also have a detrimental affect on the ozone layer.

Functional groups

We have seen that when a hydrogen atom has been removed from an alkane the resulting group has the name methyl-, ethyl-, propyl- etc. This is known as an **alkyl group**, and can be represented in general terms by the letter R-. An alkyl group cannot exist on its own, but will be bonded to another atom or group of atoms. This other atom or group of atoms is known as a **functional group**. Compounds with different functional groups will form their own homologous series and have different physical and chemical properties. Some examples of compounds with different functional groups, together with the way they are named, are as follows:

Alcohols

General formula: R-**OH**
Naming: The -e at the end of the alkane is replaced by -ol.
Examples: see Figure 15.

1 There are five structural isomers of hexane, C_6H_{14}. Four of them are: hexane, 2-methylpentane, 3-methylpentane and 2,2-dimethylbutane. Name and draw the fifth isomer.

2 Describe the mechanism for the reaction of ethane with bromine in the presence of ultraviolet light. Name the product, and draw the structural formula of the organic compound formed, when 1 mol of ethane reacts with 1 mol of bromine under these conditions.

3 When the use of CFCs was banned by the Montreal Protocol, alternatives had to be found. One of the most used alternatives is a hydrofluorocarbon, HFC, which has the commercial name R-134a. The chemical name of R-134a is 1,1,1,2-tetrafluoroethane. Draw its structural formula.

195

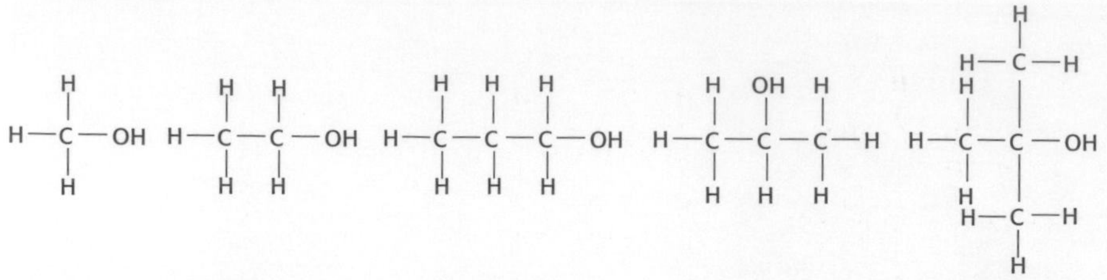

CH₃OH
methanol

C₂H₅OH
ethanol

C₃H₇OH
propan-1-ol

CH₃CH(OH)CH₃
propan-2-ol

C(CH₃)₃OH
2-methylpropan-2-ol

Figure 15 Some alcohols.

Halogenoalkanes

General formula: R-**X** (where **X** represents F, Cl, Br or I).
Naming: fluoro-, chloro-, bromo- or iodo- is placed in front of the alkane.
Examples: see Figure 16.

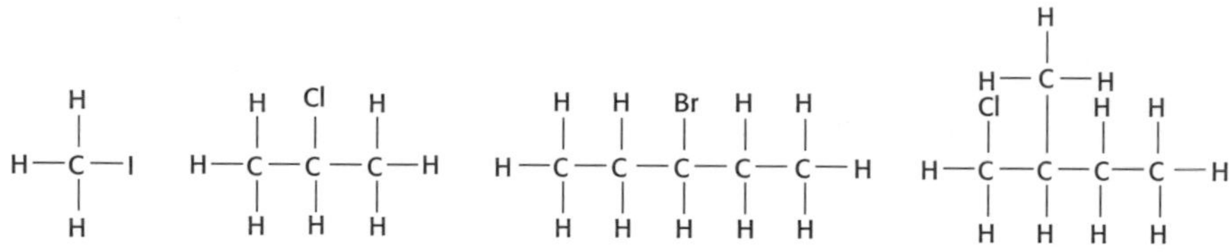

CH₃I
iodomethane

CH₃CHClCH₃
2-chloropropane

CH₃CH₂CHBrCH₂CH₃
3-bromopentane

CH₂ClCH(CH₃)CH₂CH₃
1-chloro-2-methylbutane

Figure 16 Some halogenoalkanes.

Aldehydes

General formula: R-**CHO** or

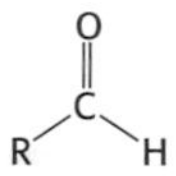

Naming: The -e at the end of the longest chain alkane is replaced by -al.
Examples: see Figure 17.

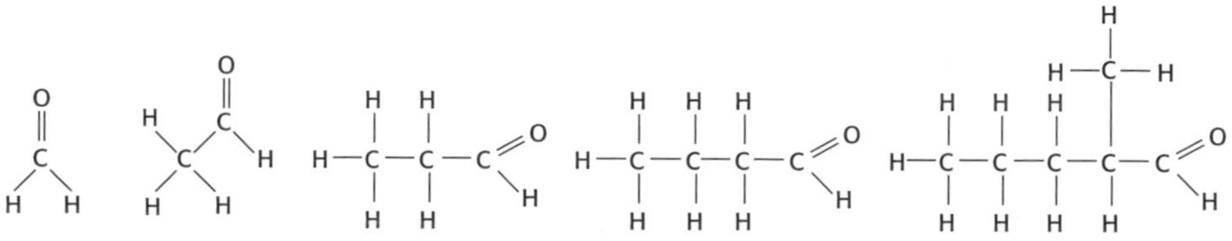

HCHO
methanal

CH₃CHO
ethanal

C₂H₅CHO
propanal

C₃H₇CHO
butanal

CH₃CH₂CH₂CH(CH₃)CHO
2-methylpentanal

Figure 17 Some aldehydes.

Ketones

General formula: R-**COR′** or 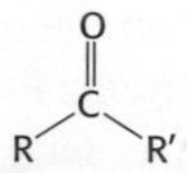 (R′ represents either the same alkyl group as R or a different alkyl group).
Naming: The -e on the end of the longest chain alkane is replaced by -one.
Examples: see Figure 18.

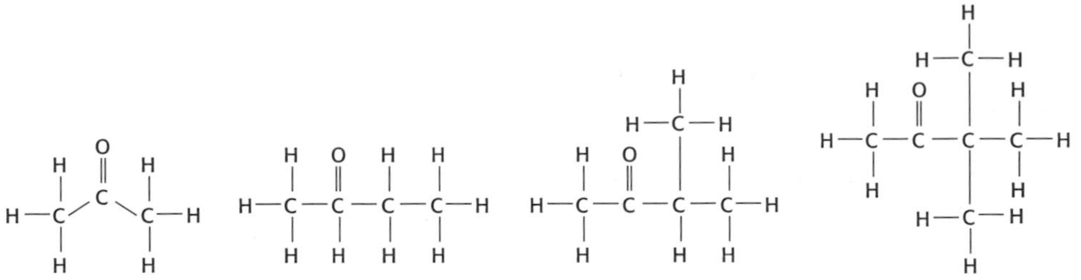

CH_3COCH_3 propanone

$CH_3COC_2H_5$ butan-2-one

$CH_3COCH(CH_3)CH_3$ 3-methylbutan-2-one

$CH_3COC(CH_3)_3$ 3,3-dimethylbutan-2-one

Figure 18 Some ketones.

Carboxylic acids

General formula: R-**COOH** or 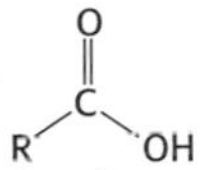

Naming: The -e on the end of the longest chain alkane is replaced by -oic acid.

Examples: see Figure 19.

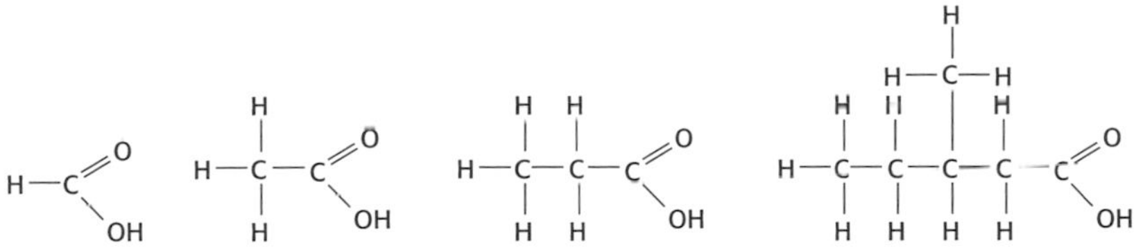

HCOOH methanoic acid

CH_3COOH ethanoic acid

C_2H_5COOH propanoic acid

$C_2H_5CH(CH_3)CH_2COOH$ 3-methylpentanoic acid

Figure 19 Some carboxylic acids.

Esters

General formula: R-**COOR′** or 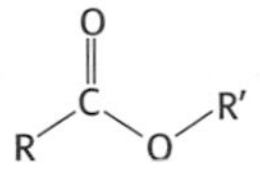

Examples: see Figure 20.

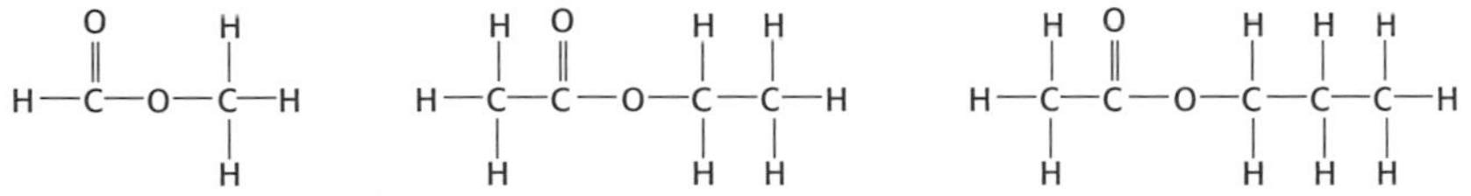

$HCOOCH_3$ methyl methanoate

$CH_3COOC_2H_5$ ethyl ethanoate

$CH_3COOC_3H_7$ propyl ethanoate

Figure 20 Some esters.

Amines

General formula: R-$\mathbf{NH_2}$

Examples: see Figure 21.

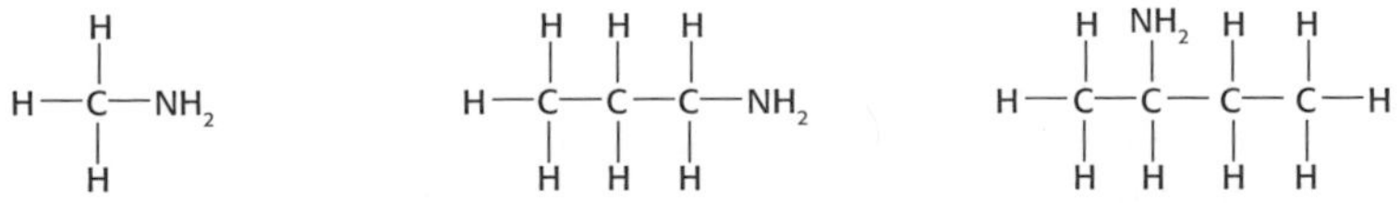

CH_3NH_2 methylamine

$C_3H_7NH_2$ propylamine

$CH_3CH(NH_2)CH_2CH_3$ 2-aminobutane

Figure 21 Some amines.

Aromatic compounds

Instead of R- representing an alkyl group, it can also represent an **aryl group**. Aryl groups contain a **benzene ring**. Benzene itself is C_6H_6, and in aryl compounds one of the hydrogen atoms has been substituted by a functional group. The simplest aryl group is the phenyl group, $-C_6H_5$.

Examples: see Figure 22.

Figure 22 Some aromatic compounds.

Physical properties of compounds with different functional groups

Alkanes are not soluble in water, and have low melting and boiling points, because they are essentially non-polar substances. However, if the functional group is polar it can have a very great effect on the physical properties of the compound, although this effect decreases as the number of carbon atoms in the non-polar R- (alkyl or aryl) group increases. The greater the polarity of the functional group, then the higher the melting and boiling points of the compounds are, and the more soluble in water they will be. We can classify the main functional groups listed in terms of the type of intermolecular forces of attraction between their molecules (Table 3).

Table 3

Group	Type of intermolecular forces	Relative boiling point	Soluble in water
alkanes	van der Waals'	very low	no
halogenoalkanes	weak dipole–dipole	low	no
aldehydes	dipole–dipole	moderate	yes
ketones	dipole–dipole	moderate	yes
alcohols	hydrogen bonding	high	yes
carboxylic acids	hydrogen bonding	high	yes

It should be stressed that the solubility in water refers only to the lowest members. As the size of the hydrocarbon chain increases, the solubility in water rapidly decreases. In addition to melting points and boiling points, the volatility of liquid compounds is often referred to. A **volatile liquid** is one that has a high vapour pressure and therefore readily evaporates at low temperature. Generally this refers to liquids that boil below 100 °C: that is, below the boiling point of water. We can illustrate Table 3 with some examples: see Table 4. To make meaningful comparisons, compounds with the same or similar molecular masses are taken. Propane and chloromethane are insoluble in water, whereas ethanal, ethanol and methanoic acid all mix freely with water.

?

1 Find the structural formula for cholesterol in the IB data booklet. Both ethanol and cholesterol contain one –OH group. Explain why ethanol is soluble in water, whereas cholesterol is insoluble in water.

2 Would you expect propan-1-ol or propan-2-ol to have the highest boiling point? Explain your reasoning, and then check your prediction.

3 The structures of two painkillers, aspirin and heroin, are given in the IB data booklet. Their structures have some large differences, but they both contain a phenyl group (benzene ring), and also have one other functional group in common. Identify this group.

Table 4

	C_3H_8 Propane	CH_3Cl Chloromethane	CH_3CHO Ethanal	C_2H_5OH Ethanol	HCOOH Methanoic acid
M_r	44.0	50.5	44.0	46.0	46.0
B.p. / °C	−42.2	−23.8	20.8	78.5	101.0

Alkenes

Alkenes contain a carbon-to-carbon double bond. They form a homologous series with the general formula C_nH_{2n}. The simplest alkene is ethene, C_2H_4 (Figure 23).

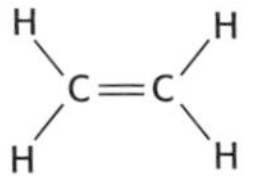

Alkenes are named by taking the longest carbon chain and then specifying where the double bond begins (Figure 24).

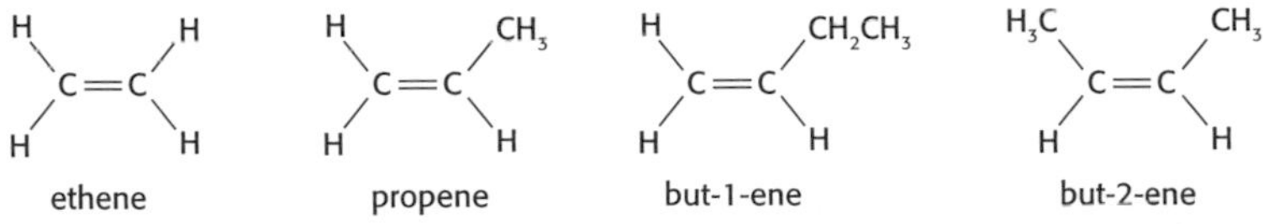

Figure 23 The structural formula of ethene.

Figure 24 Some alkenes.

The physical properties of alkenes are similar to those of alkanes, as they are either non-polar or have very low polarity. Like alkanes they have low boiling points, which increase as the molar mass increases (Table 5), and they are insoluble in polar solvents such as water.

Like all hydrocarbons, alkenes burn in excess oxygen to give carbon dioxide and water as the products of complete combustion.

$$C_nH_{2n} + (n + n/2)O_2(g) \rightarrow nCO_2(g) + nH_2O(l)$$

In air where the amount of oxygen is limited they burn with a more yellow and sooty flame than alkanes, as more incomplete combustion occurs.

Apart from combustion, because of the presence of the double bond in alkenes their other chemical properties are very different from those of alkanes.

Table 5 Boiling points of alkenes

Alkene	Formula	Boiling point / °C
ethene	$CH_2=CH_2$	−104.0
propene	$CH_3CH=CH_2$	−47.7
but-1-ene	$CH_3CH_2CH=CH_2$	−6.2
pent-1-ene	$CH_3(CH_2)_2CH=CH_2$	30.0
hex-1-ene	$CH_3(CH_2)_3CH=CH_2$	63.9

Addition reactions of alkenes

Alkenes are said to be **unsaturated**, because they can undergo addition reactions across the double bond. When this happens the product is said to be **saturated**, because both carbon atoms will then have four single bonds to other atoms, and no further addition can occur (Figure 25).

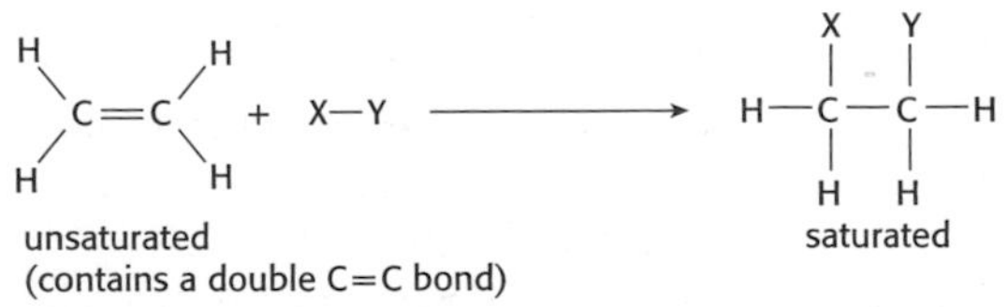

Figure 25 Addition reaction of ethene.

- *With hydrogen (hydrogenation)*
  Hydrogen will add to alkenes in the presence of a nickel catalyst at a temperature of 180 °C. The product will be the corresponding alkane. For example, ethene will produce ethane, and both but-1-ene and but-2-ene will produce butane:

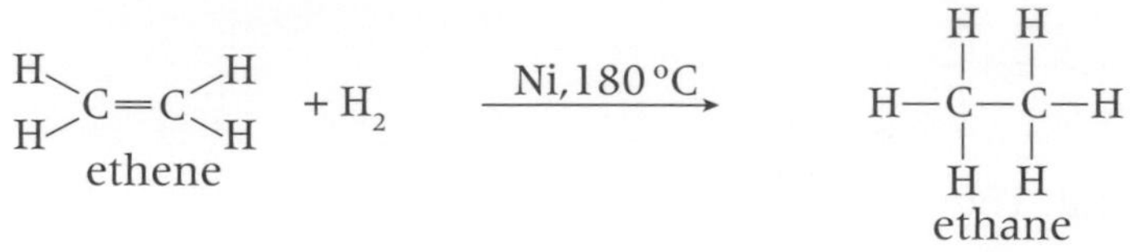

H_2C=$CHCH_2CH_3$ + H_2 $\xrightarrow{\text{Ni, 180 °C}}$ $CH_3CH_2CH_2CH_3$
but-1-ene   butane

CH_3CH=$CHCH_3$ + H_2 $\xrightarrow{\text{Ni, 180 °C}}$ $CH_3CH_2CH_2CH_3$
but-2-ene   butane

- *With halogens (halogenation)*
  Alkenes react readily with halogens to produce the di-substituted product. For example, ethene reacts with bromine in a non-polar solvent to produce 1,2-dibromoethane, and propene will react with chlorine in a non-polar solvent to produce 1,2-dichloropropane:

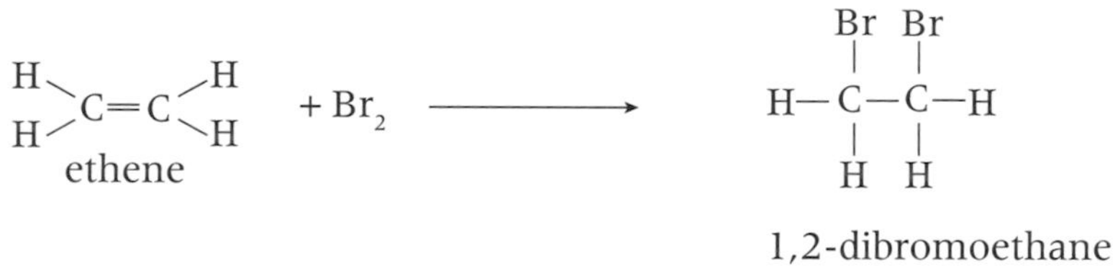

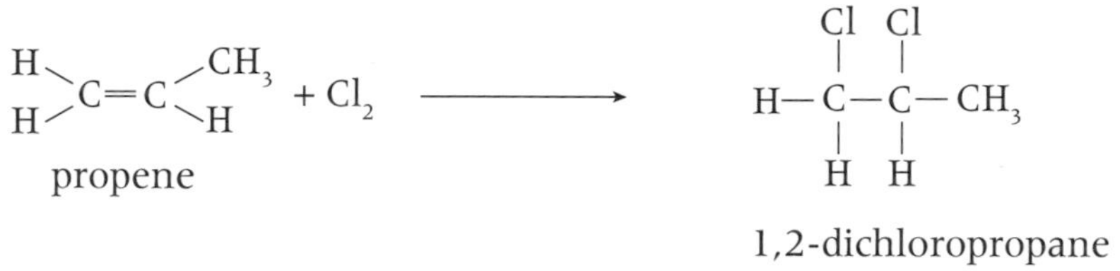

Test for unsaturation

The reaction with bromine provides a useful way to distinguish in practice between saturated and unsaturated compounds in the laboratory. Bromine water is a yellow/orange-coloured solution. If a few drops of bromine water are added to an excess of the alkene and the mixture is shaken, then the colour will rapidly disappear, because the organic product is colourless (Figure 26).

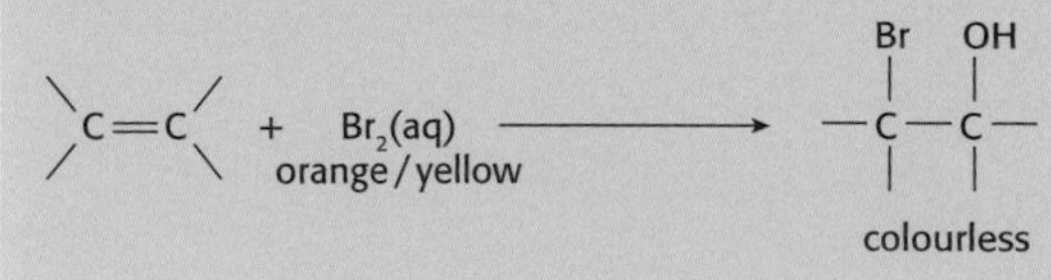

Figure 26 Test for unsaturation.

It is worth noting that, in the presence of water, the product is not actually the dibromo-addition product, because one of the bromine atoms is replaced by a hydroxyl group to form an alcohol. For example:

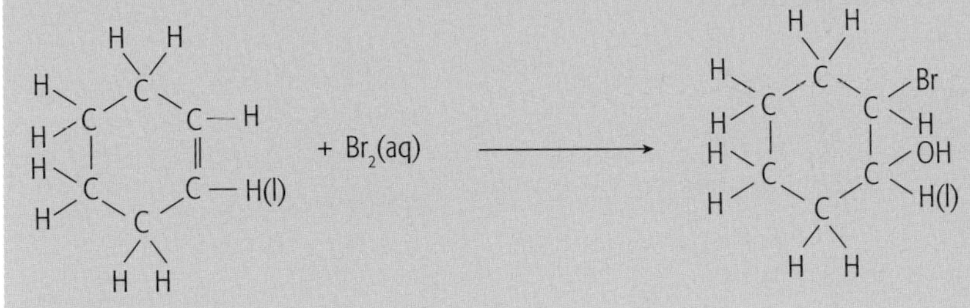

cyclohexene   + Br$_2$(aq) →   2-bromocyclohexanol

- *With hydrogen halides*
  Hydrogen halides such as hydrogen chloride and hydrogen bromide react readily with alkenes to form the addition product. For example, ethene reacts with hydrogen chloride to form chloroethane, and but-2-ene reacts with hydrogen bromide to form 2-bromobutane:

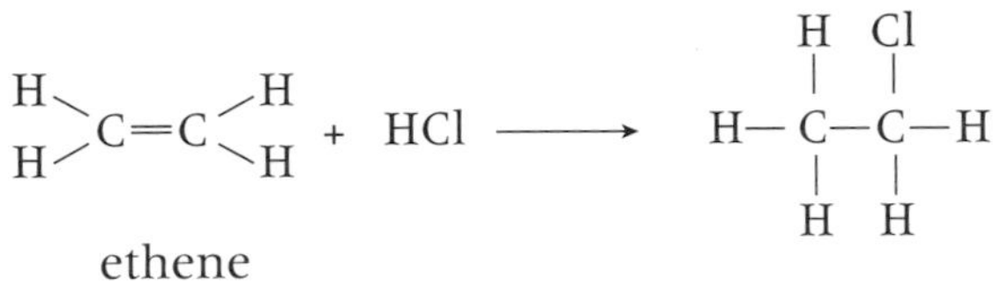

ethene   + HCl →   chloroethane

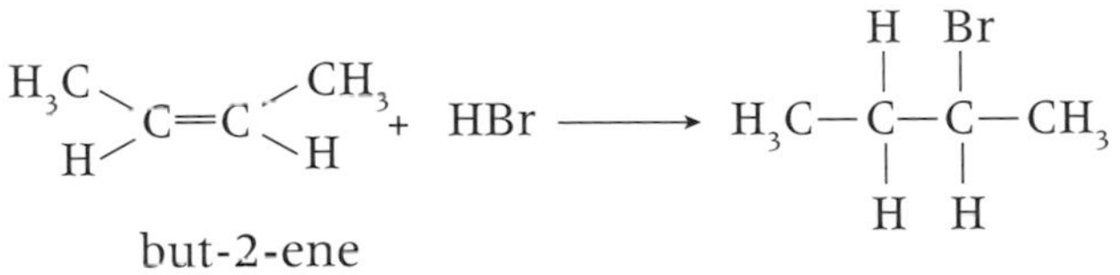

but-2-ene   + HBr →   2-bromobutane

- *With water*
  Water does not react directly with alkenes under normal conditions. However, if concentrated sulfuric acid is used, then the hydrogen sulfate addition product is formed first. This can then react with water to give an alcohol and regenerate the sulfuric acid. For example, the reaction of ethene with concentrated sulfuric acid will produce ethanol:

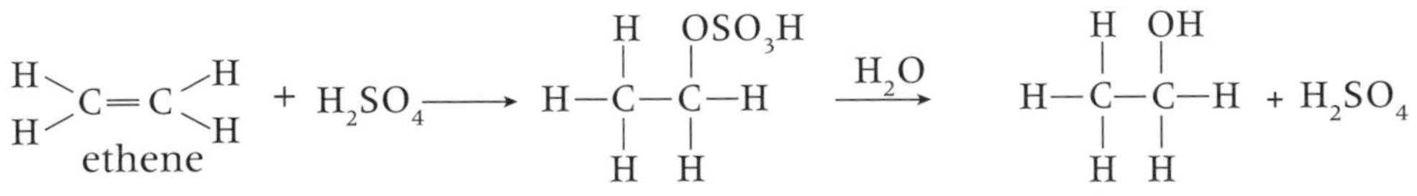

ethene   + H$_2$SO$_4$ →   ethyl hydrogensulfate $\xrightarrow{H_2O}$ ethanol   + H$_2$SO$_4$

- *With themselves (addition polymerization)*
  Under certain conditions alkenes can also add to themselves.
  For example, at high pressure ethene can polymerize to form
  poly(ethene), which is more usually known as polythene:

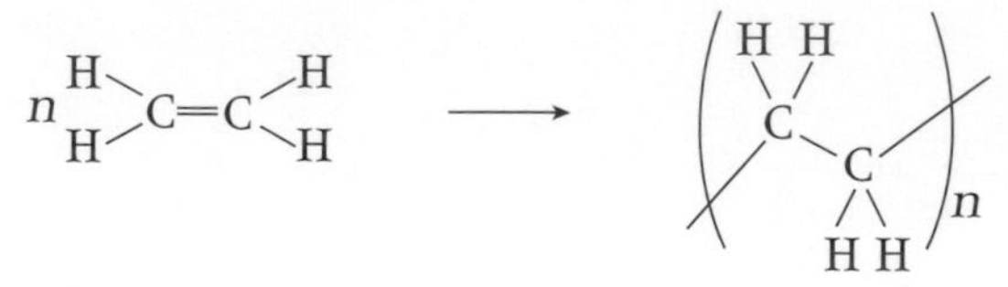

ethene monomer   repeating unit of poly(ethene)

This last type of reaction is known as **addition polymerization.**
A whole variety of addition polymers can be formed from
different alkenes. Three other common addition polymers
are: poly(chloroethene), also known as polyvinylchloride,
PVC; poly(phenylethene), also known as polystyrene; and
poly(tetrafluoroethene), PTFE, also known as Teflon® and used in
"non-stick" kitchenware.

chloroethene monomer   poly(chloroethene)
(old name: vinyl chloride)   old name: polyvinylchloride, PVC

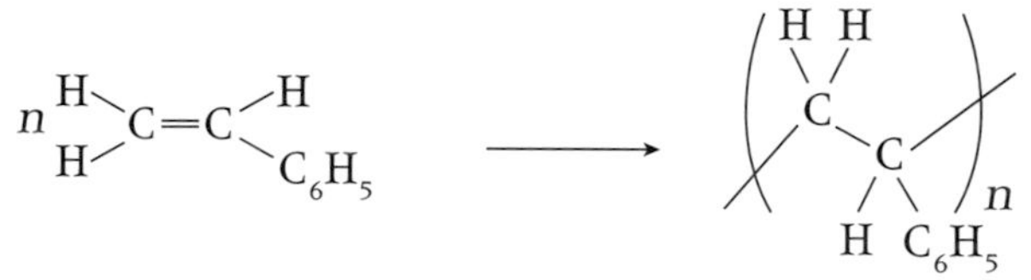

phenylethene monomer   poly(phenylethene)
(old name: styrene)   old name: polystyrene

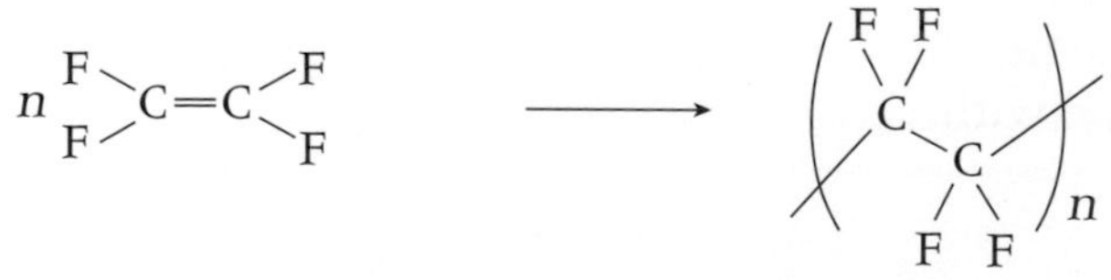

tetrafluoroethene   poly(tetrafluoroethene), PTFE

The addition reactions of ethene are summarized in Figure 27.

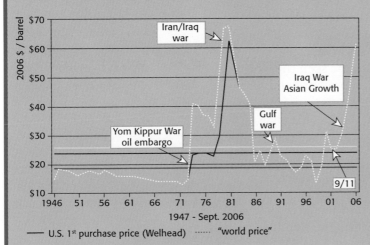

Figure 27 showing the addition reactions of ethene structures:

H—C—C—H with H H H H (alkane)

H₂

H—C—C—H with H Br H H (halogenoalkane)

HBr

C=C (alkene)

Br₂

H—C—C—H with Br Br H H (dihalogenoalkane)

H₂O (H₂SO₄ catalyst)

H—C—C—H with H OH H H (alcohol)

$H_2C{=}CH_2$

$-\!\!\left(CH_2\!-\!CH_2\right)_n$ (addition polymer)

Figure 27 The addition reactions of ethene.

The economic importance of alkenes

Extension material

The price of crude oil

Although simple alkenes do occur naturally (for example, ethene is produced when bananas ripen), they do not occur in any great quantities. Crude oil is mainly a mixture of many different alkanes. After the fractional distillation of crude oil the alkanes are grouped into different boiling ranges, known as **fractions**. The main source of alkenes is from the cracking of the higher-boiling fractions to give smaller, more useful molecules. For example, decane can be broken down ("cracked") in the presence of a catalyst at a high temperature to form octane and ethene:

$$C_{10}H_{22}(g) \rightarrow C_8H_{18}(g) + C_2H_4(g)$$

About 10% of crude oil is converted into alkenes, which are then used as a chemical feedstock. It may well be that future generations may question why we burned most of our natural resources of oil and natural gas to cause pollution and global warming, rather than use them to make other important products such as plastics, pesticides, food additives, pharmaceuticals, detergents, cosmetics, dyes and solvents. Because these products are oil-based, their price is very dependent on the price of crude oil (Figure 28).

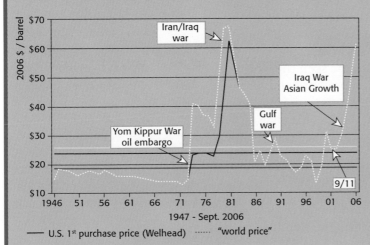

— U.S. 1ˢᵗ purchase price (Welhead)   ⋯⋯ "world price"

Figure 28 Crude oil prices in 2006 dollars.

Because much of the world's oil originates from the Middle East, events there have a large bearing on the price of crude oil. Eventually crude oil reservoirs will be so depleted that chemists will need to consider other sources of carbon for essential feedstock. At the moment the two best sources are coal, of which there are still large reserves, and biomass.

Three key areas for the use of alkenes are the hydrogenation of polyunsaturated liquid lipids to make margarine, the production of ethanol, and the production of addition polymers.

Margarine

Liquid lipids (oils) are triesters (triglycerides) made from propane-1,2,3-triol (glycerol) and carboxylic acids (fatty acids) containing at least one carbon-to-carbon double bond. The structure shown in Figure 29 is for an oil made from combining 3 mol of oleic acid, $CH_3(CH_2)_7CH=CH(CH_2)_7COOH$, with 1 mol of propane-1,2,3-triol, which is found in olive oil.

Figure 30 Margarine containing hydrogenated vegetable oils. The wrapping is aluminium foil coated with polythene.

Figure 29 A triglyceride (triester) made from oleic acid and propan-1,2,3-triol.

To enable a mixture of vegetable oils to solidify for use as margarine (Figure 30), some of the carbon-to-carbon double bonds are hydrogenated by using a nickel catalyst under pressure. This enables the long-chain molecules to pack more closely, which increases the van der Waals' attractive forces between them and so increases the melting point.

Ethanol

Ethanol has many industrial uses, principally as a solvent, as a fuel, and as an intermediate in the production of other organic chemicals such as esters and ethanolamines. For alcoholic drinks it is obtained by fermenting the sugar in grapes or other fruits, but for industrial use it is obtained mainly by hydrating ethene.

Ethene and steam are passed over a catalyst of phosphoric(V) acid, H_3PO_4, adsorbed on to a surface of silicon dioxide:

$$C_2H_4(g) + H_2O(g) \rightleftharpoons C_2H_5OH(g) \ \Delta H^\ominus = -46 \ kJ \ mol^{-1}$$

Using the knowledge gained from Chapter 7 it can be seen that the position of equilibrium will be favoured by a high pressure and a low temperature. In practice, a pressure of about 60–70 atmospheres is used, with an optimum temperature of about 570 K. This gives about 5% conversion, but the unreacted ethene can be separated and recycled.

Addition polymers

The range of addition polymers is vast, and much of modern society relies heavily on polymers. Some examples of addition polymers are listed in Table 6.

?

1 Propene can be polymerized to form poly(propene), which can be drawn into fibres to make clothes and carpets. Write the equation representing n moles of propene polymerizing to form poly(propene), and include the repeating unit for poly(propene).

2 Throughout history we have had "ages". Many of these have been based on the materials we used: for example the Stone Age, the Iron Age, the Bronze Age. Do you think that our current era will one day be known as the Plastic Age?

3 Plastics pose considerable environmental problems when they are discarded. Find out what these problems are, and use the Internet to find out how research into biodegradable plastics is progressing.

Table 6 Some examples of common addition polymers

Polymer	Common name	Repeating unit	Examples of uses
poly(ethene)	polythene	$\begin{array}{cc} H & H \\ | & | \\ -C-C- \\ | & | \\ H & H \end{array}$	wrapping food, bags, containers, electrical insulation
poly(propene)	polypropylene	$\begin{array}{cc} H & CH_3 \\ | & | \\ -C-C- \\ | & | \\ H & H \end{array}$	washing-up bowls, car fenders, packaging
poly(phenylethene)	polystyrene	$\begin{array}{cc} H & C_6H_5 \\ | & | \\ -C-C- \\ | & | \\ H & H \end{array}$	thermal insulation, household equipment, e.g. telephones
poly(2-methylpropenoate)	Perspex	$\begin{array}{cc} H & COOC_2H_5 \\ | & | \\ -C-C- \\ | & | \\ H & H \end{array}$	cast objects, e.g. domestic baths, replacement for glass
poly(tetrafluoroethene)	PTFE, Teflon	$\begin{array}{cc} F & F \\ | & | \\ -C-C- \\ | & | \\ F & F \end{array}$	reactor or plant equipment linings, medical, non-stick pans

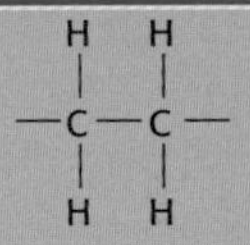

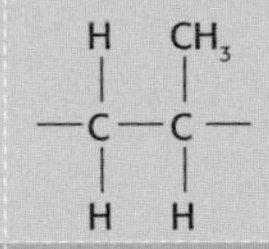

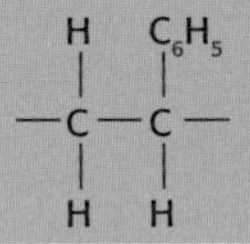

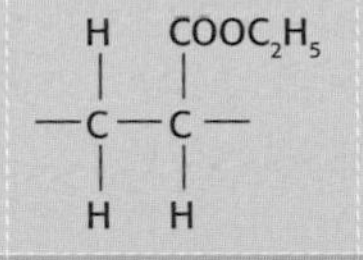

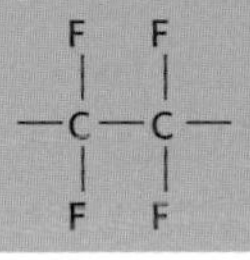

Even with just one chemical monomer the product can be altered to produce polymers with very different properties. For example, polystyrene can be modified to produce a tougher and more chemically resistant polymer than rubber. This can be moulded for casings for telephones, radios and other household goods. However, if pentane is injected during the manufacture of polystyrene it volatilizes and causes the polystyrene to expand to give a very lightweight material used for packaging and thermal insulation.

Alcohols

Combustion of alcohols

All alcohols burn readily in a plentiful supply of oxygen to form carbon dioxide and water. In some countries (e.g. Brazil), ethanol is combined with gasoline (petrol) to produce a fuel for cars called gasohol. Although this makes the country less reliant on the supply and cost of pure gasoline, weight for weight it will not produce as much energy.

This is because, unlike gasoline, alcohols are already partially oxidized, as they contain an oxygen atom. This can be verified by comparing the standard enthalpies of combustion of ethanol and octane:

$$C_2H_5OH(l) + 3O_2(g) \rightarrow 2CO_2(g) + 3H_2O(l) \quad \Delta H^\ominus = -1371 \text{ kJ mol}^{-1}$$

$$C_8H_{18}(l) + 12\tfrac{1}{2}O_2(g) \rightarrow 8CO_2(g) + 9H_2O(l) \quad \Delta H^\ominus = -5512 \text{ kJ mol}^{-1}$$

As the values are given per mole, this means that 46 g of ethanol produces 1371 kJ of energy, and 114 g of octane produce 5512 kJ of energy. Thus 1 g of ethanol produces 29.8 kJ of energy when it is

completely combusted, whereas the same mass of octane produces
48.4 kJ of energy—an increase of over 60%. Even so, if the price of
crude oil remains high, it can be economical to use ethanol as a fuel.
As it can be produced by fermentation, there is no reliance on fossil
fuels, which makes ethanol produced by this method a **carbon-
neutral** fuel.

Oxidation of ethanol

The combustion of alcohols is clearly an oxidation process, but
normally when we refer to the oxidation of organic compounds
we tend to restrict this to the oxidation of the functional group
and assume that the carbon chain remains intact. Alcohols such
as ethanol can be readily oxidized by warming with an acidified
aqueous solution of potassium dichromate(VI), $K_2Cr_2O_7$ (aq), as
the oxidizing agent. The dichromate(VI) solution is acidified using
dilute sulfuric acid. The product is ethanal, an aldehyde. During the
process the orange colour of the acidified potassium dichromate(VI)
solution changes to green, owing to the formation of chromium(III)
ions, Cr^{3+}(aq), and the smell of ethanal (which smells like apples)
can be detected. We could write the full equation for this reaction by
first writing the relevant half-equations and then combining them,
so that the electrons cancel.

$$C_2H_5OH(aq) \rightarrow CH_3CHO(aq) + 2H^+(aq) + 2e^-$$

$$Cr_2O_7^{2-}(aq) + 14H^+(aq) + 6e^- \rightarrow 2Cr^{3+}(aq) + 7H_2O(l)$$

Overall:

$$3C_2H_5OH(aq) + Cr_2O_7^{2-}(aq) + 8H^+(aq)$$
$$\rightarrow 3CH_3CHO(aq) + 2Cr^{3+}(aq) + 7H_2O(l)$$

This overall equation shows the need to acidify the potassium
dichromate solution in order to produce the necessary hydrogen ions.
However, we are only really concerned here with what happens to
the alcohol, so often a simplified oxidation equation is written, using
[O] to represent the oxygen supplied by the oxidizing agent:

In practice the reaction can proceed further than this, because the
ethanal is also readily oxidized to ethanoic acid:

ethanol ("wine") → ethanal → ethanoic acid ("vinegar")

The oxidation of ethanol to ethanoic acid is one of the oldest
chemical reactions known to man; it is essentially what happens
when wine is left exposed to the air and turns into vinegar as a
result of bacterial action (Figure 31).

Figure 31 Vinegar made by oxidizing
white wine. (photo by Geoff Neuss)

This reaction was also used in the past by the police to determine whether a car driver might have drunk too much alcohol and exceeded the legal limit of alcohol in the blood. Drivers were required to blow into a bag through a tube that contained crystals of potassium dichromate(VI) in an acidic medium (Figure 32). If the orange crystals turned green it was indicative of a high level of alcohol in the breath, and the driver was then required to provide a blood or urine sample to obtain a more accurate measurement. Modern breathalysers or intoximeters can now provide an accurate reading directly. These work using either fuel cell technology or infrared absorption.

Figure 32 A potassium dichromate(VI) breathalyser being used in the early 1990s.

If we compare the physical properties of ethanol, ethanal and ethanoic acid we can devise a way of obtaining a good yield of ethanal before it can be further oxidized to ethanoic acid. Both ethanol and ethanoic acid contain a hydrogen atom bonded directly to an electronegative oxygen atom. The intermolecular attractive forces in these two compounds are therefore due principally to hydrogen bonding. However, in ethanal there are no hydrogen atoms bonded to oxygen (the oxygen atom is bonded to a carbon atom), so there are only dipole:dipole intermolecular forces. This means that the boiling point of ethanal will be much lower than the boiling point of either ethanol or ethanoic acid (Figure 33).

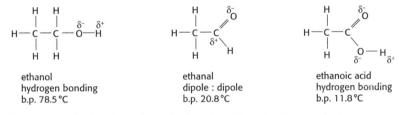

ethanol
hydrogen bonding
b.p. 78.5 °C

ethanal
dipole : dipole
b.p. 20.8 °C

ethanoic acid
hydrogen bonding
b.p. 11.8 °C

Figure 33 The boiling points of ethanol and its oxidation products

By arranging the apparatus so that the lower boiling ethanal can be distilled from the mixture, a good yield of ethanal can be obtained. If a high yield of ethanoic acid is required, then the apparatus can be arranged for reflux so that none of the ethanal can escape, ensuring that most of it will be further oxidized to the carboxylic acid. See Figure 34.

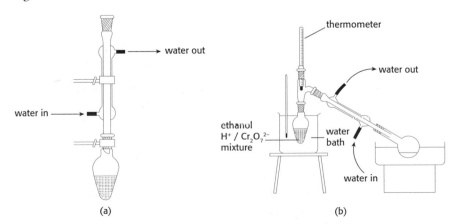

(a)

(b)

Figure 34 (a) Reflux apparatus to produce ethanoic acid from the oxidization of ethanol; (b) distillation apparatus to produce ethanal from the oxidization of ethanol.

Oxidation of primary, secondary and tertiary alcohols

Ethanol is an example of a primary alcohol. **Primary alcohols** contain two hydrogen atoms, and one alkyl group bonded to the carbon atom that contains the alcohol group. **Secondary alcohols** contain one hydrogen atom, and two alkyl groups bonded directly to the carbon containing the alcohol group. **Tertiary alcohols** contain three alkyl groups, and no hydrogen atoms bonded to the carbon atom containing the alcohol group. See Figure 35.

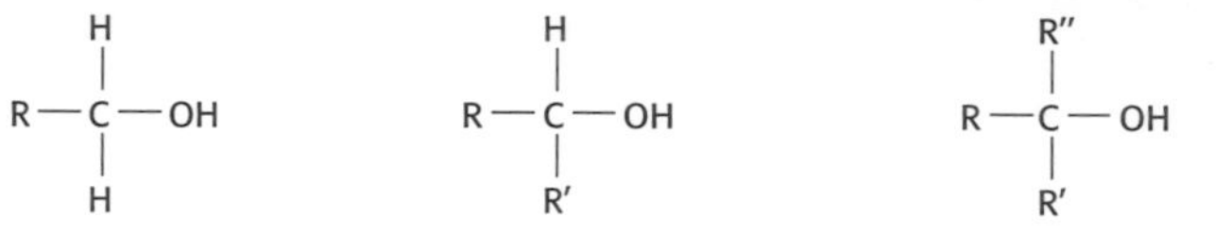

H | R—C—OH | H	H | R—C—OH | R'	R" | R—C—OH | R'
primary alcohol	secondary alcohol	tertiary alcohol

Figure 35 Primary, secondary and tertiary alcohols. R represents an alkyl group. R' and R", also alkyl groups, may be the same as or different from R.

All primary alcohols can be oxidized by an acidified aqueous solution of potassium dichromate(VI) to form, first, the corresponding aldehyde, and then oxidized further to the carboxylic acid:

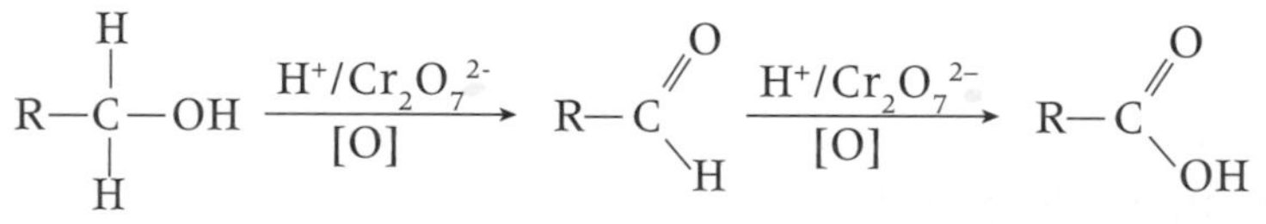

$$\underset{\text{primary alcohol}}{R-\overset{\displaystyle H}{\underset{\displaystyle H}{C}}-OH} \xrightarrow[\ [O]\ ]{H^+/Cr_2O_7^{2-}} \underset{\text{aldehyde}}{R-C\overset{\displaystyle O}{\underset{\displaystyle H}{\diagdown}}} \xrightarrow[\ [O]\ ]{H^+/Cr_2O_7^{2-}} \underset{\text{carboxylic acid}}{R-C\overset{\displaystyle O}{\underset{\displaystyle OH}{\diagdown}}}$$

Secondary alcohols can also be readily oxidized by an aqueous acidified dichromate(VI) solution. However, now the product is a ketone. Unlike an aldehyde, the carbon atom bonded to the oxygen atom in a ketone contains no hydrogen atom, so ketones cannot be oxidized further. The simplest example of this is the oxidation of propan-2-ol to form propanone:

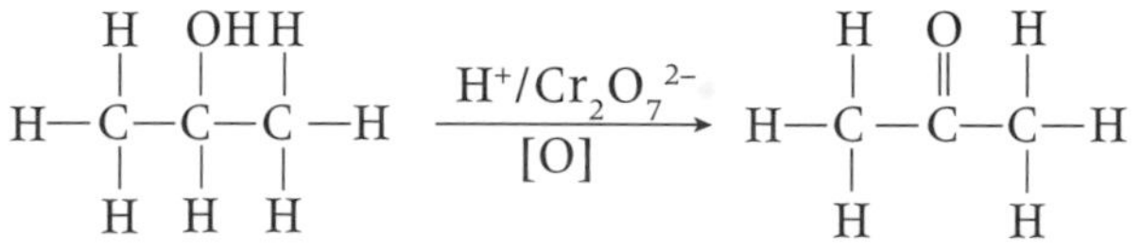

$$\underset{\substack{\text{propan-2-ol}\\\text{(secondary alcohol)}}}{H-C-C-C-H} \xrightarrow[\ [O]\ ]{H^+/Cr_2O_7^{2-}} \underset{\substack{\text{propanone}\\\text{(ketone)}}}{H-C-C-C-H}$$

Some textbooks state that tertiary alcohols cannot be readily oxidized. This is not strictly true, because they burn in air when ignited. However, they cannot be oxidized without disrupting the carbon chain, and no reaction occurs with an acidified aqueous solution of potassium dichromate(VI):

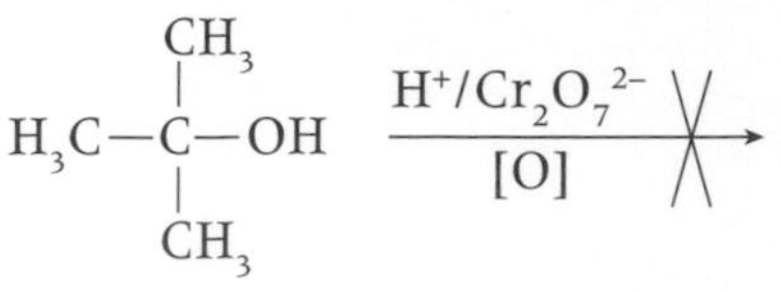

$$\underset{}{H_3C-\overset{\displaystyle CH_3}{\underset{\displaystyle CH_3}{C}}-OH} \xrightarrow[\ [O]\ ]{H^+/Cr_2O_7^{2-}} \Big/\!\!\!\!\times \quad \begin{array}{l}\text{no oxidation possible as}\\\text{no hydrogen atom on carbon}\\\text{atom bonded to the alcohol group}\end{array}$$

Halogenoalkanes

Halogenoalkanes contain fluorine, chlorine, bromine or iodine bonded to an alkyl group (Figure 36).

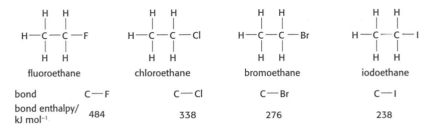

bond	C—F	C—Cl	C—Br	C—I
bond enthalpy/ kJ mol⁻¹	484	338	276	238

Figure 36 Some halogenoalkanes.

Because the carbon-to-fluorine bond is so strong, fluorocarbons are extremely unreactive. In fact they are so unreactive that one of their major uses used to be as fire extinguishers. Halon 1211 (bromochlorodifluoromethane) was used for extinguishing electrical fires and for dealing with major disasters such as aircraft fires. However, since the ratification of the Montreal Protocol its use has been phased out in most countries because it is a chlorofluorocarbon, CFC, which can deplete the ozone layer.

The main reaction of halogenoalkanes is substitution. The carbon atom bonded to the halogen has a small positive charge (δ^+) because of the greater electronegativity of the halogen atom. Negative ions or molecules that contain a non-bonding pair of electrons are attracted to the carbon atom and displace the halogen. Iodoalkanes are the most reactive, because the weak carbon-to-iodine bond breaks more readily than the stronger carbon-to-bromine and carbon-to-chlorine bonds.

Typical of these substitution reactions is the reaction between bromoethane and a warm aqueous solution of sodium hydroxide to form ethanol:

$$C_2H_5Br(aq) + NaOH(aq) \rightarrow C_2H_5OH(aq) + NaBr(aq)$$

In this reaction it is the hydroxide ion that is attracted to the carbon atom with the small positive charge. Negative ions or molecules with a non-bonding pair of electrons that substitute the halogen are known as **nucleophiles**. In these reactions they act as Lewis bases, because they donate a pair of electrons. To show the movement of a pair of electrons, chemists have adopted a convention known as the "**curly arrow**". The end of the arrow indicates the initial position of the electron pair and the head of the arrow shows where they move to. So the reaction for a halogenoalkane with hydroxide ions can be written as shown in Figure 37.

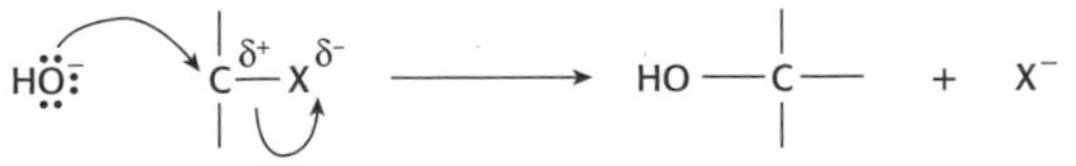

Figure 37 Nucleophilic substitution of halogenoalkanes with aqueous hydroxide ions, where X represents a chlorine, bromine or iodine atom. The carbon-to-fluorine bond is too strong for fluoroalkanes to undergo nucleophilic substitution.

The mechanisms for these substitution reactions are interesting because they differ, depending on the type of halogenoalkane that is reacting.

Primary halogenoalkanes

Bromoethane is a typical primary halogenoalkane, because the carbon atom bonded to the bromine atom has two hydrogen atoms and one alkyl group, R-, bonded to it.

Mechanisms cannot be determined from the stoichiometric equation:

$$C_2H_5Br(aq) + OH^-(aq) \rightarrow C_2H_5OH(aq) + Br^-(aq)$$

As we saw in Chapter 6 on kinetics, a valid mechanism can be proposed only if it is consistent with the experimentally determined rate equation. Experimentally, the rate of the reaction has been shown to be proportional both to the concentration of the halogenoalkane and to the concentration of the hydroxide ions.

$$Rate \propto [C_2H_5Br(aq)] \times [OH^-(aq)]$$

This implies that both the alcohol and the hydroxide ions are involved in the rate-determining step. Because two species are involved in the rate determining step the reaction is said to be **bimolecular**. The proposed mechanism is that the hydroxide ions (the nucleophile) are attracted to the carbon atom and form a transition state, with the hydroxide ions beginning to bond to the carbon as the bromine atom is beginning to leave. This involves five different atoms or groups around the central carbon atom in the transition state. The transition state is actually an ion, because it contains a negative charge. The bromine atom then leaves as a bromide ion, and the alcohol is produced (Figure 38).

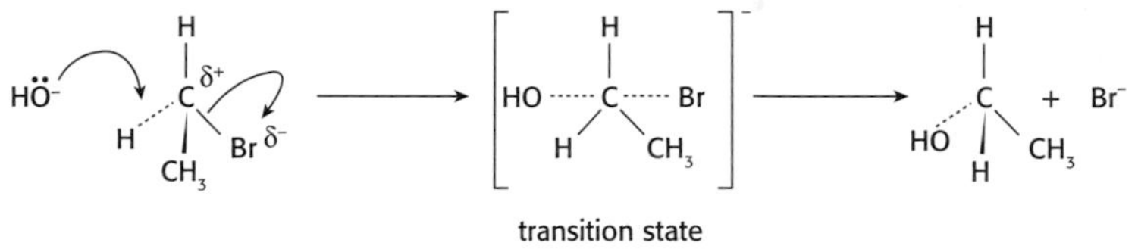

transition state

Figure 38 The S_N2 mechanism.

This mechanism, which involves two reacting species in the rate-determining step, is known as an **S_N2 mechanism**. S_N2 stands for *substitution nucleophilic bimolecular*.

Tertiary halogenoalkanes

The simplest tertiary bromoalkane is 2-bromo-2-methylpropane. It reacts with warm dilute aqueous sodium hydroxide solution to form the tertiary alcohol 2-methylpropan-2-ol:

$$C(CH_3)_3Br(aq) + OH^-(aq) \rightarrow C(CH_3)_3OH(aq) + Br^-(aq)$$

The experimentally determined rate equation now shows that the rate depends only on the concentration of the halogenoalkane, and is independent of the hydroxide ion concentration:

$$Rate \propto [C(CH_3)_3Br(aq)]$$

The proposed mechanism consistent with this rate equation is that the first step involves the breaking of the carbon-to-bromine bond to form a carbocation intermediate. This first step will be the slow step or rate-determining step. Once the carbocation intermediate is formed, it will react with a hydroxide ion to form the alcohol. This second step will be a fast step. See Figure 39.

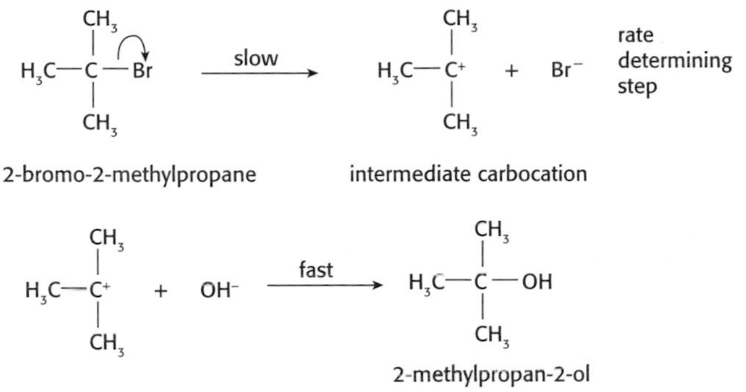

Figure 39 The S$_N$1 mechanism.

This two-step mechanism, which involves only one reacting species in the rate-determining step, is known as an **S$_N$1 mechanism**. S$_N$1 stands for *substitution nucleophilic unimolecular.*

The reasons for the different mechanisms, S$_N$2 and S$_N$1

Why do primary halogenoalkane substitution reactions proceed by an S$_N$2 mechanism whereas tertiary halogenoalkanes follow an S$_N$1 mechanism? To answer this question it is actually easier to ask it the other way round. Tertiary halogenoalkanes cannot proceed via an S$_N$2 mechanism, because this would involve five relatively bulky groups around the central carbon atom in the transition state. There is simply not room for these five groups to fit around the relatively small carbon atom—a phenomenon known as **steric hindrance** (Figure 40).

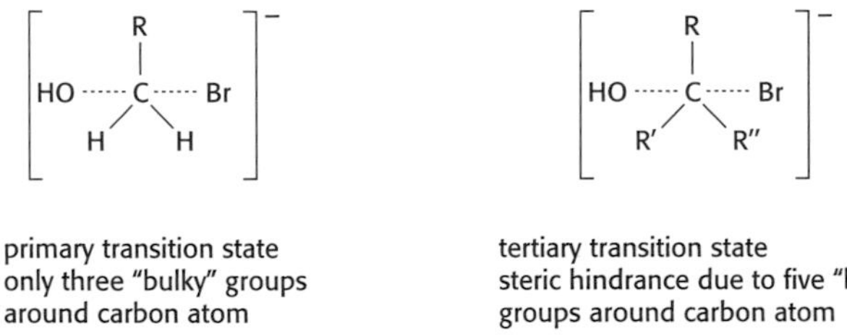

primary transition state
only three "bulky" groups
around carbon atom

tertiary transition state
steric hindrance due to five "bulky"
groups around carbon atom

Figure 40 Steric hindrance.

Primary halogenoalkanes are less likely to proceed via an S$_N$1 mechanism, because this would involve the formation of a primary carbocation intermediate. Alkyl groups tend to push the bonding pair of electrons towards the carbon they are bonded to. This is known as the **positive inductive effect**. The result is that the positive charge can be spread over more atoms, and this increases the energetic stability of the carbocation, which means it is more likely to be formed. The more alkyl groups there are surrounding the positive carbon ion, then the stronger the positive inductive effect will be, so that tertiary carbocations are much more stable and more likely to be formed than primary carbocations (Figure 41).

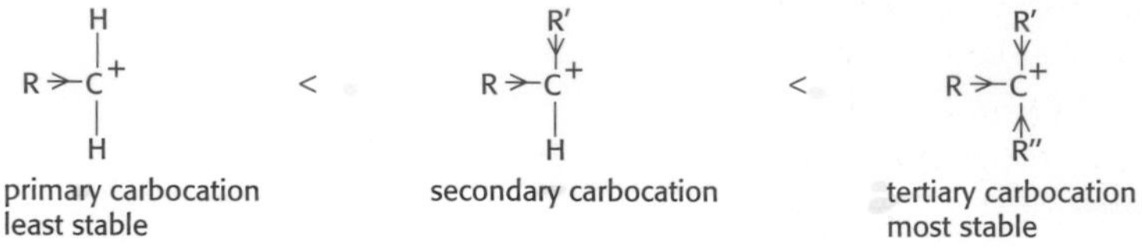

primary carbocation
least stable

secondary carbocation

tertiary carbocation
most stable

Figure 41 Stability of carbocations

Secondary halogenoalkanes proceed by a mixture of S_N1 and S_N2 mechanisms. The precise mechanism depends to some extent on the nature of the alkyl groups. Whatever the mechanism, the breaking of the carbon-to-halogen bond for primary, secondary and tertiary halogenoalkanes involves heterolytic fission as the halogen atom always leaves as the halide ion.

Reaction pathways (1)

The raw materials for many organic compounds originate from coal, crude oil or natural gas. The challenge for synthetic chemists is to devise reaction pathways to make new compounds using as few steps as possible, each with the highest yield possible. The more steps there are in an organic synthesis then the lower the final yield is likely to be, because some material will be lost during each step.

For example, but-2-ene can be obtained by cracking some of the higher-boiling fractions of crude oil. One way of obtaining butanone from but-2-ene in a two-step process would be, first, to hydrate the but-2-ene by heating with steam in the presence of a phosphoric acid catalyst to produce butan-2-ol. This secondary alcohol can then be oxidized with an acidified aqueous solution of potassium dichromate(VI):

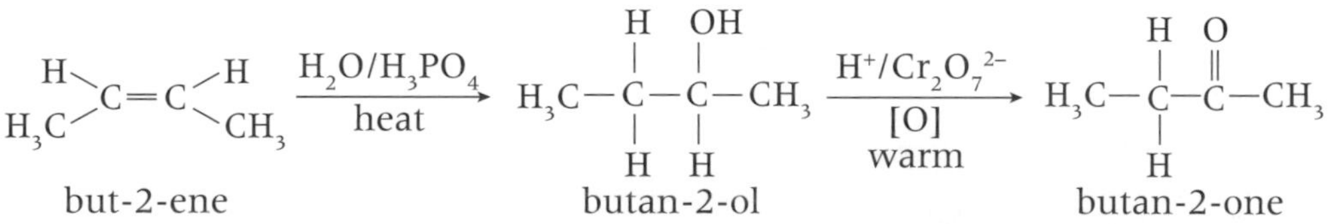

1 Devise two-step syntheses of the following products from the starting material listed. Include any necessary experimental conditions and an equation for each step.
   (a) Ethanoic acid from ethene
   (b) Butan-1-ol from butane
   (c) Hexan-3-one from 3-bromohexane
   (d) 1,2,3-trichloropropane from propene
   (e) Propanal from 1-bromopropane
2 The synthesis of a particular organic product involves seven separate stages. If each stage produces a yield of 70%, calculate the percentage yield of the final product.

H₃C, H C=C H, CH₃   →(H₂O/H₃PO₄, heat)   H₃C—C(H)(OH)—C(H)(H)—CH₃   →(H⁺/Cr₂O₇²⁻, [O], warm)   H₃C—C(H)(H)—C(O)—CH₃

but-2-ene   butan-2-ol   butan-2-one

A good knowledge of organic chemistry will be a great help in devising suitable syntheses. Rather than looking at the specific compounds it is often helpful to look at the functional groups involved and consider the reactions of these groups. For some syntheses it is helpful to work forwards from the reactant, but it can be equally profitable to work backwards from the product. In the above example butanone is a ketone, and ketones are formed from oxidizing secondary alcohols. It can then be seen that alcohols are made by hydrating alkenes.

HL
More functional groups

The structures and naming (of compounds containing up to six carbon atoms) of alkanes, alkenes, alcohols, aldehydes, ketones, carboxylic acids and halogenoalkanes have already been covered. In addition, the structures of some compounds containing the amino group, -NH$_2$, the benzene ring (phenyl group), -C$_6$H$_5$, and esters, RCOOR', have also been mentioned. For higher level you should also be able to name compounds containing the amino group and esters in addition to compounds containing two other functional groups, amides and nitriles.

Amines (R–NH$_2$)

There are several different ways to name amines that are accepted by IUPAC. Most amines are prefixed by the word amino-, with the location of the NH$_2$- group being indicated: for example, 2-aminopentane and 1,6-diaminohexane (Figure 42). It is also correct to call them by the longest alkane with the suffix -amine: for example, pentan-2-amine. However, when the number of carbon atoms is small (one, two or three), the old names of methylamine, ethylamine and propylamine tend to be used rather than aminomethane, aminoethane and aminopropane. IUPAC accepts 1-butylamine, 1-butanamine and 1-aminobutane for CH$_3$CH$_2$CH$_2$CH$_2$NH$_2$.

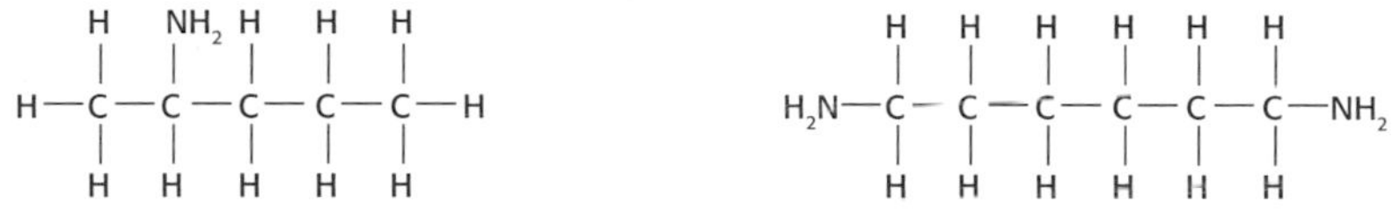

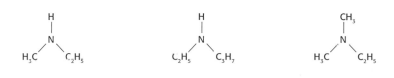

2-aminopentane
or pentan-2-amine

or H$_2$N$+$CH$_2$$\rightarrow_6$ NH$_2$
1,6-diaminohexane

Figure 42 Amines.

In the case of secondary amines the main name of the amine is taken from the longest carbon chain attached to the nitrogen atom. The other chain is prefixed as an alkyl group, with the location prefix given as an italic N. Examples include N-methylethanamine and N-ethylpropanamine. For tertiary amines there are two prefixes with an italic N: for example, CH$_3$CH$_2$N(CH$_3$)$_2$ is N,N-dimethylethanamine (Figure 43).

N-methylethanamine   N-ethylpropanamine   N,N-dimethylethanamine

Figure 43 Secondary and tertiary amines.

Esters (R–COO–R′)

Esters are named from the acid and alcohol from which they are derived. The first part of the ester is named after the R- group from the alcohol. There is then a space, followed by the name for the acid anion: for example, methyl ethanoate, ethyl propanoate and propyl methanoate (Figure 44).

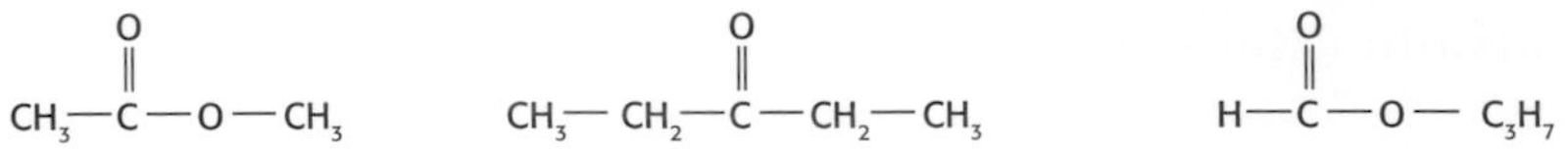

methyl ethanoate   ethyl propanoate   propyl methanoate

Figure 44 Esters.

Amides (R–CO–NH₂)

Amides are named after the longest carbon chain, followed by -amide: for example, ethanamide and 2-methylpropanamide (Figure 45). Remember that, as in carboxylic acids, the carbon atom in the functional group also contributes to the longest chain. Secondary amides are named rather like amines, in that the other alkyl group attached to the nitrogen atom is prefixed by an *N*: for example, *N*-methylethanamide

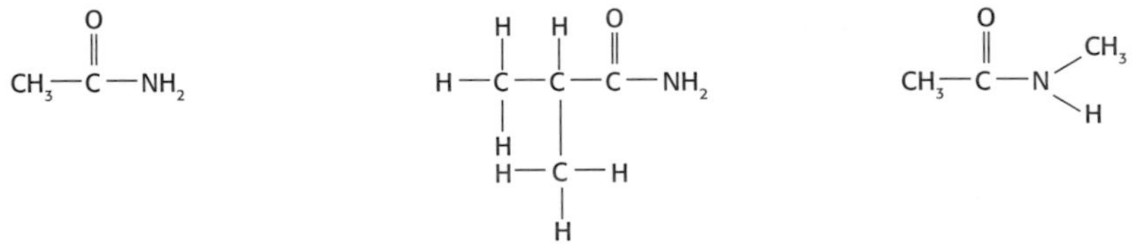

ethanamide   2-methylpropanamide   *N*-methylethanamide

Figure 45 Amides.

Nitriles (R–CN)

Nitriles used to be called cyanides, so that C_2H_5CN was known as ethyl cyanide. The IUPAC way of naming nitriles is to consider the acid from which they are derived, as the -COOH group has been replaced by a -CN group. The suffix -nitrile is added to the hydrocarbon forming the basis of the acid. For example, C_2H_5CN now becomes propanenitrile, and ethanenitrile has the formula CH_3CN (Figure 46).

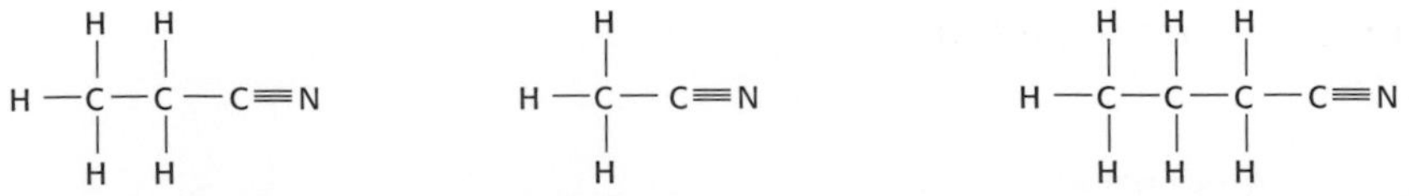

propanenitrile   ethanenitrile   butanenitrile

Figure 46 Nitriles.

❓

1 Draw the structures of:
 (a) 2-hydroxypropanenitrile
 (b) propyl propanoate
 (c) *N,N*-dimethylethanamide

2 Name the following:

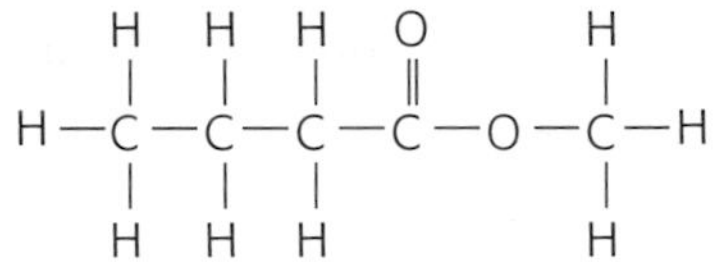

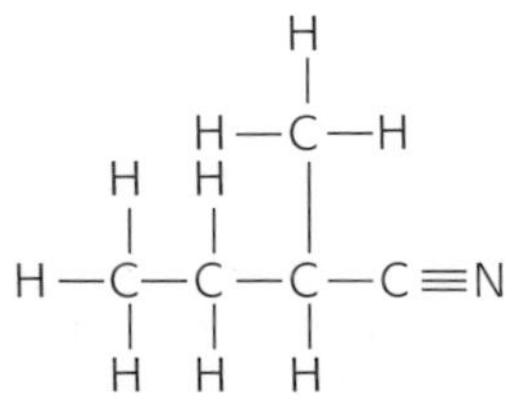

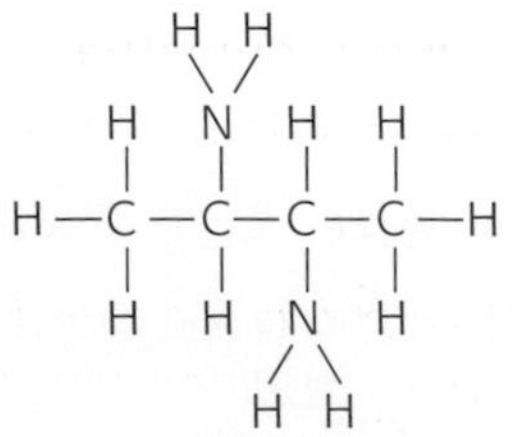

Nucleophilic substitution reactions

The hydrolysis of halogenoalkanes by a warm aqueous solution of sodium hydroxide is the classic example of a nucleophilic substitution reaction.

Apart from hydroxide ions, other common nucleophiles are water, ammonia, amines and cyanide ions. Each one reacts to replace the halogen atom, which leaves as the halide ion.

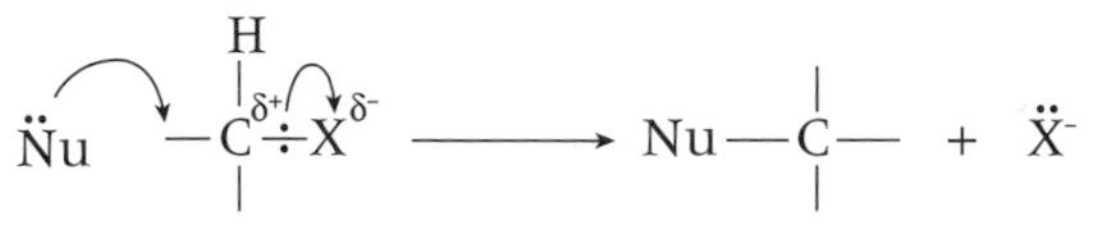

Here Nu represents a nucleophile (e.g. OH^-, CN^-, NH_3 or H_2O), and X represents a halogen atom (Cl, Br or I).

Ammonia reacts with bromoethane by an S_N2 mechanism to form ethylamine and hydrogen bromide:

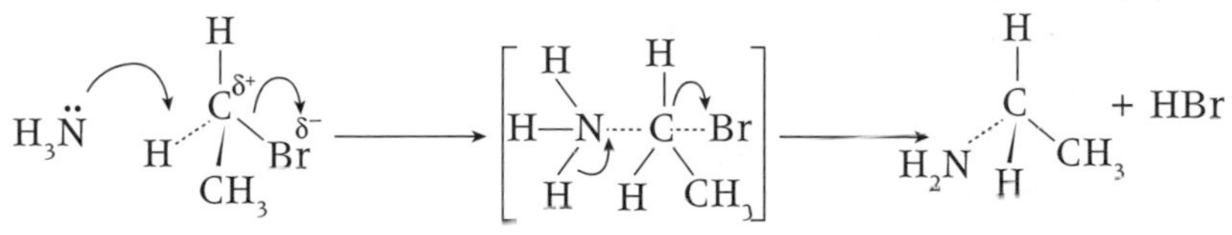

However, the product ethylamine is also a nucleophile, because the nitrogen atom still contains a non-bonding pair of electrons. The reaction can proceed further until all four pairs of electrons around the nitrogen atom have bonded to ethyl groups to give an ammonium salt that can be classified as a quaternary amine (Figure 47).

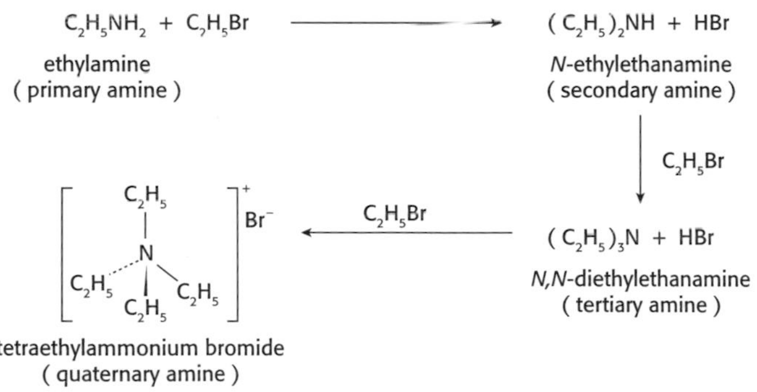

Figure 47 Nucleophilic substitution reactions with amines

Nucleophilic substitution reactions are important in organic synthesis, because the halogen atom on halogenoalkanes can be replaced by other functional groups. The reaction with potassium cyanide illustrates this well. The cyanide ion reacts to form a nitrile. For example, bromoethane reacts by an S_N2 mechanism with potassium cyanide to form propanenitrile:

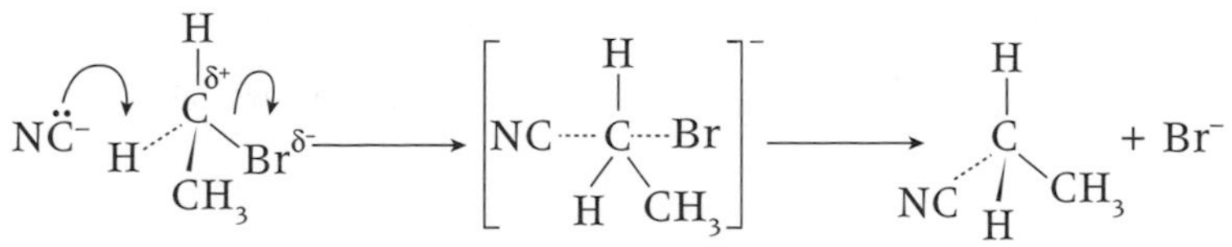

bromoethane   transition state   propanenitrile

This provides a useful way of increasing by one the number of carbon atoms in the carbon chain. Nitriles can be reduced using hydrogen and a nickel catalyst to form amines, or they can be hydrolysed in acid solution to form carboxylic acids, which can then be reacted to form many other compounds:

CH_3CH_2CN propanenitrile
$\xrightarrow{H_2/Ni}$ $CH_3CH_2CH_2NH_2$ propanamine (propylamine)
$\xrightarrow{H^+/H_2O}$ $CH_3CH_2COOH + NH_4^+$ propanoic acid

Factors affecting the rate of nucleophilic substitution

The nucleophile itself

As the nucleophile is attracted to the partially positive, δ+, carbon atom bonded to the electronegative halogen atom, the more dense the negative charge on the ion or molecule the better the nucleophile. Anions tend to be more reactive than the corresponding neutral species. For example, the rate of substitution with hydroxide ions is faster than with water. Among species with the same charge a less electronegative atom carrying a non-bonded pair of electrons is a better nucleophile than a more electronegative one. Thus ammonia is a better nucleophile than water. This is because the less electronegative atom can more easily donate its pair of electrons, as they are held less strongly. Cyanide ions are the best nucleophile of all. The strength of nucleophiles thus follows the order:

$$CN^- > OH^- > NH_3 > H_2O$$

The identity of the halogen

We have seen that fluorocarbons are very unreactive because of the strength of the carbon-to-fluorine bond. This raises an interesting point, as fluorine is the most electronegative halogen, and therefore the carbon to fluorine bond will be the most polar: that is, will have the most ionic character. As nucleophiles are attracted to the polar carbon atom, we might expect that fluoroalkanes would in fact be the most reactive. But as they are so unreactive it would appear that the polarity of the molecule is not nearly as important as the strength of the carbon-to-halogen bond. This is borne out by the fact that iodoalkanes, which are the least polar, are the most reactive, as the carbon-to-iodine bond is the weakest. The reactivity of halogenoalkanes thus depends on the ability of the carbon-to-halogen bond to break heterolytically and for the halogen to then leave. For this reason the halogen is sometimes known as the **leaving group**. Iodine is the best leaving group, and fluorine is the worst (Figure 48).

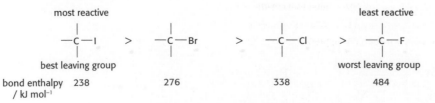

Figure 48 The reactivity of halogenoalkanes with nucleophiles.

The nature of the halogenoalkane

Primary halogenoalkanes proceed by an S_N2 mechanism. This reaction is first order with respect to the halogenoalkane, and first order with respect to the nucleophile. Both species are involved in the rate-determining step to form the transition state. Tertiary halogenoalkanes proceed by an S_N1 mechanism. The reaction is first order with respect to the halogenoalkane, and is independent of the concentration of the nucleophile. This is a two-step reaction, with an intermediate carbocation being formed. Experimentally, S_N1 reactions are generally found to be faster than S_N2 reactions. This may be because the activation energy required to form the tertiary carbocation intermediate is less than the activation energy required to form the transition state of the S_N2 reaction.

Laboratory work with halogenoalkanes

The relative rates of reaction of halogenoalkanes with nucleophiles can be elegantly demonstrated by a simple experiment (Figure 49).

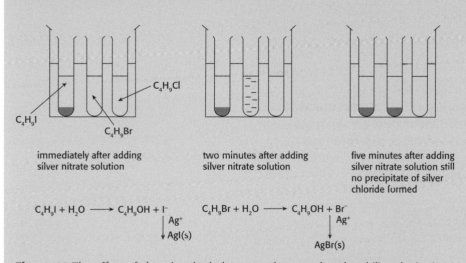

immediately after adding silver nitrate solution

two minutes after adding silver nitrate solution

five minutes after adding silver nitrate solution still no precipitate of silver chloride formed

$$C_4H_9I + H_2O \longrightarrow C_4H_9OH + I^-$$
$$\downarrow Ag^+$$
$$\downarrow AgI(s)$$

$$C_4H_9Br + H_2O \longrightarrow C_4H_9OH + Br^-$$
$$\downarrow Ag^+$$
$$\downarrow AgBr(s)$$

Figure 49 The effect of changing the halogen on the rate of nucleophilic substitution

Into each of three test-tubes place 1 cm³ of ethanol. Using separate teat pipettes, place two drops of 1-chlorobutane in the first test-tube, two drops of 1-bromobutane in the second, and two drops of 1-iodobutane in the third. Stand all three test-tubes in a beaker of water at about 60–65 °C, and place another test-tube containing about 5 cm³ of 0.05 mol dm⁻³ silver nitrate solution in the warm water. Wait until the contents of all the test-tubes have reached approximately 60 °C, then place 1 cm³ of the silver nitrate solution into each of the other test-tubes, and quickly shake each tube to mix the contents. Note and time carefully what you observe throughout the next 5 minutes.

The silver nitrate is added to precipitate any halide ions that are formed, as the silver salt. What is observed is that almost as soon as the silver nitrate is added a yellow precipitate of silver iodide is seen in the tube with the 1-iodobutane. During the next 5 minutes a cream-coloured precipitate of silver bromide appears in the tube with the 1-bromobutane, but even after 5 minutes the reaction with 1-chlorobutane is so slow that no visible precipitate of silver chloride is formed.

The experiment is then repeated. This time two drops of 1-bromobutane are added to the ethanol in the first test-tube, two drops of 2-bromobutane in the second, and three drops of 2-bromo-2-methylpropane in the third. This time, when the silver nitrate is added, an immediate cream precipitate of silver bromide is obtained with the tube containing the tertiary halogenoalkane, 2-bromo-2-methylpropane. This is quite quickly followed by the secondary halogenoalkane, 2-bromobutane, and by the end of the 5 minutes the primary halogenoalkane, 1-bromobutane, will also have produced a precipitate of silver bromide (Figure 50).

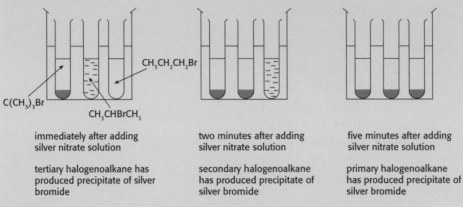

immediately after adding silver nitrate solution

two minutes after adding silver nitrate solution

five minutes after adding silver nitrate solution

tertiary halogenoalkane has produced precipitate of silver bromide

secondary halogenoalkane has produced precipitate of silver bromide

primary halogenoalkane has produced precipitate of silver bromide

Figure 50 The effect of changing the nature of the halogenoalkane on the rate of nucleophilic substitution

Elimination reactions

When a warm dilute aqueous solution of sodium hydroxide hydrolyses halogenoalkanes to form an alcohol the hydroxide ion acts as a nucleophile. However, under different conditions the hydroxide ion can act as a base (proton acceptor), and elimination of hydrogen bromide occurs to produce an alkene.

$$-\overset{\displaystyle H}{\underset{\displaystyle |}{C}}-\overset{\displaystyle Br}{\underset{\displaystyle |}{C}}- + OH^- \longrightarrow >C=C< + H_2O + Br^-$$

The conditions required are that sodium hydroxide dissolved in hot ethanol is added to the halogenoalkane, and that the mixture is heated under reflux. This provides a good example of how the same reactants can produce completely different products if the conditions are changed.

The elimination of hydrogen bromide can proceed in one of two ways:

- by a carbocation intermediate, known as an E1 mechanism because the first step involves the heterolytic fission of the carbon-to-halogen bond in a similar way to S_N1:

- or in a concerted process known as an E2 mechanism, because both the hydroxide base and the halogenoalkane are involved in the rate-determining step:

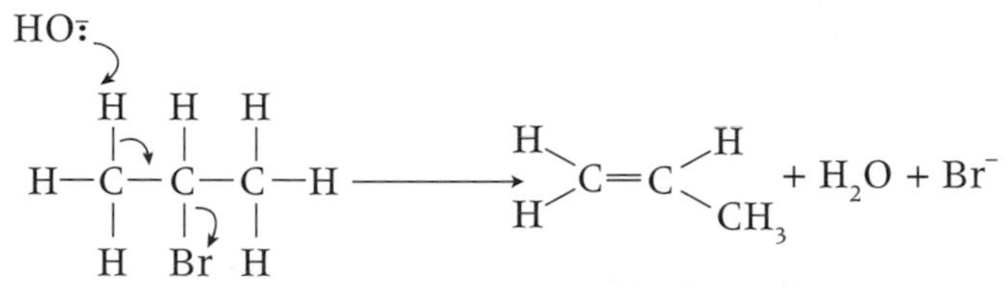

It is possible that the actual base involved in the elimination is the ethoxide ion rather than the hydroxide ion. Ethanol is an even weaker Lowry–Brønsted acid than water. Hence its conjugate base, the ethoxide ion, $C_2H_5O^-$, will be an even stronger base than hydroxide ions. Although the dissociation of the ethanol is slight, and most ethoxide ions will attract a proton from water, it is possible that enough ethoxide ions will be present to remove the proton from the halogenoalkane:

$$HO^- + C_2H_5OH \rightleftharpoons H_2O + C_2H_5O^-$$

Condensation reactions

A condensation reaction involves two molecules reacting together to produce a larger molecule with the elimination of a small molecule such as water or hydrogen chloride.

Esters

One of the best examples of a condensation reaction is **esterification**. Carboxylic acids react with alcohols when warmed in the presence of a small amount of concentrated sulfuric acid acting as a catalyst to form an ester and water. This is an equilibrium reaction, and the sulfuric acid also acts to shift the position of equilibrium towards the products by "removing" the water.

For example, ethanol reacts with ethanoic acid to produce ethyl ethanoate and water. From studies done using isotopic labelling it can be shown that it is the carbon-to-oxygen bond on the acid that breaks, not the carbon-to-oxygen bond on the alcohol:

Esters have very distinctive and individual smells. Generally these are quite fruity and pleasant, so they are used in perfumes and as flavourings in food. For example, ethyl methanoate has the flavour of rum and ethyl butanoate the flavour of pineapple (Figure 51).

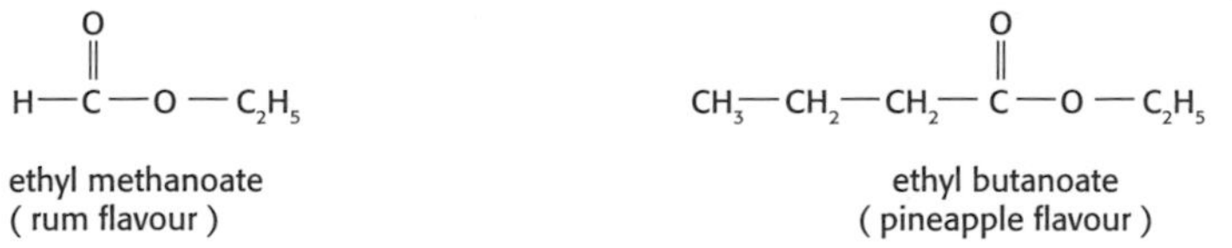

ethyl methanoate
(rum flavour)

ethyl butanoate
(pineapple flavour)

Figure 51 Two typical esters.

Esters can also be used as solvents (e.g. adhesives). Many painkillers (for example Aspirin and heroin) also contain an ester functional group (Figure 52).

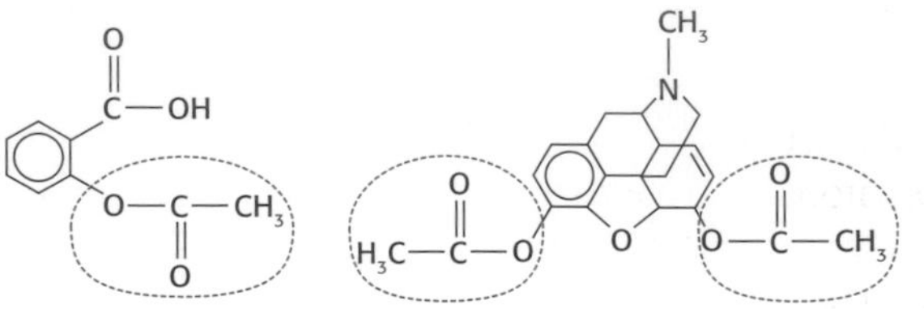

Figure 52 The structures of Aspirin and heroin with the ester functional group(s) circled

Design experiment with esters

Esters provide a useful example for a design experiment. For example, you may simply be given the instruction "Investigate an aspect of esterification". You may wish to devise a focused problem that concentrates on the rate of formation of esters, or on the value of the equilibrium constant for different esters. However, it is not easy to follow the rate of the reaction, and it takes several days for the equilibrium to be established, so these investigations may not be so simple and quick to perform in practice.

One interesting factor associated with esters is smell. Smell is a chemical property, and many chemists make their living producing perfumes and fragrances, but—unlike colour—smell cannot be described precisely and objectively. In fact smells are usually described by comparing them with other known smells.

What does smell depend on? One well-accepted theory is that it depends on the shape of the molecule as it fits into receptors located inside the nose. If this is the case, then changing the alcohols and acids that make up the ester may alter the smell by lesser or greater amounts.

Esters can be prepared easily in small amounts by warming a carboxylic acid and an alcohol for a few minutes in the presence of a few drops of concentrated sulfuric acid as a catalyst. If the mixture is then added to an aqueous solution of sodium carbonate the excess acid is neutralized and the ester floats on the surface, so it is easy to smell. Alongside are some of the genuine research questions investigated by IB Diploma Programme chemistry students who have been given this design problem.

1 To determine whether changing the type of alcohol from primary to secondary to tertiary has any effect on the smell of the ester.
2 How do the different natures of the acids and isomerism within the alcohols affect the smell of esters?
3 Is it the nature of the carboxylic acid or the alcohol that influences the smell? Or both?
4 To determine the different physical properties of esters produced when *cis*-but-2-ene-1,4-dioic acid and *trans*-but-2-ene-1,4-dioic acid react with butan-1-ol. That is, does geometric isomerism affect the smell of an ester?
5 To investigate whether there is a trend in the smells of the esters produced between 2-hydroxybenzoic acid (salicylic acid) and a homologous series of alcohols.
6 Does the smell of an ester depend on the total number of carbon atoms in it?

Amides

Carboxylic acids react with amines to form amides, $-CO-NH-$. The $-OH$ of the acid group reacts with one of the hydrogen atoms attached to the nitrogen atom to form water, and the carbon from the acid group bonds directly to the nitrogen atom:

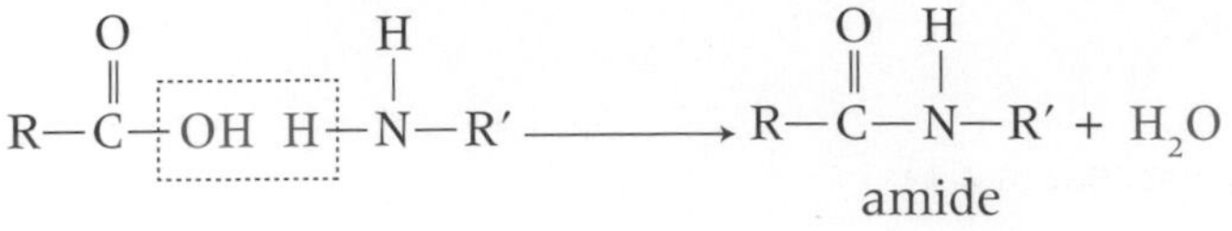

This reaction is particularly important in biological systems, because amino acids contain an amine group and a carboxylic acid group, so that two amino acids can condense together in the presence of enzymes to form an amide.

> ### Amino acids
>
> Amino acids contain both an amine functional group and a carboxylic acid functional group. When they are both attached to the same carbon atom they are known as 2-amino acids (or α-amino acids) (Figure 53).
>
> There are about 20 naturally occurring 2-amino acids. They are the monomers or basic "building blocks" of proteins in the body. The structural formulas of these naturally occurring amino acids are given in the IB data booklet, together with their common names.
>
> $$H_2N-\overset{\overset{\displaystyle H}{|}}{\underset{\underset{\displaystyle R}{|}}{C}}-COOH$$
>
> **Figure 53** 2-amino acid.

For example, 2-aminopropanoic acid (alanine) and 2-aminoethanoic acid (glycine) can react together in two different ways to form two different organic condensation products, both of which are amides:

$$\underset{\text{2-aminoethanoic acid}}{H_2N-\overset{\overset{\displaystyle H}{|}}{\underset{\underset{\displaystyle H}{|}}{C}}-\overset{\overset{\displaystyle O}{\|}}{C}-OH} \quad \underset{\text{2-aminopropanoic acid}}{H-\overset{\overset{\displaystyle H}{|}}{N}-\overset{\overset{\displaystyle H}{|}}{\underset{\underset{\displaystyle CH_3}{|}}{C}}-COOH} \longrightarrow \underset{}{H_2N-\overset{\overset{\displaystyle H}{|}}{\underset{\underset{\displaystyle H}{|}}{C}}-\overset{\overset{\displaystyle O}{\|}}{C}-\overset{\overset{\displaystyle H}{|}}{N}-\overset{\overset{\displaystyle H}{|}}{\underset{\underset{\displaystyle CH_3}{|}}{C}}-COOH + H_2O}$$

$$\underset{\text{2-aminopropanoic acid}}{H_2N-\overset{\overset{\displaystyle H}{|}}{\underset{\underset{\displaystyle CH_3}{|}}{C}}-\overset{\overset{\displaystyle O}{\|}}{C}-OH +} \quad \underset{\text{2-aminoethanoic acid}}{H-\overset{\overset{\displaystyle H}{|}}{N}-\overset{\overset{\displaystyle H}{|}}{\underset{\underset{\displaystyle H}{|}}{C}}-COOH} \longrightarrow \underset{}{H_2N-\overset{\overset{\displaystyle H}{|}}{\underset{\underset{\displaystyle CH_3}{|}}{C}}-\overset{\overset{\displaystyle O}{\|}}{C}-\overset{\overset{\displaystyle H}{|}}{N}-\overset{\overset{\displaystyle H}{|}}{\underset{\underset{\displaystyle H}{|}}{C}}-COOH + H_2O}$$

The amide group in condensation products from amino acids is also known as a **peptide bond**, and the condensation product itself is known as a **dipeptide**.

Condensation polymerization

The two dipeptides formed above both still contain two reactive functional groups. On one end of the molecule there is a carboxylic acid group, which in the presence of enzymes can react further with the amine group from another amino acid. On the other end of the dipeptide there is an amine group, which can react further with the carboxylic acid group of another amino acid (Figure 54). Provided the monomers each contain at least two reactive functional groups then condensation polymerization can occur.

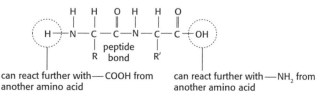

can react further with —COOH from another amino acid   can react further with —NH₂ from another amino acid

Figure 54 The presence of two reactive functional groups leading to condensation polymerization.

❓
1 Find the structures of caffeine, nicotine, diazepam (Valium), penicillin and paracetamol (acetaminophen) in the IB data booklet. Which of the above contain at least one amide group?

2 Use the structure of the hallucinogenic drug LSD given in the IB data booklet to explain what the "D" stands for in LSD.

Primary structure of proteins

Proteins consist of a long chain of amino acid residues formed by condensing the amino acids together in the presence of enzymes to form peptide bonds. The three-dimensional structure of proteins is complex, but their primary structure is simply the sequence of amino acid residues from the N terminus to the C terminus. One such protein is lysozyme. This is an enzyme, found in many organisms, which breaks down the cell walls of certain types of bacteria.

Lysozyme consists of a chain of 129 amino acid residues (Figure 55).

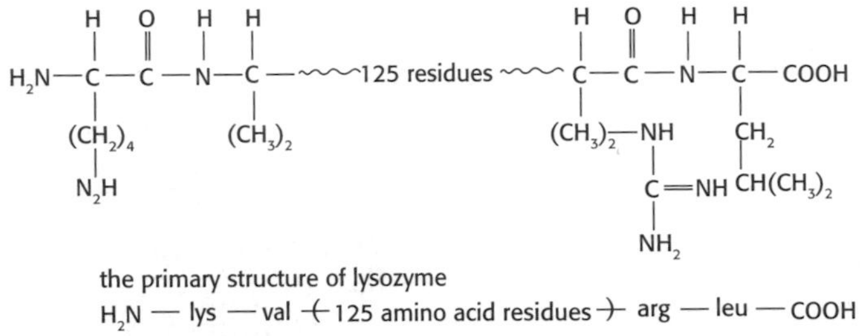

the primary structure of lysozyme

H_2N — lys — val —(125 amino acid residues)— arg — leu — COOH

Figure 55 The primary structure of lysozyme.

Polyamides

Proteins are examples of natural polyamides. Polyamides can be made in the laboratory and in industry by condensing molecules containing two amine groups with dicarboxylic acids. The classic example is the condensation of 1,6-diaminohexane acid with hexane-1,6-dioic acid to form nylon 6,6 (Figure 56). The repeating unit is the smallest part of the molecule that repeats itself.

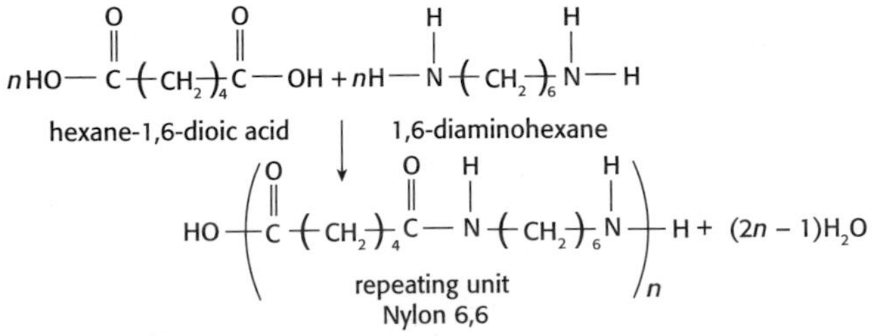

Figure 56 Nylon 6,6.

The inventor of nylon

The American chemist Wallace Carothers (1896-1937) is credited with the invention of nylon. Having been part of the team that created neoprene, the first synthetic rubber, in 1930, he then went on to create and patent nylon in 1935. Wallace Carothers suffered from mental illness, and took his own life in 1937, two years before nylon fibres were first used commercially.

Polyesters

Polyesters can be made by condensing diols (compounds containing two alcohol functional groups) with dicarboxylic acids. One example of a polyester is polyethene terephthalate (Figure 57). This polyester is known as Terylene in the UK and as Dacron in the USA.

nHO—C(=O)—⟨benzene ring⟩—C(=O)—OH   + nH—O—CH$_2$—CH$_2$—OH

benzene-1,4-dicarboxylic acid   ethane-1,2-diol

↓

HO—[C(=O)—⟨benzene ring⟩—C(=O)—O—CH$_2$—CH$_2$—O]$_n$—H   + $(2n-1)$ H$_2$O

repeating unit

Figure 57 Polyethene terephthalate.

The economic importance of condensation reactions

Polyamides and polyesters are just two examples of condensation polymers. Currently the world production of nylon alone exceeds 5 million tonnes. Both polyesters and polyamides are used to make clothes and carpets. However, there are many other uses for condensation polymers. Nylon has high strength, resists abrasion, and is easy to dye. The fibres are used to make climbing ropes for mountaineering (Figure 58), and one of the main uses of nylon is in engineering. The structure can be altered to give specific properties by using fillers, pigments, glass fibre and toughening agents. The chemical and physical properties can also be altered by changing the number of carbon atoms in the two condensed carbon chains. Thus nylon 6,10 has the repeating unit $-NH(CH_2)_6NHCO(CH_2)_8CO-$.

Stereoisomerism

Structural isomers are compounds that possess the same molecular formula but have a different structural formula. That is, the atoms are bonded in a different order: for example propan-1-ol and propan-2-ol.

Stereoisomers have the same molecular formula *and* structural formula, but their atoms are arranged differently in space. In stereoisomers each atom is bonded to the same atoms, but the way in which they are bonded is different. There are two types of stereoisomerism, geometrical isomerism and optical isomerism.

Geometrical isomerism

Geometrical isomerism occurs when bonds are unable to rotate freely. This is known as **restricted rotation**. This type of isomerism occurs in alkenes when the two atoms attached to the carbon atoms constituting the double bond are different. Consider but-1-ene and but-2-ene (Figure 59). Compared with each other they are structural isomers, because the double bond is in a different place. If we could rotate the double bond in but-1-ene, then the outcome would be the same: the two hydrogen atoms on the end carbon atom simply exchange places. So there is only one compound of

Figure 58 Two IB teachers roped together with a nylon climbing rope on the summit of the Wilde Spitze in the Austrian Alps (photo by Geoff Neuss)

1 Part of the primary structure of lysozyme contains the three amino acids alanine, serine and valine condensed together to form a tripeptide. In lysozyme they are joined in the sequence H$_2$N–ala–ser–val–COOH.
   (a) Find the structures of these three amino acids in the IB data booklet and draw out the full structure of this sequence, showing the two peptide bonds.
   (b) How many other tripeptides can be formed by condensing these three amino acids together, assuming that each amino acid is only used once in each tripeptide?
2 Give the repeating unit for the polyester that could be formed by reacting ethane-1,2-dioic acid with propane-1,3-diol.

but-1-ene. However, if we could rotate the double bond in but-2-ene, the outcome would be different. In one case the two methyl groups are on the same side of the molecule, whereas in the other case they are on opposite sides, and clearly the distance between the end two carbon atoms is different. When they are on the same side the compound is known as the *cis-* isomer, and when they are on opposite sides (across the molecule) the compound is known as the *trans-* isomer.

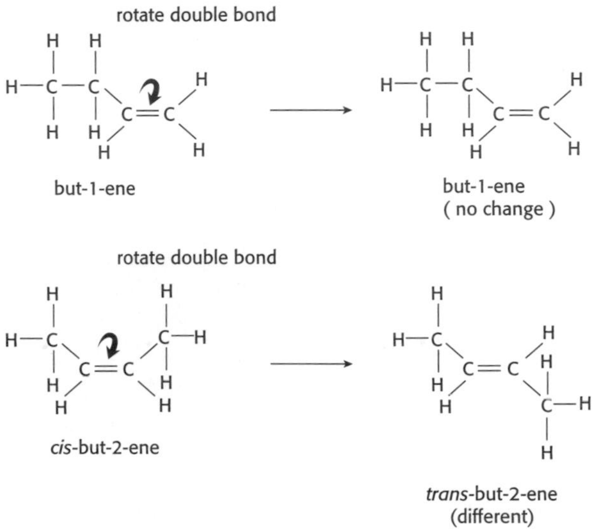

Figure 59 Geometrical isomerism.

The *cis-* and *trans-* isomers exist separately, because the double bond cannot be rotated. The carbon atoms on either side of the double bond are sp² hybridized. One of the bonds between the two carbon atoms is a sigma (σ) bond formed by the overlap of two sp² hybrid orbitals, but the other bond is a pi (π) bond formed by the sideways overlap of the p atomic orbitals on each carbon atom. For the two p orbitals to overlap they must both be in the same plane: that is, they must both be p_z orbitals or p_y orbitals (Figure 60). Any attempt to rotate one of the carbon atoms relative to the other moves the p orbitals out of the same plane, and thereby causes the bond to break. As this would require a considerable amount of energy the *cis-* and *trans-* isomers exist independently: they are not easily interconvertible, thanks to this high activation energy barrier.

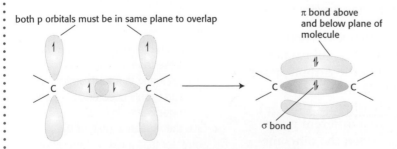

Figure 60 Why the double bond in alkenes cannot rotate

The physical and chemical properties of *cis-* and *trans-* isomers can sometimes be similar, but often they are very different. If there is a difference in melting points, for example, this may be due to the way in which they can pack together. It is not easy to generalize about which isomer will have the higher melting point. For example, *cis-*but-2-ene melts at −139 °C, which is lower than the melting point of *trans-*but-2-ene (−106 °C). By contrast, *cis-*1,2-dichloroethene melts at 60.3 °C, which is higher than the melting point of *trans-*1,2-dichloroethene (47.5 °C).

Sometimes the differences can be even more marked, thanks to the nature of the functional groups attached to the double bond. The melting points of *cis-*but-2-ene-1,4-dioic acid and *trans-*but-2-ene-1,4-dioic acid are very different. In the *trans-* isomer there is strong intermolecular hydrogen bonding between different molecules, because of the polarity of the carboxylic acid groups. In the *cis-* isomer much of this hydrogen bonding occurs internally between the two acid groups. This means the attraction between two different molecules is less, so that the melting point is much lower. See Figure 61.

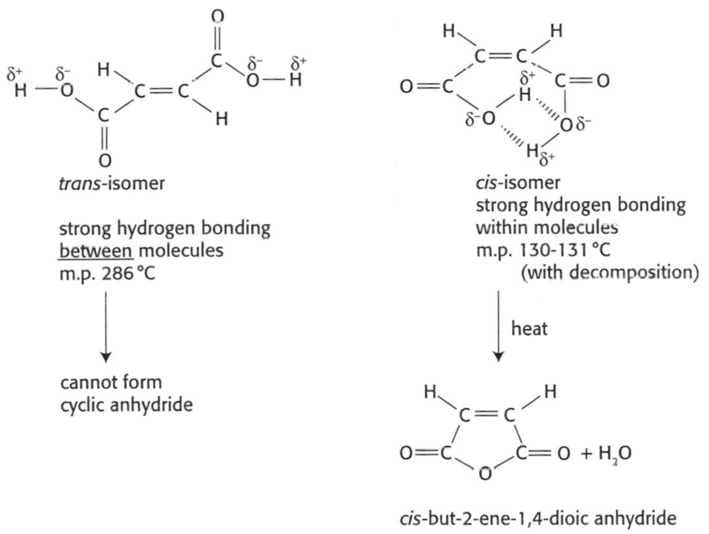

*trans-*isomer

strong hydrogen bonding <u>between</u> molecules
m.p. 286 °C

cannot form cyclic anhydride

*cis-*isomer
strong hydrogen bonding within molecules
m.p. 130-131 °C
(with decomposition)

heat

*cis-*but-2-ene-1,4-dioic anhydride

Figure 61 Physical and chemical properties of *cis-* and *trans-*but-2-ene-1,4-dioic acid

The two geometric isomers of but-2-ene-1,4 dioic acid are also very different chemically. In the *cis-* isomer the two carboxylic acid functional groups are close enough together to react, so that when the isomer is heated water can be expelled, and the cyclic acid anhydride is formed (Figure 59). No such reaction is possible with the *trans-* isomer.

Extension material

How we see

The cells of the retina at the back of our eyes contain a molecule called **retinal**. When light falls on the retina it converts molecules of retinal from the *cis-* form to the *trans-* form (Figure 62).

It is the energy in the light that enables this photochemical reaction to occur. The energy is sufficient to cause the homolytic fission of the pi bond to form radicals. The remaining single sigma bond can rotate, and then the pi bond is re-formed. It is thought that as this happens a nerve impulse is transmitted via an interaction with a protein to send a signal along the optic nerve to the brain (Figure 63).

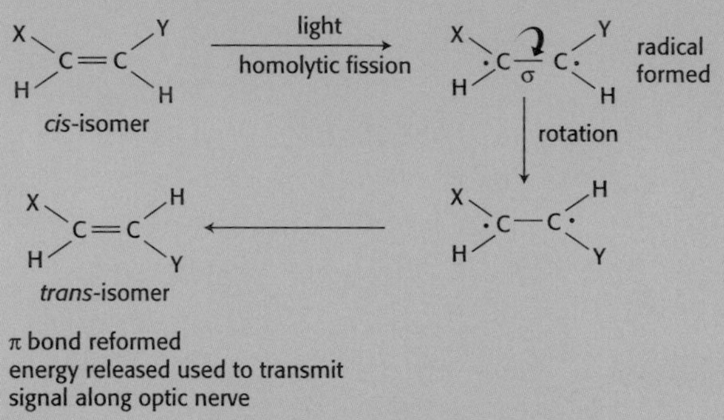

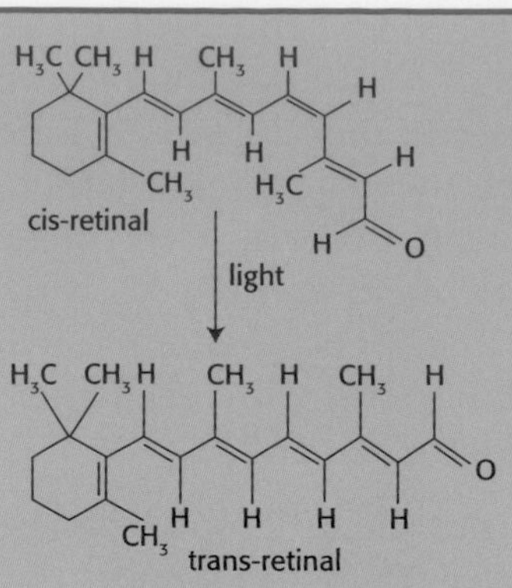

Figure 62 The effect of light on *cis*-retinal

π bond reformed
energy released used to transmit
signal along optic nerve

Figure 63 How nerve impulses in the eye are transmitted to the brain

Geometrical isomerism is also possible in cyclic compounds. Although there are only single bonds between the carbon atoms, the rigid structure of the ring prevents free rotation. Thus 1,2-dichloropropane, for example, can exist as *cis*- and *trans*- isomers (Figure 64).

cis-1,2-dichlorocyclopropane   *trans*-1,2-dichlorocyclopropane

Figure 64 Isomers of 1,2-dichloropropane.

There are four isomers of dichlorocyclobutane. 1,2-dichlorocyclobutane and 1,3-dichlorocyclobutane are structural isomers, and each one can have two geometric isomers (Figure 65).

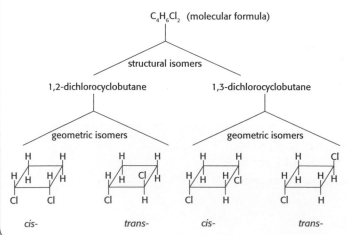

$C_4H_6Cl_2$ (molecular formula)

structural isomers

1,2-dichlorocyclobutane   1,3-dichlorocyclobutane

geometric isomers   geometric isomers

cis-   *trans-*   *cis-*   *trans-*   **Figure 65** Isomers of dichlorocyclobutane

Optical isomerism

Optical isomers occur when there are four different atoms or groups attached to a single carbon atom. This is known as an **asymmetric** carbon atom. Molecules containing an asymmetric carbon atom are said to be **chiral** molecules, although the word "chiral" is also sometimes used instead of "asymmetric" to just describe the carbon atom itself that contains the four different atoms or groups attached to it.

It is worth making two three-dimensional models of an asymmetric carbon atom. For the first model place four different groups around the carbon atom, and then for the other just interchange two of the groups. You will see that, no matter how you try, it is impossible to superimpose one upon the other without breaking and remaking bonds. The only way in which one can become the other is by reflection in a mirror: they are mirror images of each other. These two mirror images are known as optical isomers or **enantiomers**. As was explained in Chapter 4 when discussing the shapes of molecules, the three-dimensional structures can be represented in two dimensions by using a dotted line for bonds going behind the plane of the paper and wedge-shaped lines for bonds coming out from the plane of the paper (Figure 66).

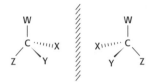

Figure 66 Mirror images (enantiomers) of an asymmetric carbon atom.

The origin of the word "chiral"

Chiral is derived from the Greek word *kheir*, which means "hand". The mirror image of the right hand is the left hand, and it cannot be superimposed on the right hand. If molecules (enantiomers) cannot be superimposed on their mirror images, they are behaving rather like left and right hands—hence the word "chiral".

Examples of simple molecules that contain an asymmetric carbon atom are butan-2-ol and 2-bromobutane. The asymmetric carbon atom within a molecule is sometimes indicated using an asterisk (Figure 67).

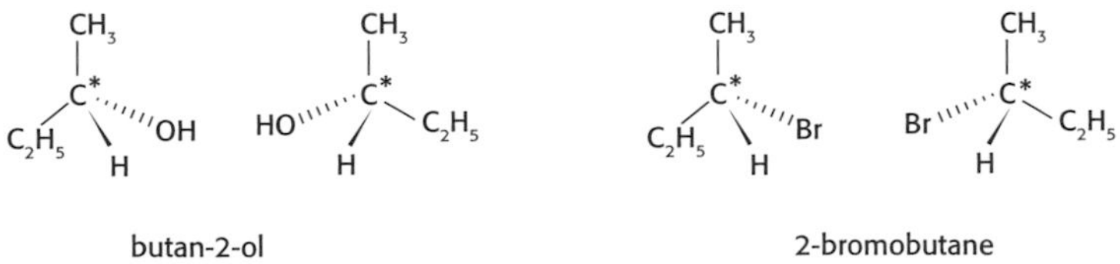

butan-2-ol   2-bromobutane

Figure 67 Enantiomers of butan-2-ol and 2-bromobutane.

Optical isomers of a compound differ in only one respect in their physical properties: they rotate the plane of plane-polarized light in opposite directions. All their other physical properties, such as density and melting point, are identical.

Plane-polarized light

Normal light is electromagnetic radiation that is oscillating in every plane. A polarizing filter acts rather like a gate. It contains many narrow and closely spaced vertical slits, so that only the light oscillating in the same vertical plane can pass through it. The light emerging is known as plane-polarized light (Figure 68).

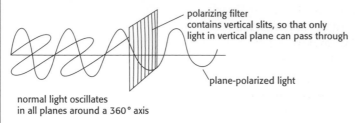

Figure 68 The formation of plane-polarized light

Our eyes are unable to distinguish between normal light and plane-polarized light. If another polarizing filter is placed in the path of plane-polarized light then, as it is rotated through 90° relative to the first filter, it will completely block the passage of all light. Polarizing filters are used in sunglasses and in camera lenses to cut down glare and reflections.

The ability of enantiomers to rotate the plane of plane-polarized light can be shown in practice by using a **polarimeter** (Figure 69). This consists of a light source, two polarizing lenses, and a tube to hold the sample of the enantiomer located between the lenses.

When light passes through the first polarizing lens (the **polarizer**) it becomes plane-polarized: that is, it vibrates in a single plane. With no sample present the observer will see the maximum intensity of light when the second polarizing lens (the **analyser**) is in the same plane. Rotating the analyser by 90° will cut out all the light. When the sample is placed between the lenses the analyser must be rotated by θ degrees, either clockwise or anticlockwise, to give light of maximum intensity. If the analyser has to be rotated clockwise, the enantiomer is said to be **dextrorotatory** (from the Latin *dexter*, meaning "right"). If it has to be rotated anticlockwise, the enantiomer is said to be **laevorotatory** (from the Latin *laevus*, meaning "left"). The two enantiomers rotate the plane of plane-polarized light by the same amount, but in opposite directions. One of the enantiomers is thus known as the **d-form** and the other as the **l-form**. If both enantiomers are present in equal amounts the two rotations cancel each other out, and the mixture appears to be optically inactive. Such a mixture is known as a **racemic mixture**.

Not only are the physical properties of the two enantiomers of a chiral compound identical; their chemical properties are also identical—except when they interact with other optically active substances. This is often the case in the body, where the different enantiomers can have completely different physiological effects.

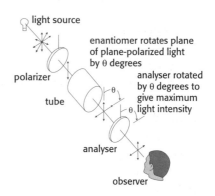

Figure 69 Using a polarimeter.

For example, one of the enantiomers of the amino acid asparagine $H_2N-CH(CH_2CONH_2)-COOH$ tastes bitter, whereas the other enantiomer tastes sweet.

Critical considerations: Thalidomide

This ability of different enantiomers to have different physiological effects became tragically apparent with one particular drug that was used in the middle of the last century. In 1958 a German pharmaceutical company launched a massive publicity campaign for a new tranquillizer to combat "morning sickness" in pregnant women. The drug was sold worldwide under brand names such as Thalidomide and Contergan. In many countries it was sold without prescription, and was marketed as completely innocuous. Reports of severe adverse side-effects began to appear in 1959, and it later transpired that as early as 1956 clinical trials by the company itself had revealed problems. Nevertheless, because it was making so much profit the company continued to market the drug heavily, and sales kept increasing until it was withdrawn in 1961. By that time many children had been born with absent or severely malformed limbs. It was later shown that one of the enantiomers of Thalidomide was active against the morning sickness, whereas the other was responsible for the fetal abnormalities (Figure 70). Thalidomide was withdrawn although it is now being prescribed again in certain cases, as it has proved to be an effective drug against leprosy and certain types of cancer.

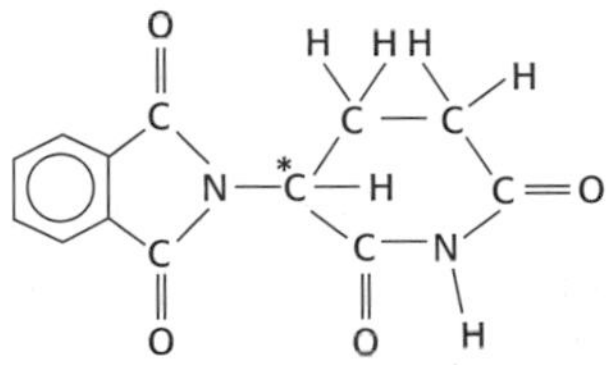

Figure 70 The structure of Thalidomide (the chiral carbon atom is marked with an asterisk).

One important group of molecules that can show optical activity is the 2-amino acids (Figure 71). Apart from the simplest, 2-aminoethanoic acid (glycine), in which the R group is a hydrogen atom, all the others can exist in both the d- and l- forms. Interestingly, in biological systems most amino acids occur as the l-form, whereas sugars, which can also show optical activity, tend to exist as the d-form. Some people have speculated that this shows that all life originated from a single ancestor. Others argue that it was simply a random event that occurred during the early stages of evolution.

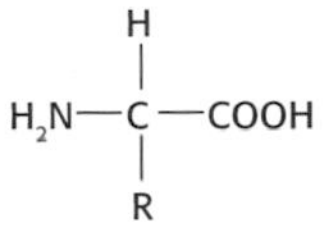

Figure 71 2-amino acid.

Thinking about science

The ability to use reasoning based on physical evidence

Although we can use STM (scanning tunnel microscopy) to gain some physical "picture" of atoms, and we can use X-ray crystallography to deduce the distances between atoms, we cannot ultimately *prove* the shapes of simple molecules or the mechanisms of organic reactions. The existence of optical isomers (enantiomers) does provide good evidence to support our picture of chemistry at the molecular level. Because we can make simple models of tetrahedrally bonded carbon atoms and see that, if they are asymmetric, then two enantiomers exist, this provides good evidence that these do represent the true shapes of these molecules.

The real strength of the ability to reason from models is the application to the S_N2 mechanism for nucleophilic substitution. This was first shown by a Latvian chemist, Paul Walden (1863–1957).

Walden reasoned that if he could find an optically active halogenoalkane that underwent nucleophilic substitution by an S$_N$2 mechanism, then the product should rotate the plane of light in the opposite direction to the starting material. The nucleophile would attack the carbon atom from one side to form the pentacoordinate carbon transition state. Then, as the halogen atom left, the bonds around the carbon atom would "flip" to form an enantiomeric product with the opposite absolute stereochemistry. The problem is that, by definition, a primary halogenoalkane will not be optically active. However, Walden inversion *does* occur in the reaction between dilute aqueous sodium hydroxide and the secondary halogenoalkane, 2-bromobutane (Figure 72). This provides strong evidence for the S$_N$2 mechanism, because, if it went by an S$_N$1 mechanism, the hydroxide ion could react on either side of the planar carbocation intermediate, and a racemic mixture would be obtained.

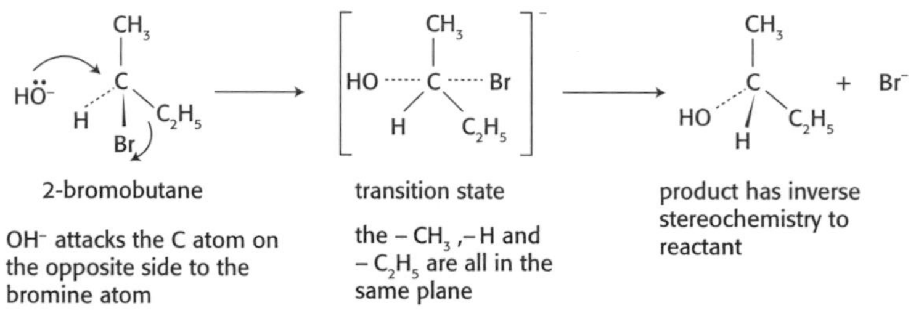

2-bromobutane	transition state	product has inverse stereochemistry to reactant
OH$^-$ attacks the C atom on the opposite side to the bromine atom	the − CH$_3$,−H and − C$_2$H$_5$ are all in the same plane	

Figure 72 An illustration of Walden inversion.

Reaction pathways (2)

Having studied the reactions of some more functional groups, you should be in a position to devise more two-step organic syntheses. As before, consider working backwards from the required product as well as forward from the reactants.

1 Devise two-step syntheses of the following products from the starting materials listed. Include any necessary experimental conditions and an equation for each step.
   (a) 1-butylamine (1-aminobutane) from 1-bromopropane.
   (b) Ethyl ethanoate from ethanol (no other organic reagents are available).
   (c) *N*-ethylethanamine from ethene.
   (d) Propanone from propene.
   (e) Cyclohexanol from cyclohexane.
2 Cyclohexanone can be synthesized from cyclohexane by the following three-step process:

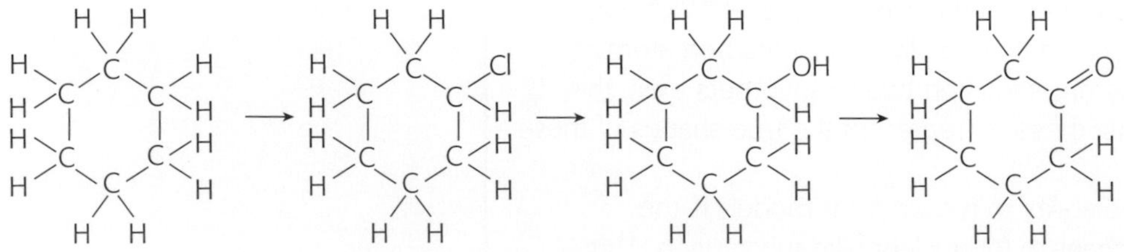

Give the necessary conditions and reagents for each step, and name the mechanisms for the first two steps.

Design experiment on homologous series

Now that the core and additional higher level material has been covered, you may be set another design experiment so that you can use your knowledge of organic chemistry. One such instruction might be: "Investigate an aspect of a homologous series." If you wanted to play safe, then an obvious dependent variable might be boiling point, and an independent variable might be the number of carbon atoms in the chain of a particular homologous series. However, you will get more satisfaction if you try to be more creative and investigate a relationship where you are less certain of the outcome. Some genuine focused research questions that IB Diploma Programme chemistry students have devised when confronted with this problem are as follows.

1 How does increasing the length of the alkyl group affect the elutive properties of an alcohol acting as a solvent in paper chromatography of black ink?

2 How long does it take the reaction to go to completion when different alcohols from the same homologous series react with sodium?

3 How does the solubility of carboxylic acids in water change when the length of the carbon chain is increased?

4 How does the enthalpy change of reaction for the formation of sodium alkoxides from sodium and the corresponding alcohol change as the carbon chain length of the alcohol increases?

5 How significantly does hydrogen bonding affect the boiling points of mixtures of propane and primary alcohols as the carbon chain length of the alcohol increases?

Uncertainty and error in measurement

Precision, accuracy, systematic errors and random errors

Quantitative chemistry involves measurement. A measurement is a method by which some quantity or property of a substance is compared with a known standard. In the past many of these standards were not fixed, but depended upon individuals: for example, a cubit was the length of the forearm from the elbow to the tip of the middle finger. In some societies this practice still remains. Nomads in Mongolia use a measurement called a *tokhoi*, which is the length between the wrist and the elbow (about 32 cm), and it is still used when buying a pair of jeans. However, most units are now carefully and precisely defined. For example a second, which used to be defined in terms related to the rotation of the Earth, is now defined as the duration of 9,192,631,770 periods of the radiation corresponding to the transition between the two hyperfine levels of the ground state of the caesium atom.

Chemists use many different instruments to make measurements, so it is sensible to question how reliable these measurements are when compared with the accepted standards. Suppose you want to know the temperature of the chemistry laboratory you are working in. The easy solution is to find a thermometer that has been in the room for a while, and simply take the reading. A problem arises if you then look at a different thermometer. The chances are that they will not both give you exactly the same reading. How can we determine which one is giving the "true" reading?

Suppose we find 10 thermometers, and the readings (all measured in degrees Celsius) are: 19.9, 22.3, 20.2, 20.0, 20.0, 20.1, 19.9, 20.3, 19.9 and 20.2. Nine of the thermometers are giving a reading between 19.9 °C and 20.3 °C, so it is reasonable to assume that the actual temperature is about 20 °C, especially as—upon closer inspection—the one that reads 22.3 °C can be seen to have an air bubble trapped in the alcohol. If we ignore the anomalous result, then we can take the average value by adding up all the other nine values and dividing by nine to give 20.06 °C, which is 20.1 °C to three significant figures (see next section). Can we now claim with certainty that the temperature in the room is 20.1 °C, even though only one of the thermometers actually gives this reading?

What we can reasonably claim from these results is that the temperature is 20.1 ±0.2 °C. We can also talk about the **precision** of the readings. Precision refers to how close several experimental measurements of the same quantity are to each other. If a single student had taken all the temperature readings they would be said to be **repeatable**, whereas if several students had taken the readings they would be said to be **reproducible**. Because we have used nine different thermometers that all give results to within ±0.2 °C we can claim they are reasonably precise results. However, we still do not know whether the true temperature of the room is 20.1 °C.

To know this we have to assume that the all the thermometers were calibrated accurately in the first place. Liquid-in-glass thermometers are calibrated by taking the melting point of an ice/water mixture as 0 °C and the boiling point of pure water at one atmosphere pressure (1.01 kPa) as 100 °C. If a mistake was made in the factory that manufactured the thermometers, then they all may be giving a biased reading.

How close a precise reading is to the true reading (or the generally accepted or literature value) is known as the **accuracy** of the measurement. If all the thermometers are reading 20 °C when the true value of the temperature is 19 °C then they are giving precise but inaccurate readings. This inaccuracy is due to a **systematic error**. Systematic errors may be due to the apparatus itself, or they may be due to the way in which the readings are taken. For example, when using a pipette to measure a known volume of an aqueous solution, the bottom of the meniscus, not the top of the meniscus, should be on the graduation line (Figure 1). If a student repeats this error when performing a titration, the results may be precise, but they will also be inaccurate.

We have not yet answered the question as to how to determine the temperature of the room accurately. Fortunately, the instrument we use—the thermometer—can provide us with some helpful information. Good measuring instruments in the chemical laboratory have not only the measuring scale clearly identified, but also the **tolerance** of the apparatus. The manufacturers may claim that the thermometers read from −10 °C to +110 °C with an uncertainty of ±0.2 °C. Although we do still have to trust the manufacturer, we can now be reasonably certain that the temperature in the room is 20.1 °C ± 0.2 °C. It is worth noting that, even if an instrument does have a systematic error, it may still be of use if it is used to measure a **difference**. For example, we could use a single thermometer with a systematic error to measure the temperature increase in a solution where an exothermic reaction was occurring. The measured temperature increase will be accurate, even though the two temperature readings are inaccurate, because the systematic error will cancel out (Figure 2).

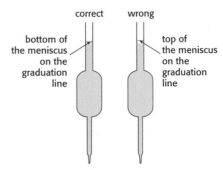

Figure 1 Measuring a volume using a pipette. Repeating measurements using the wrong method introduces a systematic error, because the pipette consistently delivers less than the stated volume.

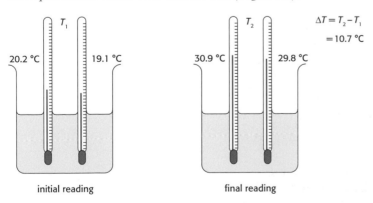

Figure 2 Using two thermometers to measure the temperature increase for an exothermic reaction in aqueous solution. Even though the thermometers give different readings, they both give the same temperature difference, because any systematic errors cancel. (This assumes that the expansion of the liquid used in the thermometers is linear.)

One other uncertainty that is less easy to control is the effect of **random errors**. A random error occurs if there is an equal probability of the reading being high or low from one measurement to the next. Random errors might include variations in the volume of glassware due to temperature fluctuations, or the decision on when an indicator changes colour during an acid–base titration. Because they are random, they can be reduced through repeated measurements, whereas repeated measurements will not reduce a systematic error. Because there is always some random uncertainty in measurements it is usual practice to report the reading from a scale as the smallest division or the last digit capable of precise measurement, even though it is understood that the last digit has been rounded up or down so that there is a random error or uncertainty of ±0.5 of the last unit.

Significant figures

Whenever a measurement of a physical quantity is taken, there will be uncertainty in the reading. The measurement quoted should include the first figure that is uncertain. This should include zero if necessary. Thus a reading of 45.310 g indicates that the mass was measured with a balance that is accurate to ±0.001 g. If a balance accurate to only ±0.01 g is used, the mass should be recorded as 45.31 g.

Zero can cause problems when determining the number of significant figures. Essentially, zero becomes significant only when it comes after a non-zero digit (1, 2, 3, 4, 5, 6, 7, 8, 9):

000123.4	0.0001234	1.0234	1.2340
zero not a significant figure		zero is a significant figure	
values quoted to 4 sig. figs		values quoted to 5 sig. figs	

Zeros after a non-zero digit but before the decimal point may or may not be significant depending on how the measurement was made. For example, 123 000 might mean exactly one hundred and twenty three thousand, or one hundred and twenty three thousand to the nearest thousand. This problem can be neatly overcome by using scientific notation:

1.23000×10^6 quoted to six significant figures
1.23×10^6 quoted to three significant figures

Calculations

When adding or subtracting, it is the **number of decimal places** that is important. Thus, when using a balance that measures to ±0.01 g the answer can also be quoted to two decimal places, which may increase or decrease the number of significant figures. For example:

7.10 g	+	3.10 g	=	10.20 g
3 sig. figs		3 sig. figs		4 sig. figs.

22.36 g	−	5.16 g	=	7.20 g
4 sig. figs		4 sig. figs		3 sig. figs

When multiplying or dividing, it is the **number of significant figures** that is important. The number with the fewest significant figures used in the calculation determines how many significant figures should be used when quoting the answer.

For example, when the temperature of 0.125 kg of water is increased by 7.2 °C:

Heat required = 0.125 kg $\times$ 7.2 °C $\times$ 4.18 kJ kg^{-1} °C^{-1} = 3.762 kJ

As the temperature was recorded to only two significant figures, the answer should strictly be given as 3.8 kJ.

Currently you may be penalized (by just one mark in the whole paper) in IB examinations if you give too few significant figures in your answers. Equally you should not give the "calculator answer", which can often be 10 or more digits.

Uncertainties in calculated results

When using measuring apparatus in the laboratory, always look to try to find the manufacturer's statement regarding the degree of tolerance. If it is not present, then you will have to estimate the uncertainty.

If you are making a single measurement with a piece of apparatus, the absolute uncertainty and the percentage uncertainty can both be stated relatively easily. For example, consider measuring 25.0 cm^3 with a 25 cm^3 pipette that measures to ± 0.1 cm^3. The absolute uncertainty is 0.1 cm^3, and the percentage uncertainty is equal to

$$\frac{0.1}{25.0} \times 100 = 0.4\%$$

If we use a burette that is claimed to measure 50 cm^3 with an uncertainty of ± 0.1 cm^3, then if we do actually use it to measure 50 cm^3 of a solution the percentage uncertainty is 0.2%. Normally, however, we use the burette to deliver a smaller volume than 50 cm^3. If the volume used from the burette is 20.0 cm^3, then the absolute uncertainty is still ± 0.1 cm^3, but the percentage uncertainty is now equal to

$$\frac{0.1}{20.0} \times 100 = 0.5\%$$

If we are simply adding or subtracting two volumes or two masses, then the absolute uncertainties can be added. For example, suppose we add two volumes of 25.0 cm^3 $\pm$ 0.1 cm^3. In one extreme case the first volume could be 24.9 cm^3 and the second volume 24.9 cm^3, which would give a total volume of 48.8 cm^3. Alternatively, the first volume might have been 25.1 cm^3, which when added to a second volume of 25.1 cm^3 gives a total volume of 50.2 cm^3. The final answer can therefore be quoted between 48.8 cm^3 and 50.2 cm^3—in other words 50.0 cm^3 $\pm$ 0.2 cm^3.

If we are using multiplication, division or powers then percentage uncertainties should be used during the calculation and then converted back into an absolute uncertainty when the final result is presented. This can be illustrated using the results obtained from a typical titration to find the molar mass of an unknown acid. Let us assume that 2.500 g of the unknown acid was dissolved in distilled water and the volume made up to 250 cm³ with distilled water in a volumetric flask. We will also assume that 25.0 cm³ of a standard solution of an alkali was pipetted into a conical flask for each titration, and the acid solution was placed in the burette. Four separate pieces of apparatus were used, each of which contributed to the overall uncertainty.

When using a balance that weighs to ±0.001 g, the uncertainty in weighing 2.500 g of the unknown substance will be equal to

$$\frac{0.001}{2.500} \times 100 = 0.04\%$$

Similarly, a pipette measures 25.00 cm³ ± 0.10 cm³. The percentage uncertainty due to the pipette is thus

$$\frac{0.01}{25.00} \times 100 = 0.04\%$$

Assuming the percentage uncertainties due to the burette and to the volumetric flask are 0.50% and 0.40% respectively, the overall percentage uncertainty is obtained by summing all the individual percentage uncertainties:

Overall uncertainty = 0.04 + 0.40 + 0.50 + 0.40 = 1.34% ≈ 1.3%

Hence if the answer for the molar mass is determined to be 129 g mol⁻¹, the uncertainty is 1.3% or 1.68 g mol⁻¹.

The answer should be given as 129 ± 2 g mol⁻¹.

If the literature value for the molar mass of the acid is 126 g mol⁻¹, then you can calculate the total error in the experiment, as this is the difference between the literature value and the experimental value, divided by the literature value, and expressed as a percentage. In this case:

$$\text{Percentage error} = \frac{(129 - 126)}{126} \times 100 = 2.4\%$$

This lies just outside the calculated uncertainty of 1.3%.

There are actually other uncertainties that you have not taken into account when calculating the total percentage uncertainty. For example, the standard solution used for the titration may have been prepared by a technician, so you have no way of knowing the uncertainty in its concentration—or indeed its purity. Similarly, there is uncertainty in exactly where the end-point was taken during the titration. These extra uncertainties should be mentioned in the evaluation section of any laboratory report, even if they are not actually quantified.

You should also be able to recognize when the uncertainty of one of the measurements is much greater than that of the others. This will then have the major effect on the uncertainty of the final result, and the approximate uncertainty can be taken as being due to that quantity alone. This may be the case whenever a thermometer is used. Compared with an analytical balance, or accurate and precise volumetric apparatus, the percentage uncertainty in the temperature readings will be much greater, particularly when the temperature difference is relatively small. Consider the experiment to measure the enthalpy change of neutralization, $\Delta H_{neutralization}$, when hydrochloric acid is neutralized with sodium hydroxide solution described in Chapter 5. The temperature rise was 6.8 °C. If the uncertainty in the thermometer scale is ±0.5 °C, and two readings were taken to measure this, then the absolute uncertainty is ±1.0 °C. This gives a percentage uncertainty of 14.7%, which is much higher than any other uncertainties inherent in the experiment.

Graphical techniques

Frequent use is made of graphs in chemistry. On a simple level they provide an instant visual representation of data. If you look at information given in Chapter 5 on the change in the amount of carbon dioxide present in the atmosphere during the past 50 years it is much easier to see the increasing trend by inspecting the graph rather than by looking at the figures. By plotting a suitable graph to give a straight line, or some other definite relationship between the variables, we can use graphs to predict unknown values. There are various methods to achieve this. They include measuring the intercept, measuring the gradient, extrapolation and interpolation. Interpolation involves determining an unknown value within the limits of the values already measured, whereas extrapolation requires extending the graph to determine an unknown value that lies outside the range of the values measured.

When plotting a graph, the convention is to plot the independent variable on the horizontal axis and the dependent variable on the vertical axis. Thus, if we are measuring the volume of gas evolved during a reaction at fixed time intervals, time will be along the x axis and the volume of gas evolved will be along the y axis. Graphs should be plotted using Système International (SI) units wherever possible. As for all graphs a suitable title should be included, the variables should have their units clearly shown, and the scale of the axes should be appropriate. Take particular care when using computer programs to plot graphs, because some tend to start all axes at zero unless you specify otherwise.

Choosing a sensible scale for the axes

Consider the following (simple) data to be plotted using Microsoft Excel:

Temperature / °C	Volume of gas / cm³
20.0	60.0
30.0	62.0
40.0	64.0
50.0	66.0
60.0	68.0

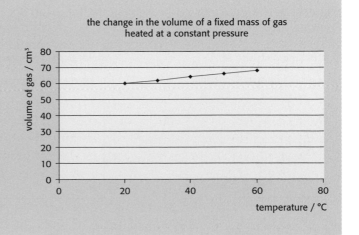

the change in the volume of a fixed mass of gas heated at a constant pressure

The first graph has both axes starting at zero, and does not use a sensible scale, as most of the graph is wasted space:

The second graph uses a sensible scale if interpolation to find the volume at 35 °C is all that is required:

The third graph expands the scale of the axes so that the graph can be extrapolated:

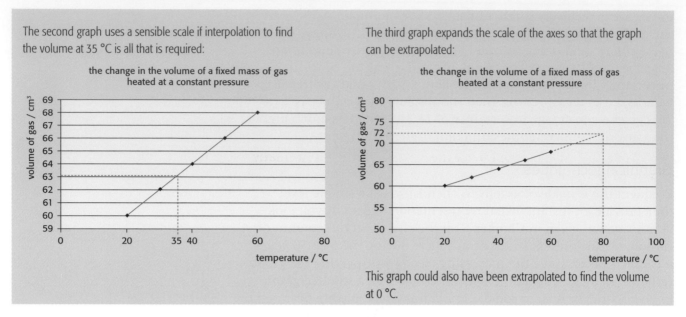

This graph could also have been extrapolated to find the volume at 0 °C.

Once the points are plotted on a graph, you will need to determine whether to join the points up directly or whether to choose the line of best fit. Occasionally the points are just joined directly. Obvious examples of this have been given in Chapter 3, where the periodicity exhibited by the elements can be seen from graphs of first ionization energies or melting points plotted against atomic number. However, it is much more usual in chemistry to find the line of best fit. This may be a straight line or a smooth curve. If you are trying to find the line of best fit you do need sufficient readings. Two readings are absolutely guaranteed to give a straight line! To be scientifically valid a minimum of five readings is usually required.

It is often sensible not to just plot the variables actually measured but to manipulate them to give a straight-line graph. For example, from the ideal gas equation it can be deduced that the pressure of a fixed mass of gas is inversely proportional to its volume at a constant temperature: that is, the pressure multiplied by the volume is equal to a constant, $PV = k$. A plot of pressure against volume will therefore give a curve. It is not easy to see the relationship from this curve. However, if a graph of pressure against the reciprocal of volume is plotted then a straight line is obtained (Figure 3). From the general equation for a straight line ($y = mx + c$, where m represents the gradient and c represents the intercept) we can see that $P = m/V$, which is equivalent to stating that $PV = $ constant.

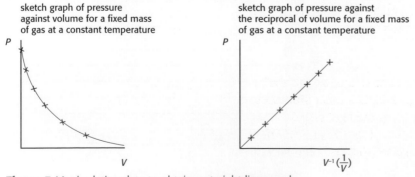

Figure 3 Manipulating data to obtain a straight-line graph

Graphs can provide other information apart from just the determination of unknown values of one of the variables either within or outside the measured range. If we have a graph of the concentration of a product, [A], as it is formed during a chemical reaction plotted against time, then by taking the gradient at a particular value of time we can obtain the rate of that reaction at that time (Figure 4). As the gradient of the slope is given by the change in concentration divided by the change in time, we can also obtain the units for the rate of the reaction, which is normally expressed as $mol\ dm^{-3}\ s^{-1}$.

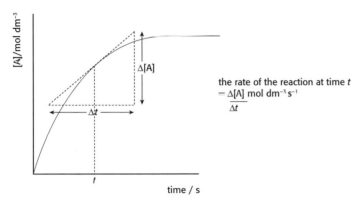

the rate of the reaction at time t
$= \dfrac{\Delta[A]}{\Delta t}$ mol $dm^{-3} s^{-1}$

Figure 4 Graph showing how the concentration of the product A changes with time.

Some graphs also show the uncertainties in the measurements by using error bars, or by drawing lines of minimum and maximum fit. These can be useful ways of expressing the degree of uncertainty in the data; they are not necessary or required by the IB when processing chemical data.

Thinking about science

One instrument we take for granted

This chapter has looked at some of the uncertainties associated with the instruments or apparatus used for taking measurements, and at how data can be presented graphically. There is one piece of apparatus that is used in every scientific experiment; it is often taken for granted, and the uncertainties associated with it are often ignored. It is the eye. Our eyes are very remarkable optical instruments, but they do have their limitations. As chemistry is an experimental science, it is useful to list some of these limitations, and also some of the limitations associated with how our brain interprets the information received through our eyes. If we are at least aware of some of these limitations we should be able to become better observers and therefore more effective chemists.

Images are inverted

It may seem obvious, but an eye is actually a convex lens, and so the image received on the retina is inverted (Figure 5). Our brain processes the nerve impulses and conveniently turns everything the right way up. If you wear special glasses that invert images you will find that, after a few days, the brain turns everything the right way up again—which is fine until you remove the glasses.

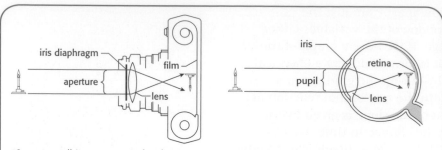

Figure 5 All images seen by the eye are inverted; the brain turns them the right way up.

Our eyes see only in the visible region of the electromagnetic spectrum

The electromagnetic spectrum covers low-energy radio waves to high-energy gamma radiation, and yet our eyes are able to see only a very narrow part of this spectrum known as the **visible region** (Figure 6). We cannot see the first series in the emission spectrum of hydrogen unaided, for example, because this occurs in the ultraviolet region of the spectrum.

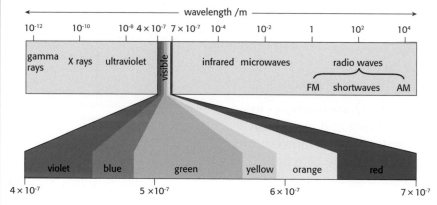

Figure 6 The eye sees only in a very small part of the electromagnetic spectrum.

The brain selects only what it thinks is important

Conjure up a picture in your mind of a particular classroom or laboratory you are very familiar with. Now ask yourself how the number six appears on the clock in the room. You will have seen this clock many times, perhaps when looking forward to the end of the lesson: is the six a 6 or a VI? I do not know the answer, because I do not know your clock. The question is: do you? There is a strong possibility that there is in fact no number six, because many clocks omit it. Our eyes are capable of recording much more information than our brain is able to record, so our brain automatically selects what it thinks is important and ignores the rest. Your brain does not think that a precise description of the six is necessary, and so is unlikely to record it. The problem is: how does your brain know what is important? Sometimes people's lives have depended on witness statements in a court of law, and yet there is no guarantee that people can accurately recall all that they saw. This can be important when performing a chemical experiment. Ask yourself what you see when you heat a test tube in a Bunsen burner flame. Although they have done it many times, most people do not say that the tube mists up on the outside, even though it is obvious once it is pointed out to you. This has serious implications if we cannot rely on observing in an unbiased way.

The brain can be trained to see things differently

We have already seen how chemists can represent the three-dimensional tetrahedral shape of methane using dotted and wedge-shaped lines, although in reality the picture is still in two dimensions. As we discussed in Chapter 4, this can cause problems with molecular modelling in computer simulations. No matter how three-dimensional the images of rotating molecules may appear, they are still in two dimensions. This is a real limitation of computer simulations. You can realize how much your brain automatically transposes two-dimensional shapes into three dimensions when you try to make sense of some of the mathematically inspired drawings of the Dutch graphic artist Maurits Escher (1898–1972) (Figure 7).

Our brain channels us along narrow pathways

Often we only look at a problem from only one angle. This can sometimes cause us to miss an obvious solution or answer. In an earlier chapter you were asked why it is difficult to hard-boil an egg on the top of Chomolungma. The normal answer given is that the pressure is so low that water boils at a much lower temperature, and so it will take much longer for the egg to cook. You were then asked whether it would be difficult to fry an egg there. Because you were thinking only in chemistry terms the answer was "no", as the oil does not have to boil. However, in reality you would have thick gloves on to prevent frostbite, the wind is likely to be blowing strongly, and psychologically most accidents on Chomolungma happen on the way down, so you would not want to spend time on the summit trying to fry an egg. Clearly it is difficult to fry an egg there, and yet many chemists will tell you the opposite.

This can be further illustrated by an experiment in school. How would you determine the molar mass of the gas in your gas taps? You could collect some of the gas in a gas syringe, measure the volume, the temperature and the pressure and use the ideal gas equation $pV = nRT$ to give a good answer. A much simpler solution would be to ring up the gas supplier or, if it is bottled gas, read the label on the bottle.

Figure 7 *Waterfall* by M.C. Escher. The drawing shows an apparent paradox whereby water from the base of a waterfall appears to run downhill before reaching the top of the waterfall. Esher is using conflicting proportions to create the paradox.

The brain accepts authority

Which photograph of Albert Einstein do you recognize—Figure 8 or Figure 9?

Figure 8 Albert Einstein aged 26.

Figure 9 Albert Einstein (somewhat older than 26).

We may have a stereotypical image of scientists as "old men", often with white hair and beards. Look on the Internet at pictures of Darwin, Newton, Mendeleyev and Galileo. All the truly great scientists who brought about a paradigm shift in scientific thinking were all actually young men at the time. Einstein was only 26 when he published his work on special relativity and the photoelectric effect, and Newton had already invented calculus and his three laws of motion by the time he was 23. There is a tendency to equate age with wisdom and authority, and perceive science wrongly as the preserve of old men. We also tend to accept what we are told or read in a textbook, and yet textbooks and sites on the Internet contain many errors. Hopefully, now that you have finished studying the core of IB Diploma Programme chemistry you will be better prepared to question everything you read and are told.

Remember: one of the aims of the IB learner profile is to encourage you to think critically. We could add to this—to observe critically too. This was elegantly summed up by the French chemist and microbiologist Louis Pasteur (1822–1895), who said, in a speech given at Lille University in 1854: *"Dans les champs de l'observation le hazard ne favorise que les esprits préparés."* This translates as: "When observing, chance favours the prepared mind."

Donald Rumsfeld, the former US Defence Secretary once said: "There are known knowns; there are things we know that we know. We also know there are known unknowns. That is to say, there are some things we do not know. But there are also unknown unknowns—the ones we don't know we don't know."

Now that you have completed the core and AHL part of IB Diploma Programme chemistry, do you think it is true to state that there are "known knowns" in chemistry? Would it be better if the IB Diploma Programme concentrated on what we do not know about chemistry rather than what we think we do know?

12 External and internal assessment

There is a continuing debate in the educational world about the purpose of education and the value of examinations. It should be clear from the IB learner profile that one of the main aims of studying for the IB Diploma is to provide you with an education for life. After the two-year course you should be able to think critically, be aware of the responsibilities associated with the possession of knowledge in a principled, balanced and caring way, and hopefully have a desire to continue enquiring and learning throughout your whole life. Equally, for many students the IB Diploma is the qualification they need in order to continue their studies in higher education at university and beyond. The assessment aims to test your knowledge and understanding in a fair and reasonable way. It aims to give you credit for what you do know and understand, rather than penalize you for errors and omissions.

It is possible to spend virtually the whole of the two years of the IB Diploma Programme chemistry course focusing almost solely on the final assessment. This might involve practising past questions from many past papers, looking at mark schemes and studying chief examiner's reports, learning by heart the answers to typical questions, and ignoring anything that is not directly on the syllabus or that cannot be tested in an examination. You may end up with a good grade by doing this, but you are unlikely to enjoy your studies or have a deep understanding and appreciation of chemistry. The aim of this book is to put chemistry in a wider context and encourage you to think critically about chemistry. If you understand and enjoy chemistry then it is much easier to learn the necessary supporting facts, and you are likely to perform much better when assessed, particularly in those areas that require a higher level of thinking. You will also be much better prepared to thrive in the challenging atmosphere of higher education. Nevertheless, it is still worth preparing yourself thoroughly so that you can meet the criteria upon which the IB Diploma Programme chemistry assessment is based and do justice to your ability and knowledge.

The assessment consists of four separate components. Three of these are tested externally in three separate examination papers, and the fourth is internally assessed but externally moderated. The three external examinations account for 76% of the final mark and the internal assessment accounts for the remaining 24%.

Objectives of assessment

Your teacher will have the IB Diploma Programme chemistry guide. It lists the aims of the course together with the content material. The objectives reflect those parts of the aims that will be assessed using the content material. There are six main objectives:

1   Demonstrate an understanding of:
    (a) scientific facts and concepts

(b) scientific methods and techniques
(c) scientific terminology
(d) methods of presenting scientific information.
2   Apply and use:
(a) scientific facts and concepts
(b) scientific methods and techniques
(c) scientific terminology to communicate effectively
(d) appropriate methods to present scientific information.
3   Construct, analyse and evaluate:
(a) hypotheses, research questions and predictions
(b) scientific methods and techniques
(c) scientific explanations.
4   Demonstrate the personal skills of cooperation, perseverance and
responsibility appropriate for effective scientific investigation
and problem solving.
5   Demonstrate the manipulative skills necessary to carry out
scientific investigations with precision and safety.

The first three objectives only are tested in the written examinations;
all the objectives, and in particular the last three, are tested during
the internal assessment. If we distil the first three into simpler
English what they are essentially testing is: (1) the knowledge of
basic information; (2) the ability to use that information to solve
straightforward problems; and (3) the higher skill of selecting the
correct knowledge and then applying it to solve more complex
problems. The detailed programme and the examinations use very
specific verbs to help indicate which objective is being assessed.
These verbs, which are known as **command terms**, have very
precise meanings within the IB context. It is worth looking at them
carefully and becoming familiar with them. They determine how
deeply you will study any particular topic, and they determine the
answer required for any specific question in the examination.

IB Command terms

Objective 1

Define   Give the precise meaning of a word, phrase or physical
             quantity.
Draw   Represent by means of pencil lines.
Label   Add labels to a diagram.
List   Give a sequence of names or other brief answers with no
             explanation.
Measure   Find a value for a quantity.
State   Give a specific name, value or other brief answer
             without explanation or calculation.

Objective 2

Annotate   Add brief notes to a diagram or graph.
Apply   Use an idea, equation, principle, theory or law in a new
           situation.
Calculate Find a numerical answer showing the relevant stages in
           the working (unless instructed not to do so).

Describe	Give a detailed account.
Distinguish	Give the differences between two or more different items.
Estimate	Find an approximate value for an unknown quantity.
Identify	Find an answer from a given number of possibilities.
Outline	Give a brief account or summary.

Objective 3

Analyse	Interpret data to reach conclusions.
Comment	Give a judgment based on a given statement or result of a calculation.
Compare	Give an account of similarities and differences between two (or more) items, referring to both (all) of them throughout.
Construct	Represent or develop in graphical form.
Deduce	Reach a conclusion from the information given.
Derive	Manipulate a mathematical relationship(s) to give a new equation or relationship.
Design	Produce a plan, object, simulation or model.
Determine	Find the only possible answer.
Discuss	Give an account including, where possible, a range of arguments, for and against the relative importance of various factors, comparisons of alternative hypotheses.
Evaluate	Assess the implications and limitations.
Explain	Give a detailed account of causes, reasons or mechanisms.
Predict	Give an expected result.
Show	Give the steps in a calculation or derivation.
Sketch	Represent by means of a graph showing a line and labelled but unscaled axes but with important features (for example, intercept) clearly indicated.
Solve	Obtain an answer using algebraic and/or numerical methods.
Suggest	Propose a hypothesis or other possible answer.

External assessment

The written examinations basically follow the same format for both higher level and standard level. The first paper consists only of multiple-choice questions, and just tests objectives 1 and 2. For this paper a Periodic Table is provided, but the use of a calculator or the IB data booklet is not allowed. At SL this paper contains 30 multiple-choice questions. It lasts for 45 minutes and tests the core material. At HL this paper contains 40 multiple-choice questions. It lasts for 1 hour and tests both the core and the Additional Higher Level, AHL, material. About 20 of the questions will be on the core and 20 on the AHL. For both SL and HL each question has four responses: A, B, C and D. You are required to select the best response, and you are not penalized if you select the wrong response.

Paper 2 tests the same material: that is, the core for SL and the core and AHL for HL. Approximately 50% of this paper tests objectives 1 and 2 and the remaining 50% tests objective 3. For this paper a calculator is required, and you will be provided with a fresh copy of the IB data booklet. The SL paper lasts for 1¼ hours, and the HL paper last for 2¼ hours. Both papers have a compulsory Section A. This consists of a data-based question and several short answer questions. In Section B on the SL paper you are required to answer one extended response question out of a choice of three. In Section B on the HL paper you must answer two extended response questions out of a choice of four. At both levels there is a 5 minute reading period before the examination starts, which provides a good time to choose which of the extended response questions to answer.

There is also a 5 minute reading time before the examination for Paper 3 starts. Paper 3 tests the options. Approximately 50% of this paper tests objectives 1 and 2 and the remaining 50% tests objective 3. The SL paper lasts for 1 hour and the HL paper lasts for 1¼ hours. In both papers there are questions on all the options. You must answer all the short answer questions in each of the two options you have studied.

The total marks from the three examination papers at both levels equal 76% of the final mark, but the weightings (given below in percentages) are different.

	Paper 1	Paper 2	Paper 3	Total
Standard level	20%	32%	24%	76%
Higher level	20%	36%	20%	76%

Answering examination questions

Examiners are looking to give you credit for your answers. To ensure you achieve to the best of your ability, you should ensure you comply with the following points.

1 Answer the question asked

The command term is crucial here. If the question asks you to describe what you would observe during the reaction between a solution of acidified potassium dichromate(VI) and excess ethanol, then your answer should include the colour change from orange to green. If, on the other hand, the question asks you to explain the same reaction, then your answer should refer to the oxidation of the alcohol to either ethanal or ethanoic acid and the reduction of the dichromate(VI) ion to the chromium(III) ion.

2 Show all your working

Show clearly how you arrive at each answer. This is important, because if you make a mistake you will only be penalized once for it. Thus, in a calculation involving several stages, if you make a mistake in the first stage then all your answers are likely to be wrong. If you can show that you have used an appropriate method correctly in the subsequent stages, then your error in the first stage will be carried forward and you will not be penalized again.

3 Use the correct number of significant figures in calculations

This has been mentioned in the previous chapter. You may be penalized once on each paper if you fail to comply with this instruction.

4 *Include the correct units (if any) in your answer*

It is not necessary to include all the units in the workings leading to an answer, but the final answer must always be given with the correct units (if any). Some questions on the paper specifically ask for the units. Sometimes you are asked to deduce the units, particularly for rate constants. If the units have not been asked for specifically, then you may be penalized once on each paper if you fail to comply with this instruction.

5 *Write your answers legibly within the allotted space*

If you do require extra space then continue your answer on a separate page and attach it to your answer booklet.

Examiner's comments on student responses

To illustrate the way in which examiners mark, and to give advice on the quality of the responses expected, the answers given by some SL students to an IB question are given below, together with comments from a Principal Examiner. The IB question was part of the SL Paper 2 examination taken in November 2005. To retain confidentiality the students' names have been changed.

(a) Two reactions in the manufacture of sulfuric acid are shown below:

reaction I $S(s) + O_2(g) \rightarrow SO_2$ $\Delta H^{\ominus} = -297$ kJ
reaction II $SO_2(g) + \frac{1}{2}O_2(g) \rightleftharpoons SO_3(g)$ $\Delta H^{\ominus} = -92$ kJ

(i) State the name of the term $\Delta H^{\ominus}$. State, with a reason, whether reaction I would be accompanied by a decrease or increase in temperature.   [3]

Karin's answer

$\Delta H^{\ominus}$ is the enthalpy of reaction. The reaction would give out heat as the sulphur is burning in oxygen.

Examiner's comment

Karin has a good idea what the question is looking for, but her answers are not precise enough to gain full marks. The superscript $^{\ominus}$ means that $\Delta H^{\ominus}$ was measured under standard conditions (298 K, 1 atmosphere pressure), and it is the standard enthalpy *change* of the reaction. The reaction does give out heat (and therefore the temperature increases), and Karin's reason is correct, but she really needed to add that the process of combustion is always exothermic. A clearer reason would be to state that, as the given value of $\Delta H^{\ominus}$ is negative, then by convention the reaction must be exothermic. It is worth noting in passing that in IB Diploma Programme chemistry sulfur is spelt with an "f", as this is the internationally recognized spelling for sulfur, but she would not be penalized for spelling it with a "ph".

(ii) At room temperature sulfur trioxide, SO_3, is a solid. Deduce, with a reason, whether the $\Delta H^{\ominus}$ value would be more negative or less negative if $SO_3(s)$ instead of $SO_3(g)$ were formed in reaction II.   [2]

Ranaivo's answer

More negative. As the SO_3 condenses to a solid it gives out heat.

Examiner's comment

Ranaivo would get both marks for this answer. It does get more negative, and he has given a reason. It would be an even better answer if he had stated that it gives out more heat because as it solidifies bonds are being formed between the molecules, and bond formation is always exothermic.

    (iii) Deduce the $\Delta H^\ominus$ value of this reaction:

$$S(s) + 1\tfrac{1}{2}O_2(g) \rightarrow SO_3(g) \qquad [1]$$

Maria's answer

Adding the equation for reaction I to the equation for reaction II gives the required equation, so the value is −389.

Examiner's comment

There is only one mark for this question. Maria has performed the calculation correctly, and so gets the mark. However, one mark is then deducted, because she has not included the units (kJ). If she omits the units in an answer again on the paper then she will not lose any more marks.

(b) (i) Define the term *average bond enthalpy*.   [3]

Mohammed's answer

The average value when bonds in similar compounds are broken or formed.

Examiner's comment

Mohammed clearly knows what is meant by the average value, but he has not answered the question fully. It is necessary to explain everything that is in italics. So he also needed to explain that bond enthalpy refers to the heat change when one mole of the covalent bond in the gaseous state is formed from its gaseous atoms under standard conditions.

    (ii) Explain why Br_2 is not suitable as an example to illustrate the term *average bond enthalpy*.   [1]

Roger's answer

There is only one compound with a Br−Br bond, so there cannot be an average value.

Examiner's comment

This is the correct answer. Other bonds with precise rather than average values include H−H, O=O and N≡N, and the other halogens such as Cl−Cl and I−I.

    (iii) Using values from Table 10 of the IB data booklet, calculate the enthalpy change for the following reaction:

$$CH_4(g) + Br_2(g) \rightarrow CH_3Br(g) + HBr(g) \qquad [3]$$

Stéphanie's answer

Energy absorbed to break bonds
$[(4 \times C-H) + Br-Br] = 1648 + 193 = 1841$ kJ
Energy evolved when bonds are formed
$[(3 \times C-H) + C-Br + H-Br] = 1236 + 276 + 366 = 1878$ kJ
Enthalpy change $= 1878 - 1841 = +37$ kJ

Examiner's comment

Stéphanie has broken all the bonds in the reactants and formed all
the bonds in the products, which is a sensible approach. She could
also have just broken one of the C−H bonds in the methane and
not remade the three C−H bonds in the CH_3Br, which would have
given her the same answer. However, she has not realized that more
energy has been given out than was put in, so the answer should be
written as $1841 - 1878$, which equals -37 kJ, not $+37$ kJ.

> (iv) Sketch an enthalpy level diagram for the reaction
> in part (b) (iii). [2]

Stéphanie's answer

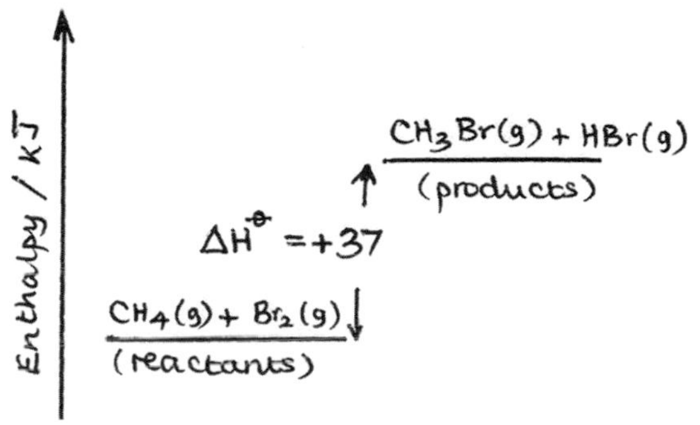

Examiner's comment

Although this answer is wrong, it is consistent with Stéphanie's
answer to part (b) (iii). Because she gave $+37$ kJ as the answer to
that question, she has implied that the reaction is endothermic:
therefore the products should have a higher enthalpy value than
the reactants. She will therefore gain full marks, as she will not
be penalized twice for the same mistake. This is an example of
what is known by examiners as "Error Carried Forward" (ECF).
As the reaction is exothermic, the correct answer should show the
products having a lower enthalpy value than the reactants. It is
worth noting that Stéphanie has put a double-headed arrow on her
diagram to show the value for $\Delta H^\ominus$. This is acceptable, because $\Delta H^\ominus$
is the difference between the two enthalpy values, and it is only by
convention that the value is positive for endothermic reactions and
negative for exothermic reactions.

Internal assessment

It is worth questioning why we do practical work for about 25%
of the IB Diploma Programme chemistry course. There are several
good reasons:
1   Chemistry is an experimental science, and is based upon observation.

2 To improve your manipulative skills in different techniques.
3 To illustrate, teach and reinforce the theoretical part of the course.
4 To develop an appreciation of the benefits and limitations of scientific methodology.
5 To have fun.

The practical scheme of work devised by your teacher should address each of these points several times during the two years. One of the real strengths of the IB practical scheme of work is that each teacher is free to determine their own set of investigations, and much of the work does not need to be assessed formally. For each practical investigation that you carry out you should ensure that you have a clear understanding of the following four areas.

The aim of the experiment

A practical investigation often has an obvious aim, such as the preparation and purification of aspirin, but it may also have a more subtle aim. In this particular example another aim might be to give you hands-on experience of the technique of recrystallization. Another experiment might have been devised to introduce you to a specific data-logging technique. If you are to benefit fully from the practical programme, try to ensure you are aware of all the aims of each experiment.

Safety

Practical chemistry is a potentially hazardous activity. However, if you are aware of the possible hazards, and perform a careful risk assessment, there is little danger of anything going seriously wrong. Essentially there are two types of hazard to be aware of: the immediate and the long term.

Immediate hazards include problems such as cutting yourself on glassware, tripping over steps etc., and problems due to particular chemicals. For example, water should never be added to concentrated sulfuric acid, and certain chemicals should be used only in a well-ventilated fume cupboard. To minimize any problems caused by immediate hazards you should always wear a lab coat and eye protection when working in the laboratory.

Long-term hazards are due to the effects caused by certain chemicals on the body; these may become apparent only after a long time has elapsed. Thirty years ago benzene was a common solvent used in all chemistry laboratories; today its use is banned, because it is potentially carcinogenic.

It is good practice to always use minimum quantities in well-ventilated spaces. Always replace stoppers on bottles after use, and clear up any spillages immediately. Look at the label on the bottle or container, and make sure you are familiar with the potential risks associated with all the chemicals you are using. To help you assess these risks, hazard warning cards for all chemical substances are available (in the UK these are known as Hazcards—see Figure 1), and you should be familiar with the international safety symbols found on containers (Figure 2).

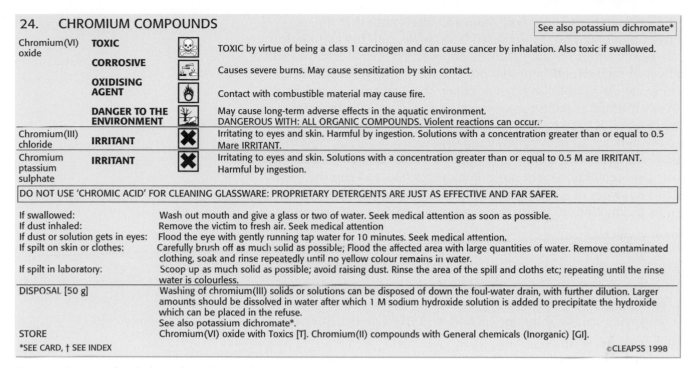

24.	CHROMIUM COMPOUNDS			See also potassium dichromate*
Chromium(VI) oxide	**TOXIC**		TOXIC by virtue of being a class 1 carcinogen and can cause cancer by inhalation. Also toxic if swallowed.
	CORROSIVE		Causes severe burns. May cause sensitization by skin contact.
	OXIDISING AGENT		Contact with combustible material may cause fire.
	DANGER TO THE ENVIRONMENT		May cause long-term adverse effects in the aquatic environment. DANGEROUS WITH: ALL ORGANIC COMPOUNDS. Violent reactions can occur.†
Chromium(III) chloride	**IRRITANT**		Irritating to eyes and skin. Harmful by ingestion. Solutions with a concentration greater than or equal to 0.5 M are IRRITANT.
Chromium ptassium sulphate	**IRRITANT**		Irritating to eyes and skin. Solutions with a concentration greater than or equal to 0.5 M are IRRITANT. Harmful by ingestion.

DO NOT USE 'CHROMIC ACID' FOR CLEANING GLASSWARE: PROPRIETARY DETERGENTS ARE JUST AS EFFECTIVE AND FAR SAFER.

If swallowed:	Wash out mouth and give a glass or two of water. Seek medical attention as soon as possible.
If dust inhaled:	Remove the victim to fresh air. Seek medical attention
If dust or solution gets in eyes:	Flood the eye with gently running tap water for 10 minutes. Seek medical attention.
If spilt on skin or clothes:	Carefully brush off as much solid as possible; Flood the affected area with large quantities of water. Remove contaminated clothing, soak and rinse repeatedly until no yellow colour remains in water.
If spilt in laboratory:	Scoop up as much solid as possible; avoid raising dust. Rinse the area of the spill and cloths etc; repeating until the rinse water is colourless.
DISPOSAL [50 g]	Washing of chromium(III) solids or solutions can be disposed of down the foul-water drain, with further dilution. Larger amounts should be dissolved in water after which 1 M sodium hydroxide solution is added to precipitate the hydroxide which can be placed in the refuse. See also potassium dichromate*.
STORE	Chromium(VI) oxide with Toxics [T]. Chromium(II) compounds with General chemicals (Inorganic) [GI].
*SEE CARD, † SEE INDEX	©CLEAPSS 1998

Figure 1 An example of a hazard warning card

SAFETY SYMBOLS:

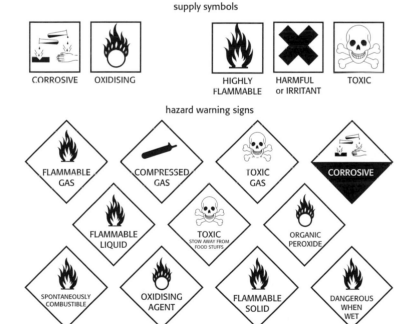

supply symbols

CORROSIVE OXIDISING HIGHLY FLAMMABLE HARMFUL or IRRITANT TOXIC

hazard warning signs

FLAMMABLE GAS COMPRESSED GAS TOXIC GAS CORROSIVE

FLAMMABLE LIQUID TOXIC STOW AWAY FROM FOOD STUFFS ORGANIC PEROXIDE

SPONTANEOUSLY COMBUSTIBLE OXIDISING AGENT FLAMMABLE SOLID DANGEROUS WHEN WET

Figure 2 International safety symbols and hazard warning signs

The effect on the environment

"The environment is not a gift from our fathers—it is a loan from our children."

(adage from the Haidi Indians of North America)

All of us have a responsibility to ensure that we respect the environment. Dealing with chemicals poses particular problems, and ultimately the only way to be absolutely sure that no environmental damage is caused is simply not do to any practical work. You should think carefully about the probable environmental effect of each experiment you do, and make every effort to keep these effects as small as possible. Keep the quantities of chemicals to a minimum, and use small-scale or micro-scale techniques wherever it is practicable. Your laboratory should have specific containers for all heavy metal and organic waste (both chlorinated and non-chlorinated); try to ensure that only harmless substances are disposed of down the sink. Remember, too, to keep the use of distilled water to a minimum, because considerable energy is used in its production; in many cases tap water will do just as well.

Assessment

Much of the practical work you do will not be assessed formally. However, during the two years some of the work will be assessed according to five different criteria. These criteria, and the method of assessment, are the same for both SL and HL students. They are:

Design: D
Data collection and processing: DCP
Conclusion and evaluation: CE
Manipulative skills: MS
Personal skills: PS

The maximum mark for each criterion is 6. The first three are assessed through your written work, and the two best marks out of six for each of these criteria count towards your final mark. The fourth criterion is assessed by your teacher as the standard you reach out of a maximum of 6 marks over the whole course. The final criterion is only assessed once, during the group 4 project. This makes a total of 48 marks. The mark you receive out of 48 is then divided by 2 to give a mark out of 24, which constitutes the IA component (maximum 24%) of your final assessment mark.

Each of the assessment criteria can be separated into three aspects as shown in the following sections. Descriptions are provided to indicate what is expected in order to meet the requirements of a given aspect completely (c) and partially (p). A description is also given for circumstances in which the requirements are not satisfied—not at all (n).

Each aspect is worth a maximum of two marks. A "complete" is awarded 2 marks, a "partial" 1 mark and a "not at all" 0 marks.

During your programme your written work should be assessed several times on the first three criteria. It is possible that all three will be assessed on the same experiment, but often only one or two of them will be assessed at the same time. Your teacher should provide you with feedback so that you can learn from earlier mistakes and hopefully improve. If you do get low marks initially, which is very likely, find out why. Remember it is the two best marks that count, not the average of all the marks you get, for each of the first three criteria. Each of the three aspects for each criterion is carefully worded, and you need to ensure that you address these aspects as completely as possible to gain maximum marks. Make sure that either you or your teacher keep all the marked work, because it may need to be sent away for moderation at the end of the course.

Design

	Aspect 1	Aspect 2	Aspect 3
Levels/marks	Defining the problem and selecting variables	Controlling variables	Developing a method for collection of data
Complete/2	Formulates a focused problem/research question and identifies the relevant variables.	Designs a method for the effective control of the variables.	Develops a method that allows for the collection of sufficient relevant data.
Partial/1	Formulates a problem/research question that is incomplete or identifies only some relevant variables.	Designs a method that makes some attempt to control the variables.	Develops a method that allows for the collection of insufficient relevant data.
Not at all/0	Does not identify a problem/research question and does not identify any relevant variables.	Designs a method that does not control the variables.	Develops a method that does not allow for any relevant data to be collected.

Four examples of design experiments have already been given in earlier chapters. You will be given minimum instructions, which will probably take the form "Investigate an aspect of …". As has been explained previously, you will need to formulate a focused research question and identify and then classify all the variables by distinguishing clearly between the controlled variables, the independent variable and the dependent variable. You will then need to design a method that allows you to control these variables and enables you to collect sufficient data. The method should be clearly described in sufficient detail so that it could be reproduced by someone else from the information given. An appropriate diagram is often helpful. The apparatus and materials used can be listed separately or mentioned in the text. Remember to give precise relevant details about all apparatus and chemicals. Although not a requirement, the use of suitable data-logging apparatus in design is perfectly acceptable as a way of collecting the raw data.

Below are two lists of the same apparatus and materials used for a particular investigation. Which one of the two lists would you award "**complete**" to?

I	II
Beakers	2×100 cm³ beakers
Balance	Analytical electronic balance (± 0.001 g)
Volumetric flask	250 cm³ volumetric flask (± 0.2 cm³)
Burette	50.0 cm³ burette (± 0.1 cm³)
Pipette	25.0 cm³ pipette (± 0.1 cm³)
Thermometer	Mercury thermometer (-10 to $+110$ °C, ± 0.2 °C)
Conical flasks	3×100 cm³ conical flasks
Clock	Electronic stop clock (± 1 s)
Ethanoic acid	1.00 mol dm⁻³ ethanoic acid, CH_3COOH(aq)
Propanoic acid	1.00 mol dm⁻³ propanoic acid, C_2H_5COOH(aq)
Sodium carbonate	Sodium carbonate decahydrate, $Na_2CO_3.10H_2O$(s)
Sodium hydroxide	0.0912 mol dm⁻³ sodium hydroxide, NaOH(aq)
Data logger	pH data logger (make: Hanna, model: pH turtle)

The definition of "sufficient relevant data" for Aspect 3 depends on the context. You should plan to be able to collect sufficient data so that your aim or research question can be suitably addressed and an evaluation of the reliability of your data can be made. If a straight-line graph is anticipated, then at least five data points will be needed. Your plan should allow for repeated measurements to calculate a mean where necessary. It may need to show the need for a trial run and repeats until consistent results are obtained as, for example, with titrimetric determinations.

Data collection and processing

Levels/marks	Aspect 1 Recording raw data	Aspect 2 Processing raw data	Aspect 3 Presenting processed data
Complete/2	Records appropriate quantitative and associated qualitative raw data, including units and uncertainties where relevant.	Processes the quantitative raw data correctly.	Presents processed data appropriately and, where relevant, includes errors and uncertainties.
Partial/1	Records appropriate quantitative and associated qualitative raw data, but with some mistakes or omissions.	Processes quantitative raw data, but with some mistakes and/or omissions.	Presents processed data appropriately, but with some mistakes and/or omissions.
Not at all/0	Does not record any appropriate quantitative raw data or raw data is incomprehensible.	No processing of quantitative raw data is carried out or major mistakes are made in processing.	Presents processed data inappropriately or incomprehensibly.

Raw data is defined as the actual data measured, and includes relevant qualitative data (observations) as well as quantitative data. All raw data have associated uncertainties, and you should attempt to quantify all the uncertainties as explained in the previous chapter. You can record raw data either by writing by hand or electronically. It is generally best to set out the results of measurements in tabular form, making sure you pay attention to significant figures, appropriate SI units and an indication of the absolute uncertainty.

Here are two lists of raw data obtained from the same experiment. The second list and comments illustrate how raw data should be recorded.

I

Time (min)	Volume of gas evolved (cm^3)
0	0
0:30	32.3
1	59
1:30	67.8
2	72
2:30	79.1
3	80

(No qualitative comments given).

II

Title (A relevant title is given to the table of results)

Time / s (±1 s)	Volume of gas evolved / cm³ (±0.2 cm³)
0	0.0
30	32.3
60	59.0
90	67.8
120	72.0
150	79.1
180	80.0

During the reaction a colourless gas was seen to be given off as soon as the reaction was started, and the flask became warm during the reaction as heat was evolved. The solution remained colourless throughout the reaction, but the white solid had completely reacted and the product dissolved by the end of the third minute, and no more gas was then evolved.

The processing of data involves, for example, combining and manipulating raw data to determine the value of a physical quantity (such as adding, subtracting, multiplying, dividing, taking reciprocals or taking logarithms), and taking the average of several measurements and transforming data into a form suitable for graphical representation. If the data you have collected can be plotted directly, then a line of best fit must be drawn and the gradient and/or intercept determined (or an extrapolation or interpolation performed) in order for the raw data to have been effectively processed. Plotting raw data (without a graph line) does not constitute processing data. It is permissible to put both the raw data and the processed data into one table. This might be particularly suitable with rate experiments, where the raw data of time might be in one column and the processed data of the reciprocal of time might be in another column to the right of the raw data columns.

When processing data, the uncertainties associated with the data must also be considered. If the data are combined and manipulated to determine the value of a physical quantity (for example, specific heat capacity), then the uncertainties in the data must be propagated in percentage form, as shown in the previous chapter. You are expected to decide upon a suitable presentation format yourself (for example, spreadsheet, table, graph, chart or flow diagram). You should provide clear, unambiguous headings for calculations, tables or graphs. Make sure that all your graphs have a title with appropriate scales, labelled axes with units, and accurately plotted data points with a suitable line of best fit or curve. You should present your data so that all the stages to the final result can be followed clearly. Your final derived quantities should be given using metric/SI units and should be expressed to the correct number of significant figures together with the calculated total uncertainty.

Conclusion and evaluation

	Aspect 1	Aspect 2	Aspect 3
Levels/marks	Concluding	Evaluating procedure(s)	Improving the investigation
Complete/2	States a conclusion, with justification, based on a reasonable interpretation of the data.	Evaluates weaknesses and limitations.	Suggests realistic improvements in respect of identified weaknesses and limitations.
Partial/1	States a conclusion based on a reasonable interpretation of the data.	Identifies some weaknesses and limitations, but the evaluation is weak or missing.	Suggests only superficial improvements.
Not at all/0	States no conclusion or the conclusion is based on an unreasonable interpretation of the data.	Identifies irrelevant weaknesses and limitations.	Suggests unrealistic improvements.

Conclusions that are supported by your data are acceptable, even if they appear to contradict accepted theories. However, make sure that your conclusion takes into account any systematic or random errors and uncertainties. You should compare the percentage error with the total estimated random error as derived from the propagation of percentage uncertainties.

In justifying your conclusion, you should discuss whether systematic error or further random errors were encountered. Include here uncertainties or errors over which you had no control, such as the preparation of solutions by a technician. You should try to appreciate the direction of any systematic errors, as exemplified in the account of the determination by gravimetric analysis of the empirical formula of magnesium oxide given in Chapter 1. Where possible you should compare the experimental value you have obtained with the textbook or literature value. You should reference fully any literature consulted.

When evaluating your procedure, comment on the design and method of the investigation as well as on the quality of the data. You should specifically look at the processes, use of equipment and management of time. When listing the weaknesses you should also show that you appreciate how significant the weaknesses are. Comments about the precision and accuracy of the measurements are relevant here.

The suggestions you make for improving the investigation should be based on the weaknesses and limitations you have already identified. Modifications to the experimental techniques and the data range can be addressed here. The modifications should address issues of precision, accuracy and reproducibility of the results. You should suggest how to reduce random error, remove systematic error, and/or obtain greater control of the variables. The modifications you propose should be realistic and clearly specified. It is not sufficient to state generally that more precise equipment and purer chemicals should be used.

Manipulative skills

This criterion will be assessed by your teacher throughout your course. The three aspects detailed below are quite self-evident. The third aspect is particularly important, because the way in which you work affects the safety of others as well as your own.

Levels/marks	Aspect 1 Following instructions	Aspect 2 Carrying out techniques	Aspect 3 Working safely
Complete/2	Follows instructions accurately, adapting to new circumstances (seeking assistance when required).	Competent and methodical in the use of a range of techniques and equipment.	Pays attention to safety issues.
Partial/1	Follows instructions but requires assistance.	Usually competent and methodical in the use of a range of techniques and equipment.	Usually pays attention to safety issues.
Not at all/0	Rarely follows instructions or requires constant supervision.	Rarely competent and methodical in the use of a range of techniques and equipment.	Rarely pays attention to safety issues.

The group 4 project

For 10 hours of your course you are required to participate in a group 4 project. The intention of the group 4 project is that students from all the different group 4 subjects work together to investigate or solve a common problem. This aim is to provide all students with the opportunity to appreciate both the implications of using science and the limitations of scientific study, in a local context. The main philosophy of the group 4 project is to emphasize interdisciplinary cooperation, and the processes involved in a scientific investigation, rather than the products of such an investigation. Cooperation on the group 4 project between schools, particularly those in different countries, is an approach that the IB really encourages.

Schools differ in the arrangements they make for the group 4 project but essentially it is separated into three distinct phases, as follows.

1 Planning

This takes about 2 hours. You will meet either as a whole IB group or perhaps in smaller groups if you are a student in a school with a large number of IB students. Ideally the students themselves in your school will have a "brainstorming" session and agree on an overall topic. This is often related either to the school itself or to the local area. Once the overall topic is agreed then, either individually or in small groups, you will plan your own particular contribution to the chosen topic. At this stage you will need to plan what equipment and chemicals, if any, you will need, and prepare a plan of action. This will include a risk assessment and an environmental assessment to address any safety issues and environmental concerns. If you are liaising with other schools you will want to inform them of your plans at this stage.

2 Action

About 6 hours is devoted to the action stage. You can investigate your topic either in mixed subject groups or in a chemistry group. Some schools spread this time out over one or two weeks; others complete it in a day or weekend. Remember the focus is on the process, not the product. You may plan to do experimental work, but it is perfectly permissible and sometimes more efficient to get your data from elsewhere. For example, many chemistry contributions to the group 4 project involve the analysis of water. Some simple practical analysis can be performed using the facilities of a normal school laboratory. However, with just one phone call to your local water provider you might be able to obtain a much more accurate and sophisticated analysis of your local water. If you do perform your own practical work then you must pay attention to safety, ethical and environmental considerations.

3 Evaluation

The time allotted to this is usually about 2 hours. The emphasis during this stage is on sharing your findings, both successes and failures, with other students. There are many ways in which you might do this. Some schools devote a morning, afternoon or evening to a symposium where all students, as individuals or as groups, give brief presentations. Other evaluations take the form of a science fair, where students circulate around posters summarizing the activities of each group—or all the students in the school could contribute their findings to a specially designed web page on the school's intranet.

Personal skills (for group 4 project assessment only)

This criterion essentially addresses objective 4, which is concerned with the personal skills of cooperation, perseverance and responsibility appropriate for effective scientific investigation and problem solving. There are various ways in which your teacher may decide to assess this. One way that can be quite useful is for you to complete a self-evaluation form, another way is peer review, but neither of these is compulsory.

	Aspect 1	Aspect 2	Aspect 3
Levels/marks	Self-motivation and perseverance	Working within a team	Self-reflection
Complete/2	Approaches the project with self-motivation and follows it through to completion.	Collaborates and communicates in a group situation and integrates the views of others.	Shows a thorough awareness of their own strengths and weaknesses and gives thoughtful consideration to their learning experience.
Partial/1	Completes the project but sometimes lacks self-motivation.	Exchanges some views but requires guidance to collaborate with others.	Shows limited awareness of their own strengths and weaknesses and gives some consideration to their learning experience.
Not at all/0	Lacks perseverance and motivation.	Makes little or no attempt to collaborate in a group situation.	Shows no awareness of their own strengths and weaknesses and gives no consideration to their learning experience.

Some examples of successful group 4 projects:
1 Food from the school cafeteria.
2 Factors accounting for poor grass growth on the school fields.
3 Saving water in and around the school.
4 How safe is the school swimming pool?
5 Producing biogas from organic waste.
6 The far-ranging implications of the local dam.
7 Safety concerns in the town parks and playgrounds.
8 Alternative energy sources for the school.
9 Environmental evaluation of the school.
10 The marine environment of the local bay.
11 Human physiology at high altitude.
12 How safe is our school?

Extended essay

An extended essay in chemistry is a substantial piece of writing of up to 4,000 words that enables you to investigate a chemical topic of special interest that you have chosen yourself.

It is a compulsory part of the IB Diploma Programme, and should take about 40 hours to complete. The purpose of the extended essay is to promote high-level research and writing skills, intellectual discovery, and creativity. It provides you with the opportunity to engage in personal research under the guidance of a teacher in your school. By its very nature the research process encourages intellectual risk-taking, and requires you to reflect extensively with an open mind. The argument you present should be balanced, honest and fair.

Many students find this one of the most challenging and also one of the most intellectually satisfying parts of their IB Diploma Programme studies. It is also highly valued by universities, because they see it as an excellent preparation for independent research and for acquiring the skills required to communicate knowledge. It may appear daunting at first, but you are not expected to produce chemical knowledge that is new to mankind—just to research a topic that is new to you. Although you are working independently, you will have a supervisor to provide you with advice and guidance in the skills of undertaking research, and to offer support and guidance in the writing of the essay.

Extended essays may be just literature based, or they may also involve individual experimental work. This means that there are basically three types of extended essay in chemistry: those where there is no individual experimental work; those where the experimental work is performed in a university or industrial research laboratory; and those where the experimental work is performed in the school chemistry laboratory or local surroundings. Although it is possible to write a good extended essay where all the data have been generated by others, it is much harder to show personal input. It is also harder to show personal input when the equipment used is the sophisticated machines and apparatus found in a research laboratory, because often such specialized equipment can be manipulated only by a trained technician. The best chemistry extended essays tend to be those performed in a school laboratory, because it is usually much easier for students to control, modify or redesign their experimental method. A school laboratory in this context can be interpreted in the widest context, and may include fieldwork in the local environment.

The research question

The key to a successful extended essay (in any subject) is the research question. It must be sharply focused, and capable of being addressed in 40 hours and 4,000 words. Your supervisor will be able to give you advice to ensure that your chosen research question is appropriate, but the best extended essays are those where you have chosen the basic topic yourself and the supervisor has just helped you to formulate the research question correctly. Find an area of chemistry that interests you, and then read as much as you can about it. Although textbooks can be useful for the background chemistry, you will need to consult original resources, such as scientific journals, newspaper articles, the Internet, and personal contacts. Gradually you should become aware of what is known about the topic and what is unknown. This should enable you to focus in on a problem to investigate.

In the box below are some of the techniques available to you in a school laboratory, each with one example of a question they could be used to investigate. Some of them are just extended essay titles, and would need to be focused more sharply to qualify as specific research questions. They are only meant as an illustration, and you are encouraged to think beyond them. Many good research questions are capable of several approaches, as the merits of the different approaches can then be evaluated. For example, the ethane-1,2-dioic acid (oxalic acid), $(COOH)_2$, content of the rhubarb plant (*Rheum rhabarbarum*), which is used in cooking and as a traditional medicine in some cultures, could be ascertained using either redox or acid–base titrations.

Some different techniques available in schools (together with an example) that can be used for chemistry extended essays

Redox titration Is the intrinsic value of a cent coin more than its face value?	**Refinement of a standard practical** How can the yield be increased in the laboratory preparation of 1,3-dinitrobenzene?
Extension of a standard practical What gas is evolved when a zinc half-cell is connected to a copper half-cell, and what factors affect its formation?	**Polarimetry** Is it possible to separate a racemic mixture in a school laboratory?
Acid–base titration How do storage time and temperature affect the vitamin C content of fruit juices?	**Theoretical** What influences the value of the C=O bond enthalpy in organic compounds?
Chromatography Do all strawberry jellies worldwide contain the same red dyes(s)?	**Data-logging probes** Does the nature of the catalyst affect the rate equation for the decomposition of H_2O_2?
Calorimetry How good a fuel is dried animal dung?	**Visible spectrometry/colorimetry** Do different underarm deodorants contain different amounts of aluminium?
pH meter Can different types of chewing gum affect the pH of the mouth and prevent tooth decay?	**Gravimetric analysis** What is the amount of caffeine in a cup of coffee prepared by different methods?
Steam distillation Can mosquito repellants be extracted from the Papua New Guinea plant genus *Ocimum*?	**Inorganic reactions** An investigation into the oxidation states of manganese: does Mn(V) exist?

Electrochemistry	**Microscale/small scale**
To find the relationship between the concentration and the ratio of $O_2:Cl_2$ evolved during the electrolysis of NaCl(aq).	How can the residues from a typical college practical programme be reduced?
Iodimetry	**Compleximetric titrations**
What is the amount of unsaturation in different fatty acids?	What is the calcium content of different types of milk?

Extended essay pathway

Some research (led by Dr John Munro at the University of Melbourne in Australia) has been carried out on how successful students manage the process of writing their IB extended essay from start to finish. The research has shown that when students choose their own research question they obtain higher marks, particularly in those areas involving such qualities as initiative and personal involvement. As a guide to students, Dr Munro has produced the pathway shown in Figure 1.

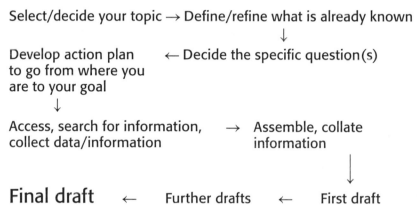

Select/decide your topic → Define/refine what is already known
↓
Develop action plan ← Decide the specific question(s)
to go from where you
are to your goal
↓
Access, search for information, → Assemble, collate
collect data/information   information
↓
Final draft ← Further drafts ← First draft

Figure 1 Pathway to an extended essay.

It can be helpful to follow this route, although it does not have to be the linear process summarized here. What is important is to remember that it is only the final written extended essay that is marked. Keep written records of everything that you do, and make a note of all references so that you can build up your footnotes and bibliography as you proceed. Question the validity of each source of information. This is particularly important for information obtained from the Internet, and you should also make a note of the date it was accessed. You have already learned how to question the uncertainties in your own experimental work. If your chosen method of experimentation works first time, then try to expand it to cover new areas of investigation. If it does not "work" (and most do not the first time), try to analyse what the problem is. Try changing some of the variables: perhaps increase the concentration of reactants, or change the temperature, or alter the pH. It may be that the equipment itself is faulty or unsuitable. Try to modify it. Use your imagination to design new equipment in order to address your particular problem.

Before you actually write the extended essay, make sure that you are familiar with the assessment criteria, and ask your supervisor to provide you with some examples of excellent extended essays written in previous years. Also ask your supervisor to provide you with the relevant sections from the IB Extended Essay Guide, particularly the part that gives advice to students.

When you are writing the essay, make sure that, throughout, you are addressing the stated research question. Make sure, too, that the essay is a chemistry extended essay. If you need supporting information that is not chemistry, then include this in the appendix. Show too that you understand all the underlying chemistry, and show clearly your personal input and engagement. Use digital photographs if they enhance the essay. Examples of photographs that might be useful are apparatus you have designed or modified yourself, and pictures of developed chromatograms. Do not include spurious photographs of bottles of chemicals used, or pictures of you performing a standard procedure, such as a titration.

You should be able to produce one or more drafts of your essay which your supervisor can comment on before the final version. Most schools set their own deadline for when the final version is due. Make sure you proofread your final version to correct any errors, and ensure that all chemical formulas and terminology are correct before you hand the essay in. Do not just rely on a spellchecker; often they will not pick up mistakes such as using "form" when you mean "from".

Finally, be absolutely certain that the extended essay is your own work. At the beginning of this book we made reference to academic honesty. Do not be tempted to copy or "cut and paste" the work of others in an attempt to pass it off as your own. When you do need to use the words, ideas or results of another person make sure you acknowledge them properly in a recognized way. Your supervisor is required to confirm that the essay is your own work, and all essays are checked using plagiarism software to ensure that they do not contain passages copied from the work of others, including past extended essays.

After you have handed in the final version your supervisor may give you a *viva voce*. This is a short interview where your supervisor will give you the opportunity to reflect and comment on the successes and difficulties you have encountered, and on what you have learned from the process. Your supervisor will write a report to accompany the essay, so the *viva voce* also gives you the opportunity to mention anything that you might wish your supervisor to include in his or her report.

Assessment criteria

All extended essays, whatever the subject, are marked according to the same assessment criteria, although each subject has its own particular interpretation of them. To gain high marks you must address these assessment criteria fully. Chemistry extended essays that are written addressing the internal assessment criteria rather than the extended essay criteria are unlikely to score highly, because they may omit certain key elements.

There are 11 separate criteria worth a total of 36 marks. For each criterion the descriptor below is given for the maximum number of marks available. The following information is taken, or adapted, from the IB Extended Essay Guide.

Criterion A: Research question (2 marks)

Descriptor: The research question is clearly stated in the introduction and sharply focused, making effective treatment possible within the word limit.

Make sure your research question is sharply focused. For instance, "What is the amount of caffeine in a cup of coffee prepared by different methods?" is not sharply focused. You would need to be more precise about the type and amount of coffee used, and the different methods of preparation and measurement. In chemistry, many research questions can be formulated as an actual question or questions. A typical example is: "What gas is evolved when zinc is added to aqueous copper(II) sulfate solution, and what factors affect its rate of formation?" However, it is also perfectly reasonable to formulate the research question as a statement or as a hypothesis rather than an actual question. "An analysis of the amount of aluminium in three different brands of underarm deodorant by visible spectroscopy" and "The kinetics of oxidation of iodide ions with hydrogen peroxide in acidic solutions" are two such examples of statements rather than questions. Whichever way the research question is formulated, it should be identified clearly as such, and you should set it out prominently in the introduction.

Criterion B: Introduction (2 marks)

Descriptor: The context of the research question is clearly demonstrated. The introduction clearly explains the significance of the topic and why it is worthy of investigation.

The purpose of the introduction is to set your research question into context: that is, to relate your research question to existing knowledge in chemistry. It is usually appropriate to also include the underlying chemical theory required to understand how your research question has arisen. Some research questions require some background knowledge unrelated to chemistry—for example, "Do the fossils found in different strata of rocks at a stated location contain different amounts of sulfur?" For the extended essay based in this question to make sense, it would be important to state the ages of the rock and give some geological background. In such cases, only the essential non-chemistry information should be provided in the introduction, because your essay will be marked on its chemical content. If it is necessary to include more non-chemistry information (geological information, for example), then the appropriate place for it is the appendix.

results are calculated from data obtained by changing one of the variables, it is generally good practice to show *one* example of the calculation. The remainder can be displayed in tabular or graphical form.

Bibliography

There are several accepted ways of listing resources in a bibliography. It does not matter which system you use, but it should be consistent. Your supervisor or school librarian will be able to give you advice on this. One method gives the author followed by the publication (in italics), the publisher, and then the date of publication. The following exemplifies this:

Bibliography

1   P. Atkins and J. de Paula, *Physical Chemistry*, Oxford University Press, 7th edition, 2002.
2   L. A. Bloomfield, *How Things Work*, John Wiley & Sons, Inc., 1997.
3   J. A. Dean, *Lange's Handbook of Chemistry*, McGraw-Hill Book Company, 13th edition, 1985.
4   D. R. Lide, *Handbook of Chemistry and Physics*, CRC Press, 72nd edition, 1991–1992.
5   G. Neuss, *Chemistry for the IB Diploma*, Oxford University Press, 2001.
6   University of Leiden, *Molecular nano-optics and spin*, http://www.monos.leidenuniv.nl, accessed 20 November 2006.
7   University of Miami, *Lecture supplemental materials*, http://www.physics.miami.edu/, accessed 20 October 2006.
8   Organic Chemistry Portal. *Microwave chemistry highlights*, http://www.organic-chemistry.org/highlights/2006/15october.shtm, accessed 19 October 2006.
9   M. R. Spiegel, J. Schiller and R. Srinivasan, *Probability and Statistics*, Schaum's Outline Series, McGraw-Hill Publications, 2nd edition, 2000.
10 J. G. Stark and H. G. Wallace, *Chemistry Data Book*, John Murray Publishers Ltd, 1991.
11 Wikipedia, *Article on ozone*, http://www.wikipedia.org/, accessed 20 November 2006.
12 H. D. Young and R. A. Freedman, *University Physics*, Addison-Wesley Publishing Company, Inc., 1996.

Criterion J: Abstract (2 marks)

Descriptor: The abstract clearly states the research question that was investigated, how the investigation was undertaken, and the conclusion(s) of the essay.

Your abstract must not exceed 300 words. It will be judged on the clarity with which it presents an overview of the research and the extended essay, not on the quality of the research question itself, nor on the quality of the argument or the conclusions.

Criterion K: Holistic judgment (4 marks)

Descriptor: The essay shows considerable evidence of intellectual initiative, depth of understanding and insight.

Qualities that are rewarded under this criterion include the following.

- *Intellectual initiative*: Ways of demonstrating this in chemistry extended essays include the choice of topic and research question, and the use of novel or innovative approaches to address the research question.
- *Insight and depth of understanding*: These are most likely to be demonstrated as a consequence of detailed research, reflection that is thorough, and by well-informed and reasoned argument that consistently and effectively addresses the research question.
- *Originality and creativity*: These will be apparent by clear evidence of a personal approach backed up by solid research and scientific reasoning.

An example of an abstract that shows the research question, the scope of the investigation and the conclusion

Abstract

Molecular polarity is an important property of a large number of chemical substances, both because of the physical and chemical properties these substances exhibit, such as strong intermolecular bonding and surface tension, and because we can utilize this property in microwave chemistry. Microwave chemistry has been shown in some cases to catalyse chemical processes, to give superior process yields and reduce reaction times noticeably. This extended essay seeks to answer the following research question:

Does there exist a positive correlation between the dipole moment of a polar liquid and the heat gained when heated in a household microwave oven?

It will investigate whether a correlation between the dipole moment of polar liquids and the heat added to the substance per unit mass, unit volume or unit amount can be found when heated in a standard household microwave oven, and if the presence of such a correlation could be used for estimating dipole moments. The experiments were done very simply by passing the same amount of microwave energy through a variety of polar liquids, including several consecutive members of the homologous series of monohydric alcohols, for a fixed period of time in a set position in a microwave oven, and then measuring the temperature increase. The experiments show that there seems to be a strong linear correlation between the heat added per unit mass/heat added per unit volume and dipole moments, if the molecular structure of the chemical heated is taken into account. In particular, cyclic compounds seem to gain a lower amount of heat energy per unit mass/unit volume than non-cyclic compounds. After this discovery, regression models from the experimental data were used to predict a dipole moment in the range of $1.0D - 1.3D$ for butan-2-ol.
(Word count: 288)

This abstract was written by Christian Bjartli, a Norwegian IB student at Atlantic College in Wales, UK.

Option A: modern analytical chemistry

Analytical techniques

The old procedure to determine the structure of an organic compound was to find both its empirical formula and relative molecular mass experimentally, and then deduce the nature of the functional groups from its chemical reactivity. However, modern well-equipped laboratories now employ a variety of instrumental techniques that, if used in combination, are able to unambiguously determine the exact structural formula. These methods are becoming ever more refined, and some of them (such as mass spectrometry and atomic absorption spectroscopy) can be used on extremely small samples. The main uses for these analytical techniques are structural determination, the analysis of the different composition of compounds, and to determine purity.

Some examples of the uses of different analytical techniques

- **Infrared spectroscopy**: organic structural determination, information on the strength of covalent bonds, information about the secondary structure of proteins, measuring the degree of unsaturation of oils and fats, and determining the level of alcohol in the breath.
- **Mass spectrometry**: organic structural determination, isotopic dating (e.g. ^{14}C dating).
- **^1H nuclear magnetic resonance**: organic structural determination, body scanning.
- **Atomic absorption spectroscopy**: determining the concentration of heavy metals in water; monitoring contamination in foods, detergents and pharmaceuticals.
- **Ultraviolet and visible spectroscopy**: assaying of metal ions, organic structural determination, and detection of drug metabolites.
- **Gas chromatography-mass spectrometry (GC-MS)**: drug testing in the blood and urine, food testing and forensic science.

Chromatography

Before analysing a sample of a mixture it is helpful to be able to separate it into its individual components. Coupled with other techniques such as mass spectrometry, chromatography can be used to separate and identify complex mixtures, both quantitatively and qualitatively. It can also be used to determine how pure a substance is.

There are several different types of chromatography. They include paper, thin layer (TLC), column (C), gas-liquid (GLC) and high-performance liquid chromatography (HPLC). In each case there are two phases: a **stationary phase** that stays fixed and a **mobile phase** that moves.

Chromatography relies upon the fact that, in a mixture, the components have different tendencies to adsorb onto a surface or dissolve in a solvent. This provides the means of separating them.

Adsorption involves a solid stationary phase with a moving liquid phase (C, HPLC and at times TLC). The rate at which the solute moves through the solid phase depends on the equilibrium between its solubility in the moving liquid phase and its adsorption to the solid phase. The more tightly the component is adsorbed, the slower it will elute. **Partition** involves a stationary liquid phase and a mobile gaseous or liquid phase (paper, TLC and GLC). It is dependent on the relative solubility of each component in the two phases, if both are liquid. If the mobile phase is gaseous then the rate of movement depends upon the volatility of the components. The more soluble or volatile the component in the mobile phase, the faster it will elute.

Paper chromatography

Paper largely consists of cellulose fibres. These contain a large number of hydroxyl groups, making the paper quite polar. Water molecules hydrogen-bond to these groups, so that a sheet of "dry" paper actually contains about 10% water. It is this water that acts as the stationary phase. The mobile phase is a solvent, either water itself or a polar organic liquid such as ethanol, propanone or ethanoic acid. Normally a small amount of the mixture is spotted onto the paper, about 1.0 cm from its base. The paper is then suspended in a small quantity of the solvent (known as the **eluent**) in a closed container (Figure 1). (The container is closed in order to saturate the atmosphere within it and prevent evaporation of the solvent from the paper to give better and faster separation.) As the solvent rises up the paper the components in the mixture partition between the two phases, depending on their relative solubility. As the solvent nears the top of the paper a mark is made to record the level, and the paper is then removed and dried. Coloured components can often be seen with the naked eye. Other components can be made visible by staining (e.g. with iodine) or by irradiating with an ultraviolet lamp.

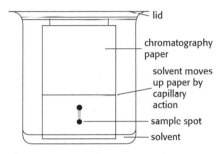

Figure 1 Paper chromatography.

As with all equilibria, the partition of a solute between the two phases is constant at a fixed temperature: thus the solute will always move the same fraction of the distance moved by the solvent. Each solute will have a particular retention factor (R_f) for a given eluent. It is obtained by measuring the distance from the original spot both to the centre of the particular component and to the solvent front:

$$R_f = \frac{\text{distance moved by solute}}{\text{distance moved by solvent (eluent)}}$$

Substances can be identified by their R_f values and by comparison with pure samples. For example, a student wished to determine the composition of the red dye(s) present in a sample of tomato ketchup. A spot of the ketchup (K) was placed on the chromatographic paper, and spots of pure samples of six known red food dyes (D_1 to D_6) were placed alongside. A suitable solvent was used as the eluent. The developed chromatogram is shown in Figure 2.

It can be seen that the solvent has moved by 9.4 cm. The ketchup sample contains three separate components with R_f values of 0.07, 0.40 and 0.63. Two of the spots have the same R_f values as two of the pure samples, D_4 and D_6. We can therefore conclude that the

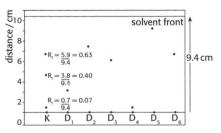

Figure 2 Separation and identification of the components in a mixture using paper chromatography.

ketchup is likely to contain food dyes D4 and D6 and one other unknown red dye.

Thin layer chromatography (TLC)

TLC is very similar to paper chromatography but uses a thin layer of a powdered solid such as alumina (Al_2O_3) or silica (SiO_2) on an inert support such as glass, aluminium or plastic. When absolutely dry it works by adsorption, but, like paper, silica and alumina have a high affinity for water: therefore the separation relies more on a partition process, with water as the stationary phase. The big advantage of TLC over paper chromatography is that the separated components can be recovered pure by scraping off the section containing the component and dissolving it in a suitable solvent. Pregnancy tests may use TLC to detect pregnanediol in urine.

Column chromatography

This technique is similar to TLC but uses larger quantities of both sample and eluent, and is usually used to separate the components of a mixture for further use rather than for identification. The stationary phase is very fine particles of alumina or silica gel. The column is set up by packing the dry stationary phase on top of a piece of glass wool in a long glass tube with a tap at the end, and then saturating it with the eluting solvent (Figure 3). The sample is added at the top, and as it moves down the column more eluent is added. After some of the components have been eluted it is possible to change the solvent to elute the more tightly held components. By evaporating the solvent from the separated components each component can be obtained pure.

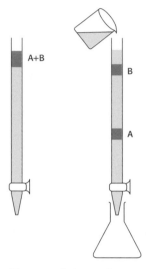

Figure 3 Column chromatography.

Principles of spectroscopy

Each type of spectroscopy depends upon a transition in energy between two states. **Emission spectroscopy** involves the analysis of light energy emitted by excited atoms, ions or molecules as they return to their ground state. The atomic emission spectrum of hydrogen (discussed in Chapter 2) is a good example of this. Many other analytical techniques involve **absorption spectroscopy**. When radiation is passed through a sample, some of the energy is absorbed by the sample to excite an atom, ion or molecule to the excited state. The spectrometer analyses the transmitted energy relative to the incident energy. As the energy levels are quantized, only radiation with a frequency corresponding to the difference in the energy levels will be absorbed. The relationship between energy and frequency is given by

$$E = hf$$

where h represents Planck's constant, 6.626×10^{-34} Js.

The greater the energy difference between the levels, the higher the frequency (and hence the shorter the wavelength) of the light absorbed. The most energetic absorptions are atomic electronic transitions that involve bond breaking and ionization. Absorptions in the ultraviolet and visible region are due to atomic and molecular transitions in which electrons become excited to higher levels.

Molecular vibrations (stretching and bending) occur in the infrared region and molecular rotations in the microwave region. The weakest transitions of all involve nuclear spin. These occur in the radio wave region and form the basis of nuclear magnetic resonance spectroscopy. See Figure 4.

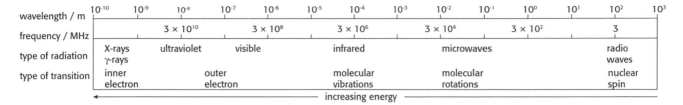

Figure 4 The electromagnetic spectrum.

Infrared (IR) spectroscopy

Covalent bonds behave rather like springs, and vibrate at a particular frequency that depends on the nature of the bond and on the type of vibration—that is, stretching or bending. The energy of these vibrations corresponds to the infrared region of the electromagnetic spectrum. When infrared radiation with a frequency equal to the frequency of the vibration is passed through the molecule it will absorb the radiation. For diatomic molecules made up of two atoms such as hydrogen chloride, HCl, there is only one form of vibration. This is stretching, where the atoms alternately move further apart and then closer together. Different molecules absorb at different frequencies, because the energy needed to excite a vibration depends on the bond enthalpy. Weaker bonds require less energy (Table 1).

Table 1 Infrared absorptions of the hydrogen halides

Molecule	Bond enthalpy / kJ mol^{-1}	Absorption / cm^{-1}
H–Cl	431	2886
H–Br	366	2559
H–I	299	2230

Infrared absorptions are usually quoted as the reciprocal of wavelength. Frequency and wavelength are related by the equation $c = \lambda f$, where c represents the velocity of light. As c is a constant, the reciprocal of wavelength is a direct measure of frequency. This reciprocal of wavelength $(1/\lambda)$ is known as the **wavenumber** and has the units cm^{-1}. Hence an absorption of 2886 cm^{-1} corresponds to a wavelength of 3.465×10^{-6} m. The longer the wavelength, the lower the energy and the smaller the value of $1/\lambda$ in cm^{-1}. Conversely the higher the wavenumber the higher the energy.

For a covalent bond to absorb infrared radiation there must be a change in the dipole moment (bond polarity) of the molecule as the vibration occurs. Thus simple diatomic gas molecules containing only one element, such as H_2, Cl_2 and O_2, are non-polar and do not absorb infrared radiation.

For more complex molecules only those vibrations that result in a dipole change will be infrared active. For example, the symmetrical stretch in carbon dioxide will be infrared inactive, whereas the asymmetric stretch and the bending are both infrared active, because they result in a dipole change. See Figure 5.

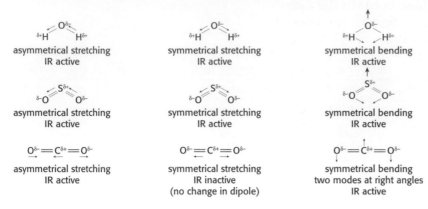

Figure 5 Fundamental vibrations of H_2O, SO_2 and CO_2.

Stretching and bending are the main modes of vibration, but bending can be subdivided into rocking, scissoring, twisting and wagging, as exemplified by the $-CH_2-$ group (Figure 6).

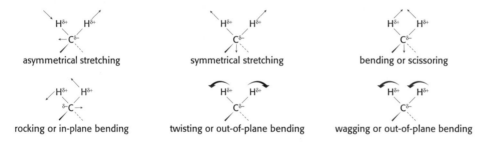

Figure 6 The different types of bending in the $-CH_2-$ group

A double-beam infrared spectrometer

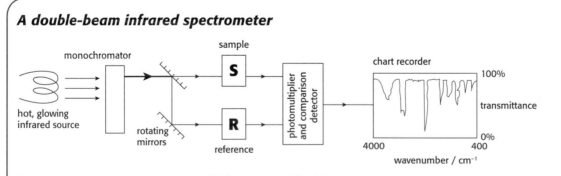

Figure 7 Schematic diagram of a double-beam infrared spectrometer.

By using a rotating mirror the beam of monochromatic radiation is alternately passed through the sample and a reference. A photomultiplier converts photons of radiation into an electrical current. The spectrum is generated by comparison of the currents produced by the sample and the reference beams. Modern spectrometers pass all the frequencies from 4000 cm⁻¹ to 400 cm⁻¹ through the sample at the same time, and a system known as Fourier transformation is used to automatically analyse the transmission at each wavelength.

Uses of infrared spectroscopy

Infrared spectroscopy is particularly useful in characterizing organic compounds, because the various bonds in the functional groups absorb at different frequencies. Also, the precise absorption depends on the neighbouring atoms. Computers can store the infrared spectra of all known compounds, so most can be identified from their "fingerprint" region. This is the complex region due to molecular vibrations between about 1400 and 400 cm^{-1}. For simple laboratory analysis there are several key absorptions that can be used to identify particular bonds, and a list of these can be found in the IB Data Booklet. The most common of these are the broad absorption due to –OH and the sharper absorption peaks due to C–H, C=O, C=C and C–O bonds (see Figure 8 and Table 2). Infrared spectroscopy can also be used quantitatively: for example, the amount of C–H stretch in ethanol present in breath forms the basis of the roadside intoximeter.

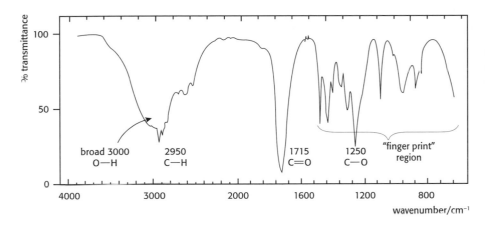

Table 2 A simplified IR correlation chart

Bond	Wavenumber / cm^{-1}
C–O	1000–1300
C=C	1610–1680
C=O	1680–1750
C≡C	2070–3300
O–H (in carboxylic acids)	2500–3300
C–H	2840–3095
O–H (in alcohols)	3230–3500

Figure 8 Infrared spectrum of propanoic acid, C_2H_5COOH.

Mass spectrometry

The workings of a mass spectrometer have already been outlined in Chapter 2. When a molecule is vaporized it will form the molecular ion M$^+$, which will give the value of its molecular mass, but it will also fragment into smaller ions. By looking at the difference in mass from the molecular ion peak it is often possible to identify particular fragments (Figure 9).

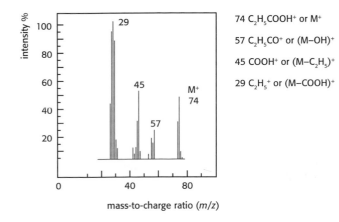

74 $C_2H_5COOH^+$ or M$^+$

57 $C_2H_5CO^+$ or (M–OH)$^+$

45 COOH$^+$ or (M–C$_2$H$_5$)$^+$

29 $C_2H_5^+$ or (M–COOH)$^+$

Figure 9 Mass spectrum of propanoic acid, C_2H_5COOH.

For example:

$(M_r - 15)^+$ due to loss of $-CH_3$

$(M_r - 17)^+$ due to loss of $-OH$

$(M_r - 29)^+$ due to loss of $-C_2H_5$ or $-CHO$

$(M_r - 31)^+$ due to loss of $-CH_3O$

$(M_r - 45)^+$ due to loss of $-COOH$

However, some care must be exercised when analysing fragmentation patterns, because not all fragments are formed. Some do not last long enough to be detected, and sometimes some unexpected fragments form.

¹H NMR spectroscopy

Whereas infrared spectroscopy gives information about the types of bond in a molecule, ¹H NMR spectroscopy provides information on the chemical environment of all the hydrogen atoms in the molecule. The nuclei of hydrogen atoms possess spin, and can exist in two possible states of equal energy. If a strong magnetic field is applied, the spin states may align themselves either with the magnetic field or against it, and there is a small energy difference between them. The nuclei can absorb energy when transferring from the lower to the higher spin state. This energy is very small, and occurs in the radio wave region of the electromagnetic spectrum. The precise energy difference depends on the chemical environment of the hydrogen atoms (Figure 10).

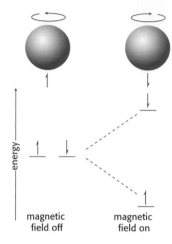

Figure 10 Nuclear spin.

The position in the ¹H NMR spectrum where the absorption occurs for each hydrogen atom in the molecule is known as the **chemical shift** from a known reference, and is measured in parts per million (ppm). The precise values depend on the solvent used, and so can vary. However, the area under each peak corresponds to the number of hydrogen atoms in that particular environment, and this can be extremely useful in determining the structure. The area under each peak is usually given by the height of the integration trace. See Table 3 and Figure 11.

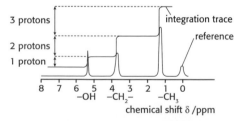

Figure 11 ¹H NMR spectrum of ethanol, CH_3CH_2OH. The additional trace integrates the area under each peak. The height of each section is proportional to the number of hydrogen atoms in each chemical environment.

Magnetic resonance imaging

In 2003 two scientists, Paul C. Lauterbur (1929–2007) from the USA and Peter Mansfield (1933–) from the United Kingdom won the Nobel Prize in Physiology or Medicine for their work in developing the use of NMR in diagnostic medicine. Because the protons in water, lipids, carbohydrates, etc. give different signals, depending upon

Table 3 A simplified ¹H NMR correlation chart

Type of proton	Chemical shift / ppm
$R-CH_3$	0.9
R_2-CH_2	1.3
R_3CH	2.0
CH_3COR	2.0
$R-C-CH_3$ (C=O)	2.1
$R-OH$	4.5
⬡–H	7.3
$R-C-H$ (C=O)	9.7

their environment, an image of the whole body can be built up by placing the patient inside the magnet of a large NMR machine. This is known as **magnetic resonance imaging** (MRI). Figure 12 shows an MRI image of the human brain. MRI images can be used to diagnose and monitor conditions such as cancer, multiple sclerosis, osteomyletisis and hydrocephalus. Because MRI uses only low-energy radio waves it is completely non-invasive and causes no harm to the patient. In addition to ^{1}H, the spectra produced by other nuclei that possess spin, such as ^{13}C and ^{31}P, can also be used in MRI.

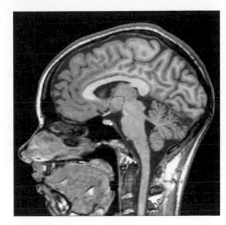

Figure 12 An MRI image of the human brain.

Atomic absorption (AA) spectroscopy

In Chapter 2 we saw that each element produces a unique emission spectrum. This explains, for example, the different flame colours for sodium (yellow) and potassium (lilac). Atomic emission spectra are obtained by giving the atoms of the element sufficient electrical or heat energy and then recording the energy of the light emitted as the electrons fall from an excited state to a state of lower energy. Atomic absorption spectroscopy is the reverse process as the electrons in atoms of the element absorb just enough energy to be promoted to a higher energy level. By measuring the amount of light absorbed the concentration of that element in a particular sample can be precisely quantified. Atomic absorption spectroscopy is extremely sensitive, and can measure concentrations as low as 1.0×10^{-6} g dm^{-3}, that is, one part per billion. For that reason it is used to determine the concentration of metals in water, blood, soils, oils and foodstuffs.

Details of how an atomic absorption spectrometer works can be found on the Internet, but modern machines employ the double beam principle already outlined for an infrared spectrometer. Essentially, for each element being analysed there is a different light source. This utilizes the emission spectrum of the element under study, so that light of a specific frequency is being passed through the sample (the **analyte**). For example, if the concentration of mercury in soil is to be determined, then the light source would consist of excited mercury atoms.

The sample to be analysed is first turned into a fine mist or aerosol in a nebulizer. This mixes with a combustion mixture made up from a fuel and an oxidizing agent. The most common combustion mixture is an ethyne (fuel)–air (oxidizing agent) mixture. The monochromatic light from the source is then passed through this vaporized sample contained inside the flame, and the amount of light absorbed by the atoms is detected by using a photomultiplier to convert it into an electrical signal (Figure 13). By changing the light source the concentration of different elements in the same sample can be determined.

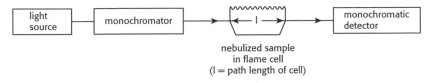

Figure 13 Simplified diagram of atomic absorption spectrometer.

Determining the concentration of an element

The concentration of each element can be determined because there is a direct relationship between the amount of light absorbed and the concentration of the element in the path of the light. The amount of light absorbed, the absorbance, is measured by comparing the logarithm of the initial intensity of the light, I_0, with the logarithm of the intensity of the emerging light, I, after it has passed through the sample. This relationship is known as the Beer–Lambert law, which applies only for dilute solutions:

$$\text{Absorbance}, A = \log_{10}\left(\frac{I_0}{I}\right) = \varepsilon c l$$

where ε represents the molar absorptivity coefficient, and is a constant for a particular element at a fixed frequency; l represents the path length through the sample cell; and c represents the concentration of the element in the sample.

As the path length can be fixed, and the molar absorptivity coefficient is constant, this means that the absorbance is directly proportional to the concentration.

In practice, samples of known concentrations are used to produce a calibration curve. The absorbance of the unknown sample is then measured, and the concentration can be read directly by interpolation from the calibration curve (Figure 14).

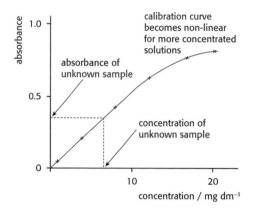

Figure 14 The use of a calibration curve to determine an unknown concentration.

Combining different techniques to aid identification

Some of the most satisfying problems to solve in chemistry consist in interpreting the data obtained for the same compound from several different analytical techniques to arrive at only one possible structural formula. Consider the two isomeric compounds methyl propanoate and propyl methanoate (Figure 15).

methyl propanoate

propyl methanoate

Figure 15 Two structural isomers of $C_4H_8O_2$

Which would be the best technique to distinguish between them? Both have the molecular formula $C_4H_8O_2$. Both would give a peak at $m/z = 88$ due to the molecular ion M^+ in their mass spectra. Both would give absorptions in the region of 2840–3095 cm^{-1}, 168–1750 cm^{-1} and 1000–1300 cm^{-1} in their infrared spectra because of the presence of C–H, C=O and C–O respectively. However, we shall find a difference if we look at their ^{1}H NMR spectra, even though both have eight hydrogen atoms. Methyl propanoate will show three separate peaks, with their areas in the ratio 3:2:3, because the hydrogen atoms are in three different chemical environments. Propyl methanoate will have four separate peaks, with their areas in the ratio 1:2:2:3, because the hydrogen atoms are in four different chemical environments.

We could confirm our findings by looking more closely at the mass spectra. Both could lose a –CH$_3$ group to give a fragment with an m/z value of 73, thanks to the remaining $C_3H_5O_2^+$ fragment, but only propyl methanoate will show a fragment at $m/z = 43$ because of the loss of –HCOO leaving $C_3H_7^+$. Similarly we would expect to find a peak at $m/z = 57$ in the mass spectrum of methyl propanoate owing to the loss of –OCH$_3$ to leave $C_2H_5CO^+$, which would be absent in the mass spectrum of propyl methanoate.

1 Identify the correct compound from the information provided.
   (a) The infrared spectrum contains a broad peak centred at 3300 cm^{-1} and no peak between 1680 cm^{-1} or 1750 cm^{-1}. Is the compound ethanol or ethanal?
   (b) The infrared spectrum contains a sharp peak at 1715 cm^{-1}. The ^{1}H NMR spectrum shows three peaks with their areas in the ratio 3:2:1. Is the compound propanone or propanal?
   (c) The mass spectrum shows peaks with m/z values of 60, 31 and 29. Is the compound ethanoic acid or methyl methanoate?
   (d) The mass spectrum shows a peak with an m/z value of 60. The infrared spectrum shows a broad peak centred at 3300 cm^{-1}. The ^{1}H NMR spectrum shows three peaks with their areas in the ratio 6:1:1. Is the compound propan-1-ol or propan-2-ol?

2 The permitted concentration of lead(II) ions, Pb^{2+}(aq), in household drinking water is 15 ppb (0.015 mg dm^{-3}). In some old houses that still have lead pipes this concentration can be exceeded. An atomic absorption spectrometer was calibrated using standards to give the following results:

Concentration of Pb^{2+}(aq) standard / mg dm^{-3}	Absorbance
0.30	0.131
0.60	0.273
1.00	0.419
1.50	0.640
2.00	0.852

650 cm^3 of drinking water from an old house was boiled until the volume was reduced to 9.62 cm^3. When this reduced volume was tested it was found to have an absorbance of 0.381. Plot the calibration curve and determine whether the water from the house was within the permitted concentration of lead(II) ions.

3 Suggest a reason why the technique used in hospitals is called magnetic resonance imaging (MRI) rather than *nuclear* magnetic resonance imaging (NMRI).

Visible and ultraviolet (UV-Vis) spectroscopy

In Chapter 3 we saw that the reason why transition metal complexes are coloured is because the d orbitals are split. As a d electron is promoted between the split d levels it gains energy, which corresponds to the wavelength of visible light, so that the colour that is transmitted is the complementary colour of the light that has been absorbed.

> ### Complementary colours
>
> If white light is shone on the solution of the transition metal complex and yellow light is absorbed, then the solution appears blue-violet because that is the colour of the remaining light that is transmitted (Figure 16).

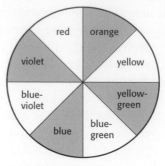

Figure 16 A colour wheel. Complementary colours are opposite to each other.

In the gaseous free transition metal ion the five d orbitals are degenerate: that is, they are of equal energy. However, as the six ligands approach to form the coordinate bonds in the octahedral complex they "see" the ligands lying along the axes first, and repel them to higher potential energy relative to the remaining three ligands, which lie between the axes (Figure 17).

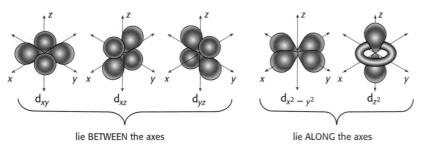

Figure 17 The shapes of the five d orbitals.

The amount by which the d orbitals are split will depend upon several factors. The more electron-dense the ligands are, the more the splitting. Ligands can be placed in order of the splitting they cause. This is known as the **spectrochemical series**:

$$I^- < Br^- < Cl^- < OH^- < H_2O < NH_3 < CN^-$$

Iodide ions cause the smallest splitting and cyanide ions the largest splitting. The energy of light absorbed increases when ammonia is substituted for water in copper(II) complexes as the splitting increases—that is, in going from $[Cu(H_2O)6]^{2+}$ to $[Cu(NH_3)_4(H_2O)_2]^{2+}$ (Figure 18). This means that the wavelength of the light absorbed decreases, and this is observed in the colour of the transmitted light, which changes from blue to blue-violet. Other factors that will affect the splitting include the nature of the transition metal itself, and the oxidation state of the transition metal ion. A fourth factor is the stereochemistry of the complex, because the splitting will be different for tetrahedral and square planar complexes than for an octahedral complex.

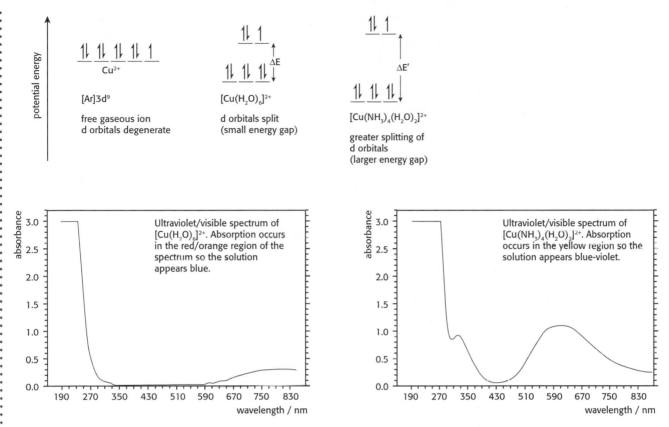

Figure 18 How changing the ligand affects the colour of copper complexes.

UV-Vis and organic molecules

Compounds containing unsaturated groups such as C=C, C=O, −N=N−, −NO₂ and the benzene ring can absorb in the ultraviolet or visible part of the spectrum. Such groups are known as **chromophores**, and the precise energy of absorption is affected to some degree by the other groups attached to the chromophore, which are known as **auxochromes**. The absorption is due to electrons in the bond being excited to an empty orbital of higher energy, usually an anti-bonding orbital. The energy involved in this process is, relatively, quite high, and most organic compounds absorb in the ultraviolet region and thus appear colourless. For example, ethene absorbs at 185 nm. However, if there is extensive conjugation of double bonds in the molecule involving the delocalization of pi electrons, then less light energy is required to excite the pi electrons and the absorption occurs in the visible region. Good examples include vitamin A (retinol), β-carotene (Figure 19), chlorophyll, phenolphthalein (mentioned in Chapter 8; Figures 20, 21) and methyl orange (Figure 22).

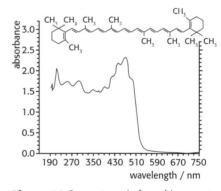

Figure 19 β-carotene is found in carrots and has a characteristic orange colour. It contains 11 double bonds and absorbs strongly in the violet-blue (400–510 nm) region. A similar compound showing extensive conjugation is retinol (vitamin A) the structure of which can be found in the IB Data Book.

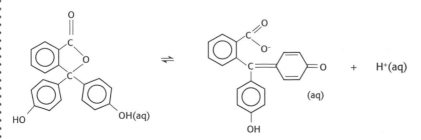

Figure 20 Phenolphthalein.

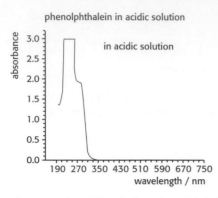

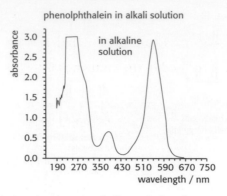

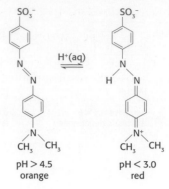

Figure 21 In acidic solution phenolphthalein is colourless. In alkaline solution the anion contains much more conjugation and appears pink.

Figure 22 Methyl orange: an acid–base indicator. The change in the amount of conjugation (delocalization) in the two different forms accounts for the change in colour.

> ### *Sunscreens*
>
> One useful application of conjugated compounds absorbing UV radiation is the application of sunscreens. A typical component of sunscreens is 4-aminobenzoic acid (Figure 23). This absorbs UV radiation in the region of 265 nm.

Figure 23 4-aminobenzoic acid (also known as PABA, *para*-aminobenzoic acid).

Using visible spectroscopy to determine the concentration of a solution

If a component of a solution is coloured then its concentration can be determined using a visible spectrometer. The Beer–Lambert Law described for atomic absorption spectroscopy also applies to visible spectroscopy. As the molar absorption coefficient ε is a constant for each absorbing substance, and the path length l can be kept constant, the concentration is directly proportional to the absorbance. A spectrum of the solution is first run to determine the wavelength of maximum absorption, λ_{max}. The absorbance at this wavelength is then measured for samples of known concentration to produce a calibration curve, from which the unknown concentration can be determined once its absorbance is known. This method can be used to determine the amount of iron in blood, or the amount of aluminium in deodorant (by first converting the aluminium ions into a coloured complex), and for investigating the reaction kinetics of a reaction involving one or more coloured species.

High-resolution ^{1}H NMR spectroscopy

The real strength of ^{1}H NMR as an analytical tool lies in the high-resolution spectra where splitting patterns can be seen. Splitting patterns are due to spin–spin coupling. If there is one adjacent proton ($n = 1$) next to a methyl group, then it will either line up with the magnetic field or be aligned against it. The effect will be that the methyl protons will thus experience one slightly stronger and one slightly weaker external magnetic field, resulting in an equal splitting of the peak. This is known as a **doublet** (Figure 24).

If there is a $-CH_2-$ group ($n = 2$) adjacent to a methyl group, then there are three possible energy states available when an external magnetic field is applied:

1 Both proton spins are aligned with the field: ↑↑
2 One is aligned with the field and one against it (two possible combinations): ↑↓ or ↓↑

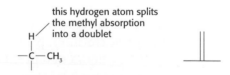

Figure 24 The splitting pattern of a methyl group caused by an adjacent hydrogen atom.

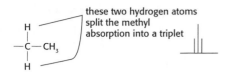

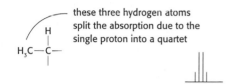

3 Both are aligned against the field: ↓↓

This results in a triplet, whose peaks are in the ratio 1:2:1 (Figure 25).

By using Pascal's triangle it is possible to predict the splitting pattern to cover all possible combinations:

Number of adjacent protons, *n*	Splitting intensities	Number of lines	Type of splitting
0	1	1	singlet
1	1 1	2	doublet
2	1 2 1	3	triplet
3	1 3 3 1	4	quartet
4	1 4 6 4 1	5	quintet

Figure 25 The splitting pattern caused by two adjacent hydrogen atoms.

The number of lines is always equal to one more than the number of adjacent hydrogen atoms. Thus a methyl group (*n* = 3) next to a proton will cause it to split into a quartet with peaks in the ratio 1:3:3:1 (Figure 26).

Figure 26 The splitting pattern caused by three adjacent hydrogen atoms.

One of the classic patterns observed in a ^{1}H NMR spectrum is a quartet together with a triplet due to the presence of an ethyl group, $-CH_2-CH_3$.

Using TMS as a reference

The advantages of using tetramethylsilane, $Si(CH_3)_4$ (Figure 27), as the standard reference are:
- All the 12 protons are in the same chemical environment, so it gives a strong single peak.
- It is not toxic, and is very unreactive (and so does not interfere with the sample).
- It absorbs upfield, well away from most other protons.
- It is volatile (has a low boiling point), and so can easily be removed from the sample.

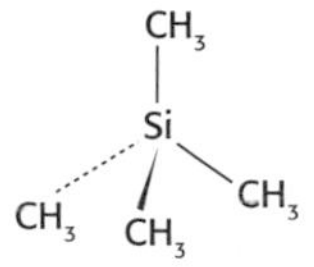

Figure 27 Tetramethylsilane as a reference sample.

Interpreting high-resolution ^{1}H NMR spectra

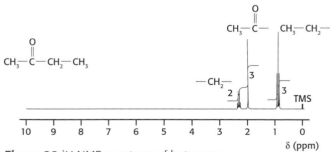

Figure 28 ^{1}H NMR spectrum of butanone.

Figure 28 shows the ^{1}H NMR spectrum of butanone.

- The three different peaks show that the hydrogen atoms within the molecule are in three different chemical environments.
- The integrated trace shows that the hydrogen atoms are in the ratio 2:3:3.
- The chemical shifts of the three peaks identify them as:

R–CH$_3$ 0.9 ppm, CH$_3$–C– 2.0 ppm and R–CH$_2$–C– 2.3 ppm.

(with O double bonds above the C in each carbonyl group)

- The –CH$_2$– group has three adjacent protons and so is split into a quartet ($n + 1 = 4$).

The CH$_3$–C– protons contain no adjacent protons, so no splitting occurs.

The CH$_3$– group next to the –CH$_2$– group is split into a triplet ($n + 1 = 3$)

Gas-liquid chromatography and high-performance liquid chromatography

Gas-liquid chromatography (known as GC or GLC) is used to separate and identify the components in mixtures of volatile liquids that do not decompose at temperatures at or near their boiling points. Particular uses include analysis of urine samples from athletes to test for the presence of banned substances, determination of the amount of ethanol present in blood in cases of drunken driving, and detection of the presence of potentially explosive or poisonous gases in underground mines.

The stationary phase consists of a liquid (e.g. a long-chain alkane) coated onto a solid support in a long, thin capillary tube (Figure 29). The mobile phase is an inert gas such as nitrogen or helium. The sample is injected through a self-sealing cap into an oven for vaporization. The sample is then carried by the inert gas into the column, which is coiled and fitted into a thermostatically controlled oven. At the end of the column the separated components exit into a detector. This is usually a flame ionization detector. The current produced is connected to a chart recorder. Each component will have a separate retention time, and the area under the peak will be proportional to the amount of component present. It is possible to programme the temperature of the oven to increase during the operation to speed up the elution of the less volatile components.

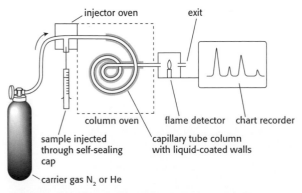

Figure 29 Schematic diagram of a gas–liquid chromatograph.

In **high-performance liquid chromatography** (HPLC) the mobile phase is forced through a column under pressure. The stationary phase is usually composed of silica particles with long-chain alkanes adsorbed onto their surface. The separation is very efficient, so long columns are not needed. The separated components are usually

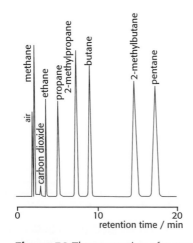

Figure 30 The separation of a sample of natural gas by GLC. The less volatile the component gas, the greater the retention time.

detected by ultraviolet spectroscopy. As in GLC, the results are recorded onto a chart showing the different retention times.

HPLC can be used for identification as well as for separation. It is particularly useful for non-volatile components, or components that decompose near their boiling point. Examples include the analysis of oil pollutants, the analysis of alcoholic drinks, antioxidants, sugars and vitamins in food and quality control of insecticides and herbicides. It can also be used to separate enantiomers (chiral separation) using columns containing optically active material.

Both GLC and HPLC are particularly powerful when connected to a mass spectrometer, as very small quantities of materials can be separated into their individual components and their identities quickly established.

1 Describe how the integration trace and the splitting patterns differ for the high-resolution ^1H NMR spectra of the following pairs of compounds:
   (a) methyl propanoate, $C_2H_5COOCH_3$, and propyl methanoate, $HCOOC_3H_7$
   (b) ethyl ethanoate, $CH_3COOC_2H_5$, and butanoic acid, C_3H_7COOH
   (c) propan-2-one, $CH_3COCH_2CH_2CH_3$, and butan-3-one, $CH_3CH_2COCH_2CH_3$.
2 You are provided with a standard solution of 1.00 mol dm^{-3} copper(II) sulfate solution, a visible spectrometer, and any other laboratory equipment you require. Design an experiment to determine the concentration of copper(II) ions in a sample of "Cuprinol", which is used to treat timber against infestation by wood-boring creatures.

Option B: human biochemistry

Energy requirements

A moderately active woman requires about 8400 kJ (2000 kcal) of energy intake per day. For an adult male undertaking physical work this increases to about 14,700 kJ (3500 kcal). Energy is provided by lipids, carbohydrates and proteins. Carbohydrates provide the main source of energy but, like proteins, they are already partially oxidized, so do not provide as much energy weight for weight as fats, which are used to store energy. The amount of energy in a dried food can be determined by combustion experiments (see Chapter 5) using a food calorimeter.

In many countries rice is the main source of dietary energy. For example, in Bangladesh 75% of the energy intake of the population comes from rice. Boiled rice has an energy content of about 150 kcal per 100 g, which means that a man undertaking manual labour needs to consume the equivalent of about 2.5 kg of rice each day.

> ### Use of a food calorimeter
>
> 1.13 g of rice raises the temperature of 525 g of water by 3.31 °C.
> Energy evolved $= 0.525 \times 4.18 \times 3.31 = 7.26$ kJ
> Energy content of rice $= 7.26 \times 100/1.13$
> $= 642$ kJ (154 kcal) per 100 g

Proteins

Proteins are also an essential part of a healthy diet. In addition to being an energy source rice contains about 7–8% protein. Other rich sources of protein include meat, fish, cheese, eggs and nuts.

Proteins have many different functions in the body. They can act as biological catalysts (enzymes) for specific reactions, and act as transporters e.g. hemoglobin in the blood. They can give structure (for example, hair and nails consist almost entirely of polypeptides coiled into proteins called **keratin** and composed of α-helices), and provide a source of energy. Some hormones are proteins or protein-based, for example insulin and FSH (follicle-stimulating hormone), which is responsible for triggering the monthly menstrual cycle in females.

2-amino acids

Proteins are large macromolecules made up of chains of 2-amino acids. About 20 2-amino acids occur naturally. These have the general formula $H_2NCHRCOOH$ and are listed in the IB Data Booklet. The structure of amino acids in aqueous solution alters at different pH values. At low pH (acidic medium) the amine group will be protonated. At high pH (alkaline medium) the carboxylic acid group will lose a proton. For each amino acid there is a unique

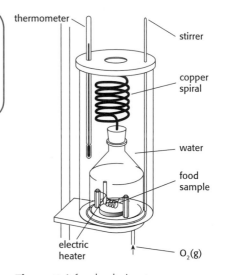

Figure 1 A food calorimeter.

pH value, known as the **isoelectric point**, where the amino acid will exists as the zwitterion (Figure 2).

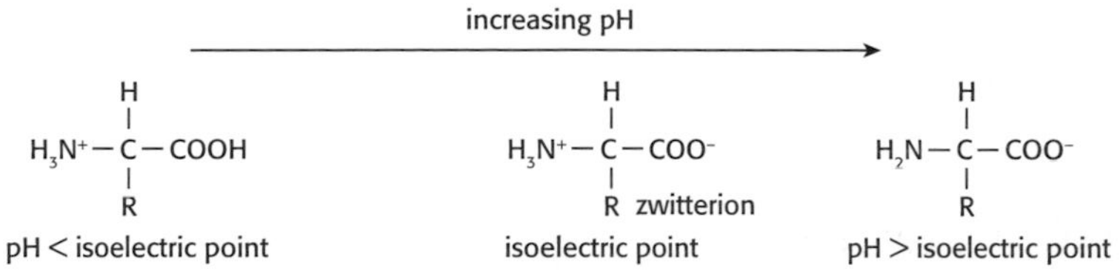

increasing pH →

$$H_3N^+ - \overset{\overset{\displaystyle H}{|}}{\underset{\underset{\displaystyle R}{|}}{C}} - COOH \qquad H_3N^+ - \overset{\overset{\displaystyle H}{|}}{\underset{\underset{\displaystyle R}{|}}{C}} - COO^- \qquad H_2N - \overset{\overset{\displaystyle H}{|}}{\underset{\underset{\displaystyle R}{|}}{C}} - COO^-$$

pH < isoelectric point   R zwitterion   pH > isoelectric point

isoelectric point

Figure 2 The structure of 2-amino acids at different pH values.

2-amino acids function as good buffers, because they can combine with either protons or hydroxide ions, thus helping to maintain the existing pH of the solution.

Two 2-amino acids can condense together to form a dipeptide with the elimination of water. Two different 2-amino acids can react to form two different dipeptides (Figure 3).

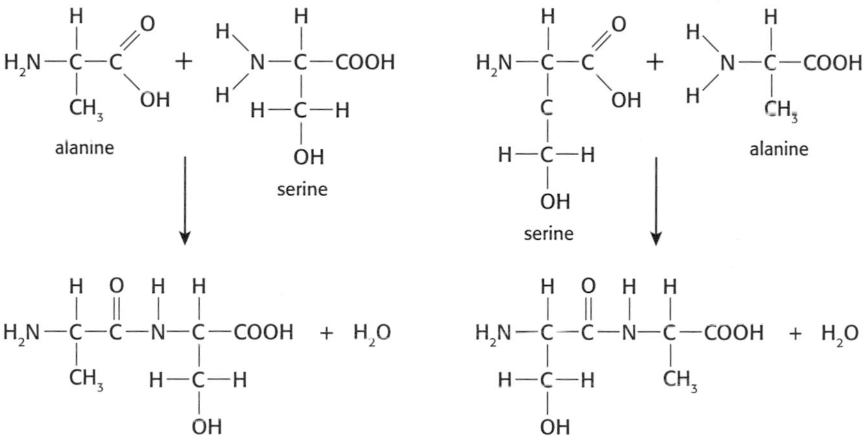

Figure 3 The condensation of alanine and serine to make two different dipeptides.

The link between them, which has the structure $-\overset{\overset{\displaystyle O}{\|}}{C}-\overset{\overset{\displaystyle H}{|}}{N}-$, is known as a **peptide bond**. Because dipeptides still contain a reactive functional group on each end they can react further in the presence of enzymes to form long chains of amino acid residues linked by peptide bonds (Figure 4). This is an example of **condensation polymerization**, and the products are known as **polypeptides**.

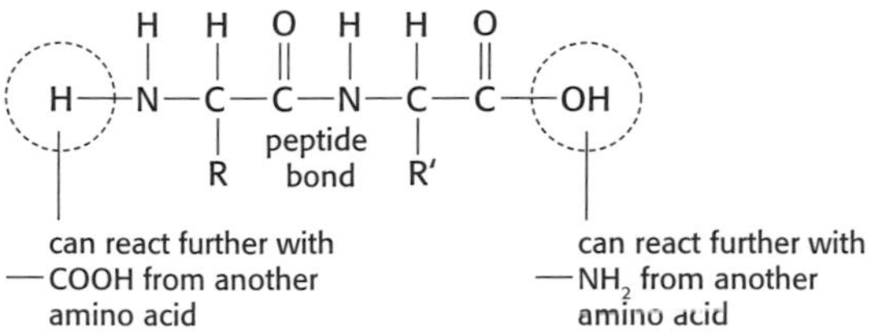

Figure 4 The ability of 2-amino acids to undergo condensation polymerization.

can react further with
—COOH from another
amino acid

can react further with
—NH$_2$ from another
amino acid

Structure of proteins

Proteins are made up from a fixed number of amino acid residues connected to each other in a unique linear sequence. This sequence, for example NH$_2$–leu–his–ala– … –val–ala–ser–COOH, is known

Find the structures of the three amino acids cysteine, phenylalanine and threonine.

1 How many tripeptides can be obtained by reacting these three different amino acids together? (Assume that each tripeptide contains the residues from all three amino acids.)

2 Give the full structural formula of one of the tripeptides.

as the **primary structure** of proteins. The **secondary structure** describes the way in which the chain of amino acids, often with repeating sequences of amino acids, folds itself because of intramolecular hydrogen bonding. The folding can either be β-helix, in which the protein chain twists in a spiralling manner, rather like a coiled spring, or β-pleated to give a sheet-like structure (Figure 5).

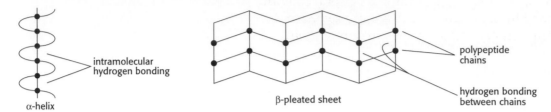

Figure 5 Secondary structure of proteins.

The **tertiary structure** describes the overall folding of the chains by interactions between distant amino acids to give the protein its three-dimensional shape. These interactions may be due to hydrogen bonds, van der Waals' attraction between non-polar side groups, or ionic attractions between polar groups. In addition, two cysteine residues can form **disulfide bridges** when their sulfur atoms undergo enzyme-catalysed oxidation (Figure 6).

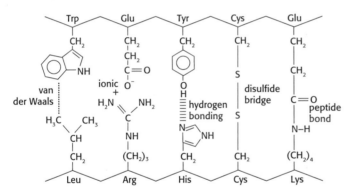

Figure 6 Examples of interactions between side groups on adjacent polypeptide chains.

Separate polypeptide chains can interact together to give a more complex structure: this is known as the **quaternary structure**. Haemoglobin has a quaternary structure that includes four protein chains (two α-chains and two β-chains) grouped together around four haem groups.

Analysis of proteins

To determine the primary structure the protein must first be completely hydrolysed by dilute hydrochloric acid to successively release the amino acids. These can then be identified by chromatography or by electrophoresis. The method for paper chromatography is described in Chapter 14. In order to observe the colourless amino acids the chromatogram is developed by spraying with ninhydrin, an organic dye that colours the amino acids (Figure 7), so that the retention factor, R_f, values of the spots can be determined and compared with known values.

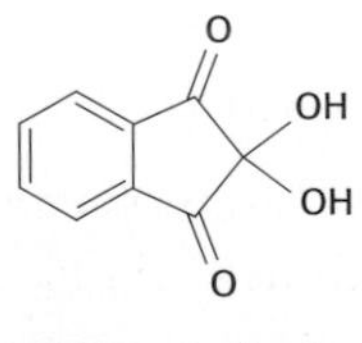

Figure 7 The structure of ninhydrin (triketohydrindane hydrate). Ninhydrin forms coloured compounds by reacting with the amine group of amino acids. It is used in forensic science to detect the residual peptides and proteins left behind in fingerprints.

In PAGE (PolyAcrylamide Gel Electrophoresis) the sample is placed in the centre of a polyacrylamide gel and a potential difference is applied across it. Depending on the pH of the buffer, the different amino acids will move at different rates towards the positive and negative electrodes. At its isoelectric point a particular amino acid will not move, because its charges are balanced (Figure 8). When separation is complete the amino acids can be sprayed with ninhydrin and identified by comparing the distance they have travelled with standard samples or from a comparison of their isoelectric points.

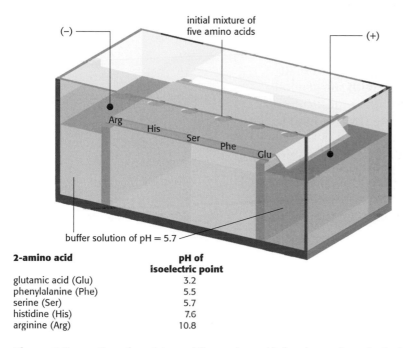

2-amino acid	pH of isoelectric point
glutamic acid (Glu)	3.2
phenylalanine (Phe)	5.5
serine (Ser)	5.7
histidine (His)	7.6
arginine (Arg)	10.8

Figure 8 Separation of a mixture of five amino acids by electrophoresis. Serine does not move, because its isoelectric point is the same pH as the buffer. Histidine and arginine contain $-\overset{+}{N}H_3$ at pH 5.7, so move towards the negative electrode. Glutamic acid and phenylalanine contain $-COO^-$ at pH 5.7, so move towards the positive electrode.

Carbohydrates

Carbohydrates are produced in plants by photosynthesis. This requires energy provided by sunlight. When taken into the human body this process is reversed, and carbohydrates are a principal source of chemical energy.

$$C_x(H_2O)_y + xO_2 \rightarrow xCO_2 + yH_2O$$

Foods such as bread, biscuits, cakes, potatoes, rice and cereals are all high in carbohydrates. The body also uses carbohydrates to store energy. In the livers and skeletal muscles of animals it is stored in the form of **glycogen**—known as animal starch. Carbohydrates are also used as precursors for other important molecules: for example, they are components of nucleic acids and play a key role in the biosynthesis of proteins. Humans also need one other form of carbohydrates, which they cannot break down. This is cellulose, hemicellulose, lignin and pectin, known collectively as **dietary fibre** or roughage. Dietary fibre is mainly plant material that cannot be hydrolysed by enzymes secreted in the human digestive tract, although it can be digested by microflora in the gut. Roughage is needed to provide good bowel movements, and a lack of dietary fibre can lead to such conditions as diverticulosis, irritable bowel syndrome, constipation, obesity, haemorrhoids, diabetes and bowel cancer.

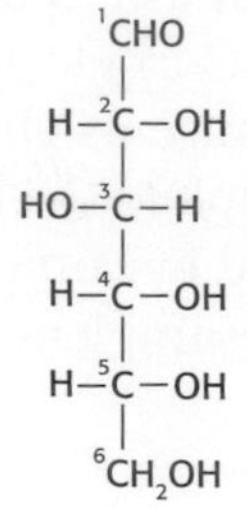

Monosaccharides

Monosaccharides have the empirical formula CH_2O and contain a carbonyl group (C=O) and at least two hydroxyl, –OH, groups. They generally have between three and six carbon atoms. Monosaccharides with the general formula $C_5H_{10}O_5$ are known as **pentoses** (e.g. ribose) and monosaccharides with the general formula $C_6H_{12}O_6$ are known as **hexoses**. One of the most common hexoses is glucose. In nature the form of glucose that is found is known as D-glucose (Figure 9).

As well as having a straight-chain formula, which contains an aldehyde group, D-glucose can also exist in a cyclic or ring structure, which has two separate crystalline forms known as α-D-glucose and β-D-glucose (Figure 10). The only difference between them is that the –OH group on the first carbon atom is inverted.

Figure 9 Straight-chain formula of D-glucose.

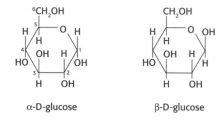

Figure 10 Cyclic (ring) structures of D-glucose.

Polysaccharides

Sugars can condense together in the presence of appropriate enzymes to form polysaccharides. **Sucrose** is a disaccharide formed from the condensation of α-D-glucose and β-D-fructose (Figure 11).

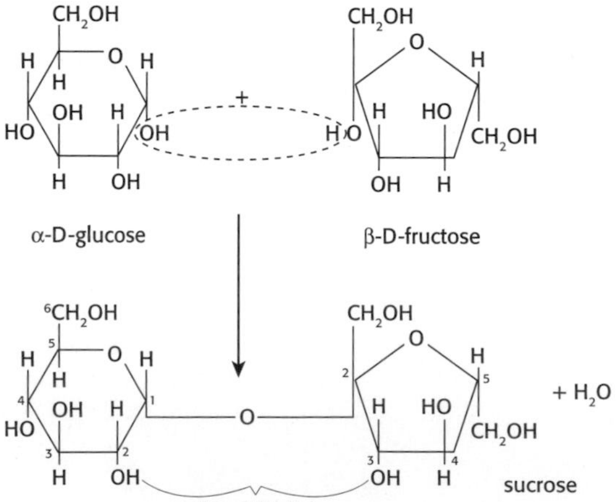

Figure 11 Formation of sucrose by a condensation reaction.

The link between the two sugars is known as a **glycosidic link**. In the case of sucrose the link is between the C-1 atom of glucose in the α configuration and the C-2 atom of fructose. The link is known as an **α-1,2 bond**.

Lactose is a disaccharide in which the β-D-galactose is linked at the C-1 atom to the C-4 atom of β-D-glucose. This is called a **β-1,4 linkage** (Figure 12).

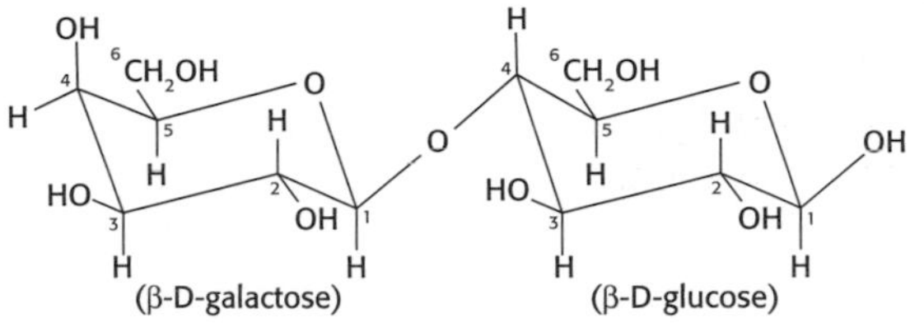

Figure 12 The structure of lactose showing the 1,4 link.

Maltose, another disaccharide, is formed when two glucose molecule condense together, forming an α-1,4 linkage.

One of the most important polysaccharides is **starch**. Starch exists in two forms: **amylose**, which is water soluble, and **amylopectin**, which is insoluble in water. Amylose is a straight-chain polymer of α-D-glucose units with α-1,4 bonds (Figure 13). Amylopectin also consists of α-D-glucose units, but it has a branched structure with both α-1,4 and α-1,6 bonds (Figure 14).

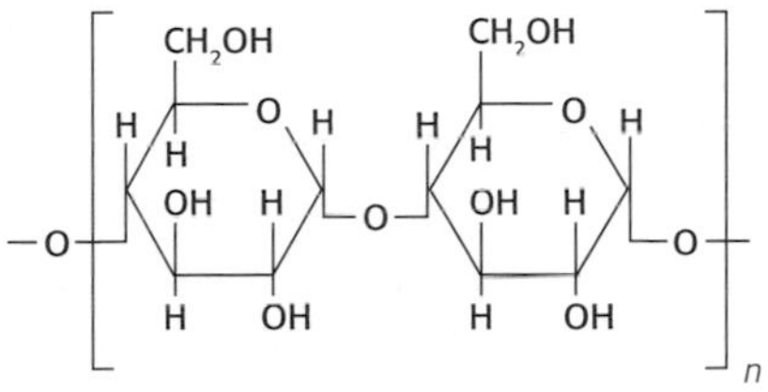

Figure 13 The structure of amylose.

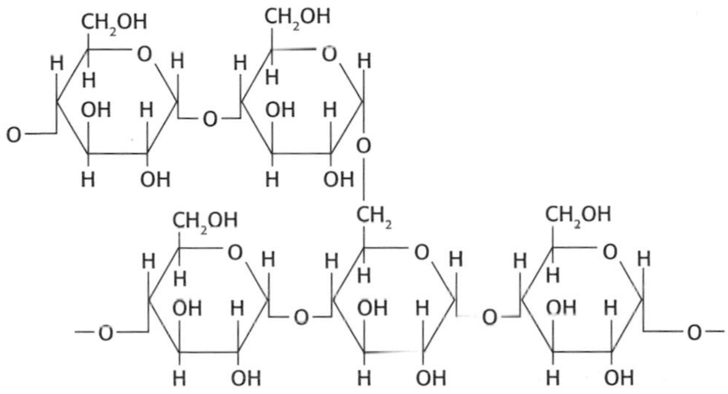

Figure 14 Structure of amylopectin.

Cellulose, a polymer of β-D-glucose, contains β-1,4 linkages (Figure 15). Cellulose, together with lignin, provides the structure to the cell walls of green plants. Most animals, including all mammals, do not have the enzyme cellulase and so are unable to digest cellulose or other dietary fibre polysaccharides.

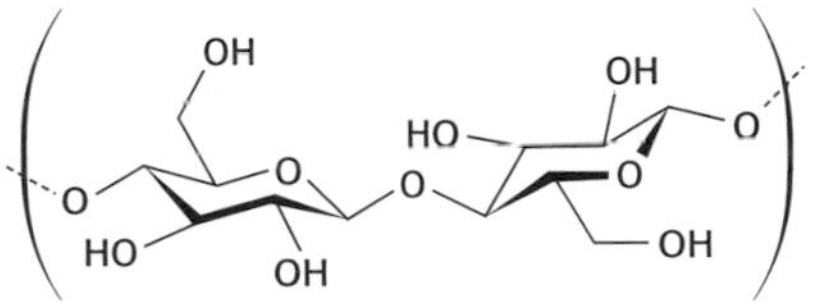

Figure 15 Repeating unit of cellulose showing β-1,4 linkages.

Lipids

Lipids are organic molecules with long hydrocarbon chains that are soluble in non-polar solvents. They are used mainly for energy storage, insulating and protecting vital organs, forming cell membranes and, in some cases, acting as hormones. Three important types of lipids are triglycerides (fats and oils), phospholipids (lecithin), and steroids (cholesterol).

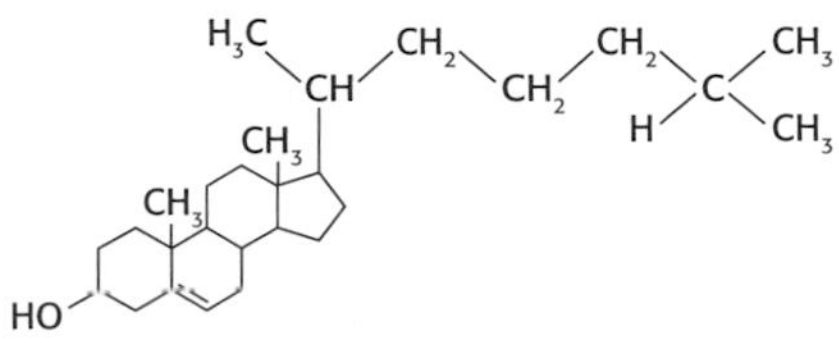

Figure 16 The structure of cholesterol.

Cholesterol

Cholesterol (Figure 16) has the characteristic four-ring structure possessed by all steroids (Figure 17). It is formed in the liver, and is found in all tissues, the blood, brain and spinal cord. It is transported by lipoproteins. **Low-density lipoproteins** (LDL) are in the order

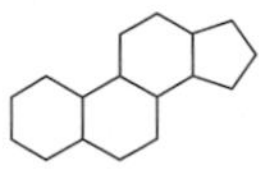

Figure 17 Steroid four-ring "backbone".

289

of 18–25 nm and transport cholesterol to the arteries, where they can line the walls of the arteries, leading to cardiovascular diseases. The major source of these low-density lipoproteins are saturated fats, and in particular those derived from lauric (C_{12}), myristic (C_{14}) and palmitic (C_{16}) acids. Smaller lipoproteins, in the order of 8–11 nm, known as **high-density lipoproteins** (HDL), can remove the cholesterol from the arteries and transport it back to the liver.

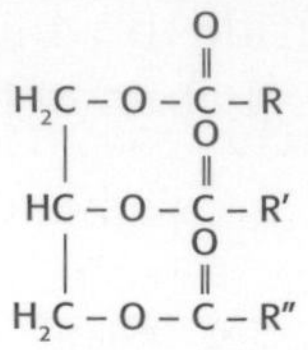

Fats and oils (triglycerides)

Fats and oils are triesters (triglycerides), formed from the condensation reaction of propane-1,2,3-triol (glycerol) with long-chain carboxylic acids (fatty acids) (Figure 18).

Fats are solid triglycerides: examples include butter, lard and tallow. Oils are liquid at room temperature, and include castor oil, olive oil and linseed oil. The essential chemical difference between them is that fats are **saturated** (that is, they do not contain carbon-to-carbon double bonds). Oils contain at least one carbon-to-carbon double bond and are said to be **unsaturated**. Most plant oils contain several carbon-to-carbon double bonds and are known as **polyunsaturated**. In the body, fats and oils are hydrolysed by enzymes, known as **lipases**, to glycerol and fatty acids. These in turn are broken down by a series of redox reactions to produce, ultimately, carbon dioxide, water and energy. Because they are essentially long-chain hydrocarbons with only two oxygen atoms each on the three carboxyl atoms, fats are in a less oxidized form than carbohydrates, so weight for weight produce more energy.

Figure 18 General formula of a fat or oil. R, R' and R" are long-chain hydrocarbons formed from carboxylic acids, which may be the same or different.

Table 1 Formulas of some fatty acids

Name	Formula	Number of C atoms per molecule	Number of C=C bonds	Melting point / °C
saturated fatty acids				
lauric acid	$CH_3(CH_2)_{10}COOH$	12	0	44.2
myristic acid	$CH_3(CH_2)_{12}COOH$	14	0	54.1
palmitic acid	$CH_3(CH_2)_{14}COOH$	16	0	62.7
stearic acid	$CH_3(CH_2)_{16}COOH$	18	0	69.6
unsaturated fatty acids				
oleic acid	$CH_3(CH_2)_7CH=CH(CH_2)_7COOH$	18	1	10.5
linoleic acid	$CH_3(CH_2)_4CH=CHCH_2CH=CH(CH_2)_7COOH$	18	2	−5.0

The body requires two essential unsaturated fatty acids that it is unable to synthesize. These are the ω-6 linoleic acid (Figure 19) and the ω-3 linolenic acid. From these the body is able to synthesize longer and more unsaturated fatty acids. Green leaves are a good source of ω-3 fatty acids, whereas most seeds and vegetable oils are a good source of ω-6 fatty acids.

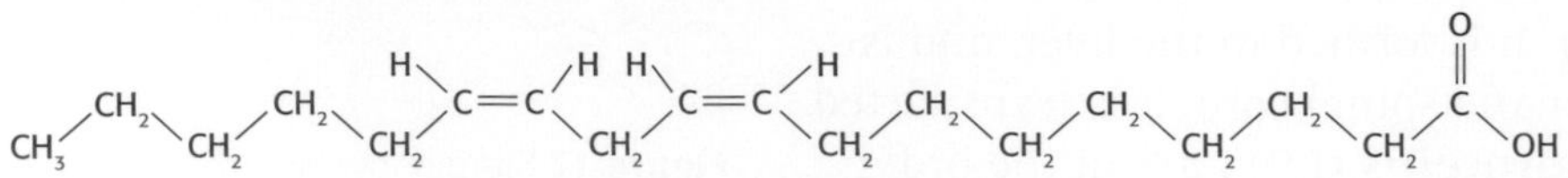

Figure 19 The structure of linoleic acid (*cis,cis*-9,12-octadecadienoic acid).

The double bonds in linoleic acid are on the ninth and twelfth carbon atoms of the acid, but it is known as an omega(ω)-6 fatty acid because the first double bond is on the sixth carbon atom from the end of the hydrocarbon chain (that is, counting from the other end). ω-3 fatty acids have the first double bond on the third carbon atom from the end of the hydrocarbon chain.

When fatty acids are made synthetically by partially hydrogenating other polyunsaturated fatty acids, then the *trans*-isomers may be formed (Figure 20). *Trans*-fatty acids are present in fried foods such as French fries and some margarines. They increase the formation of LDL cholesterol and thus increase the risk of heart disease.

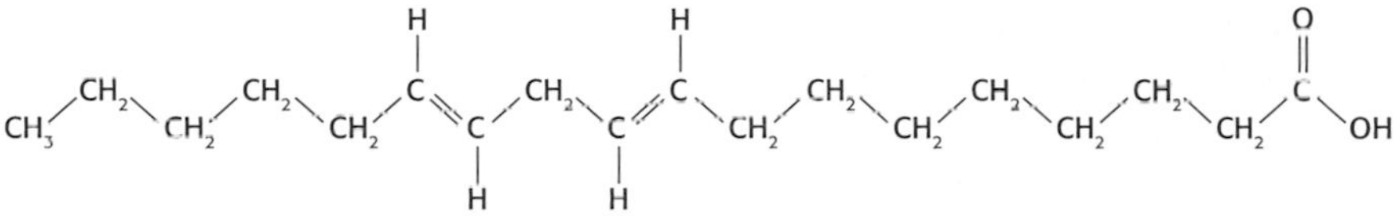

Figure 20 The structure of the *trans,trans-* form of linoleic acid.

The amount of unsaturation in an unsaturated fat or oil can be determined by titrating with iodine solution. One mole of iodine molecules will add quantitatively to 1 mol of carbon-to-carbon double bonds in the fat or oil. As the iodine is added, the purple colour of the iodine will disappear as the addition reaction takes place. The **iodine number** of a fat or oil is the number of grams of iodine that add to 100 g of the fat or oil.

Micro- and macro-nutrients

Good health requires a balanced diet. Micro-nutrients are required only in trace amounts (measured in mg or µg), which make up less than 0.005% of body mass. They include vitamins and minerals (Fe, Cu, Zn, I, Se, Mn, Mo, Cr, Co and B). Their main function is to act as a co-factor with enzymes. Macro-nutrients are required in larger quantities (> 0.005% body mass). These include carbohydrate (about 60%), protein (20–30%) and lipids (10–20%). They also include some minerals (Na, Mg, K, Ca, P, S and Cl).

Vitamins

Apart from vitamin D the body is not capable of synthesising vitamins, but it cannot function correctly without them, so they must be obtained from the diet. Vitamins can be classified as fat-soluble or water-soluble. The structure of fat-soluble vitamins is characterised by long, non-polar hydrocarbon chains or rings (Figure 21). They include vitamins A, D, E, F and K. They can accumulate in the fatty tissues of the body. The molecules of water-soluble vitamins, such as vitamin C (Figure 22) and the eight B-group vitamins, contain hydrogen attached directly to electronegative oxygen or nitrogen atoms that can hydrogen-bond with water molecules. They do not accumulate in the body, so a regular intake is required.

❓

1 Linolenic acid contains 18 carbon atoms. It has three double bonds, all in the *cis* position. Its IUPAC name is *cis,cis,cis*-9,12,15-octadecatrienoic acid. Draw the structure of linolenic acid.

2 The iodine number of a fatty acid (M_r = 278) is 274. Determine the number of carbon-to-carbon double bonds in the fatty acid.

3 Explain why eating large amounts of fatty food and taking little physical exercise may lead to obesity.

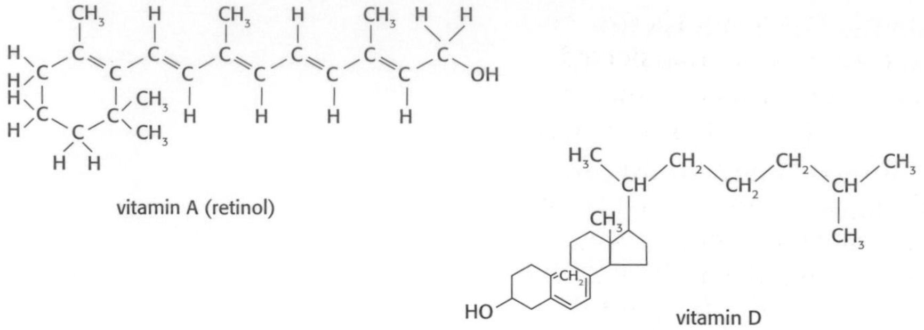

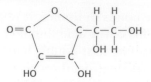

vitamin A (retinol)

vitamin D

Figure 21 Examples of fat-soluble vitamins.

Table 2 Sources of nutrients and associated deficiency diseases

Nutrient	Some sources	Deficiency disease
iron, Fe	red meat, fish, lentils, beans	anaemia
iodine, I	sea fish, shellfish, (enhanced) salt	goitre
vitamin A, retinol	cod liver oil, green vegetables, fruit	xerophthalmia (night blindness)
vitamin B$_3$, niacin	fish, chicken, brewer's yeast	pellagra
vitamin B$_1$, thiamin	brewer's yeast, oatmeal	beriberi
vitamin C, ascorbic acid	fruit and vegetables	scurvy
vitamin D, calciferol	fish oils, egg yolk and sunlight	rickets
protein	meat, fish, cheese, egg, nuts	marasmus and kwashiorkor

Hormones

Hormones are chemicals produced in endocrine glands and transported to the site of action by the bloodstream. The endocrine glands themselves are controlled mainly by the pituitary gland, which in turn is controlled by the hypothalamus in the brain. Hormones act as chemical messengers, and perform a variety of different functions. Examples of specific hormones include epinephrine (adrenaline), thyroxin, insulin and the sex hormones.

Epinephrine (adrenaline) is produced in the adrenal glands—two small organs located above the kidneys. It is a stimulant closely related to the amphetamine drugs (see Chapter 16). It is released in times of excitement; it causes a rapid dilation of the pupils and airways, and increases heartbeat and the rate of release of glucose into the bloodstream. It is sometimes known as the "fight or flight hormone".

Thyroxin (Figure 23) is produced in the thyroid gland located in the neck. It is unusual in that it contains iodine. A lack of iodine in the diet can cause the thyroid gland to swell to produce the condition known as goitre (Figure 24). Thyroxin regulates the body's metabolism. Low levels of thyroxin cause hypothyroidism, characterized by lethargy as well as sensitivity to cold and a dry skin. An overactive thyroid gland can cause the opposite effect. This is known as hyperthyroidism, with the symptoms of anxiety, weight loss, intolerance to heat and protruding eyes.

Human **insulin** is a protein containing 51 amino acid residues. It is formed in the pancreas—an organ located at the back of the

Figure 22 Vitamin C is water-soluble, because the many hydroxyl (–OH) groups in the small molecule can form hydrogen bonds with water.

Consider the problems of malnutrition that occur in your country. Include overnourishment as well as undernourishment. Which of the solutions listed below might be applicable to solve the problems?
A Providing food rations containing fresh foods rich in vitamins and minerals
B Adding nutrient supplements
C Genetic modification of food
D Adding nutrients directly to foodstuffs
E Educating the population in healthy eating
F State subsidies for healthy living
G Any other solutions

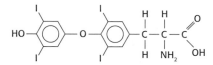

Figure 23 Thyroxin.

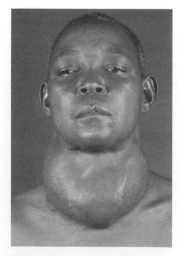

Figure 24 A patient suffering from goitre.

abdomen—and regulates blood sugar levels. In most diabetics the levels of insulin are low or absent, and glucose is not transferred sufficiently from the bloodstream to the tissues. This is known as **hyperglycemia** and results in thirst, weight loss, lethargy, coma and circulation problems. Long-term sufferers from diabetes can suffer blindness or kidney failure, and may need limbs amputated because of poor circulation. Diabetes is treated by reducing sugar intake and taking daily insulin injections. Too much insulin can cause **hypoglycemia**, where the blood sugar level falls, resulting in dizziness and fainting.

Sex hormones

The male sex hormones are produced in the testes, and comprise mainly testosterone and androsterone (Figure 25). They are anabolic—encouraging tissue, muscle and bone growth—and androgenic—conferring the male sexual characteristics. The female sex hormones are structurally very similar, with just small changes in the functional groups attached to the steroid framework (Figure 26). They are produced in the ovaries from puberty until the menopause. The two main female sex hormones are oestradiol and progesterone. They are responsible for sexual development, and for the menstrual and reproductive cycles in women.

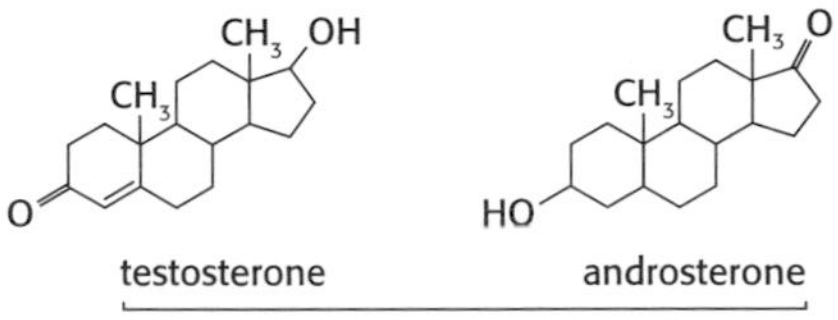

Figure 25 Male sex hormones. 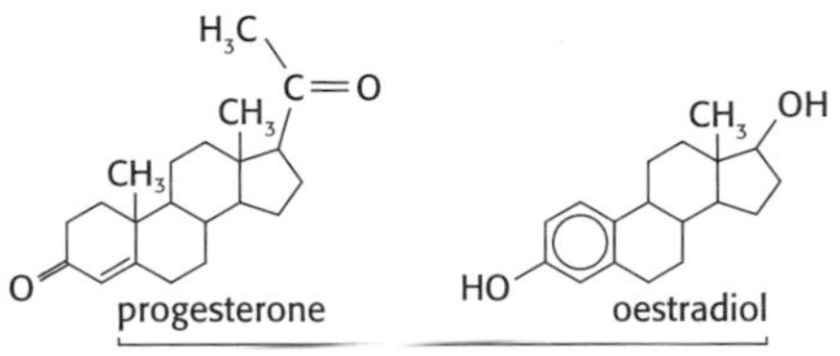   **Figure 26** Female sex hormones.

Oral contraceptives

At the beginning of the menstrual cycle the pituitary releases the follicle-stimulating hormone (FSH). FSH travels to the ovaries, causing the release of oestradiol, which prepares for the release of the ovum or egg and the build-up of the uterine wall. After about two weeks a negative feedback system stops the release of FSH and triggers the release of luteinizing hormone (LH). This travels to the ovaries and releases progesterone. The progesterone causes the egg to be transported to the uterus, as well as continuing to build up the uterine wall. If the egg is fertilized the egg embeds itself in the uterine wall and hormone levels rise dramatically; otherwise hormone levels fall, and menstruation begins.

The most common "pill" contains a mixture of oestradiol and progesterone, and mimics pregnancy by intentionally keeping the hormones at a high level so that no more eggs are released. It is usual to take the pill for 21 days and then a placebo for 7 days so that a mild period will result, but without the risk that the hormone levels will fall and allow the unexpected release of an egg.

Oestradiol and progesterone may also be given to post-menopausal women as hormone replacement therapy (HRT), partly to prevent brittle bone disease (osteoporosis).

1 Compare the structures of testosterone and progesterone. Identify the steroid backbone present in both compounds and two other functional groups that are common to both. Identify one functional group that is present in testosterone but absent in progesterone.

2 Some athletes cheat by taking anabolic steroids. Find out the structures of some anabolic steroids, and explain why they are used both legally and illegally. How can they be detected in the urine or blood?

3 Guidelines for healthy eating include:
- Eat a variety of foods
- Maintain a healthy weight
- Eat a diet low in fat, saturated fat and cholesterol
- Include plenty of fresh fruit and vegetables
- Use salt and sugar sparingly
- Moderate the intake of alcohol

Explain the chemistry behind why each of these bullet points is important.

HL ⋮ Enzymes

Enzymes are protein molecules that catalyse biological reactions. Each enzyme is highly specific for a particular reaction, and extremely efficient, often being able to increase the rate of reaction by more than one hundred million times. Like all catalysts, enzymes work by providing an alternative pathway for the reaction with a lower activation energy, by creating an environment in which the transition state is stabilized.

The specificity of enzymes depends on their particular shape. This is determined by their secondary, tertiary and quaternary structure. The part of an enzyme that reacts with the substrate is known as the **active site**. This is the part of the enzyme where the substrate will bind and undergo chemical reaction. The active site is not necessarily rigid, but can alter its shape to allow for a better fit—known as the **induced fit theory**.

Enzyme kinetics

At low substrate concentrations the rate of reaction is proportional to the concentration of the substrate. However, at higher concentrations the rate reaches a maximum, known as V_{max}. This can be explained in terms of enzyme saturation. At low substrate concentrations there are enough active sites present for the substrate to bind to and react. Once all the sites are used up the enzyme cannot work any faster.

V_{max} was first identified by a German-born American biochemist, Leonor Michaelis (1875–1949), and a Canadian doctor, Maud Menten (1879–1960). They also identified a constant K_m, known as the Michaelis–Menten constant, which is the substrate concentration when the rate of the reaction is $\frac{1}{2}V_{max}$. K_m for a particular enzyme with a particular substrate will always be the same. It indicates whether the enzyme functions efficiently at low substrate concentrations or whether high substrate concentrations are necessary for the reaction to be catalysed efficiently.

Figure 27 Maud Menten (1879–1960). Born in Ontario, Maud Menten was one of the first Canadian women to gain a medical doctorate, and went on to become Professor of Medicine at Pittsburgh University. As well as being the co-discoverer of the Michaelis–Menten equation in 1913 she is also credited with being the first to separate proteins by electrophoresis. She was an accomplished painter, and held several exhibitions of her work.

Inhibition of enzymes

Inhibitors are substances that decrease the rate of enzyme-catalysed reactions. **Competitive inhibitors** resemble the substrate in shape but cannot react. They slow down the reaction because they can occupy the active site on the enzyme, thus making it less accessible to the substrate. **Non-competitive inhibitors** also bind to the enzyme, but not on the active site. This is thought to cause the enzyme to change its shape so that the substrate cannot bind. As the substrate concentration is increased, the effect of competitive inhibitors will be diminished, because there is increased competition for the active sites by the substrates. With non-competitive inhibitors increasing, the substrate concentration will not increase the effectiveness of the enzyme, because the enzyme's shape still remains altered by the non-competitive inhibitor. See Figure 28.

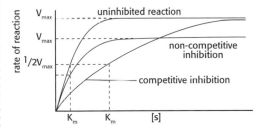

Figure 28 Effect of substrate concentration on enzyme inhibitors. For non-competitive inhibitors V_{max} is lower, but K_m is the same. For competitive inhibitors V_{max} is the same, but K_m is increased.

Factors affecting enzyme activity

Several factors can affect the efficient functioning of enzymes. The catalytic action of an enzyme clearly depends on its specific shape. Increasing the **temperature** will initially increase the rate of enzyme-catalysed reactions, because more of the reactants will possess the minimum activation energy (Figure 29). The optimum temperature for most enzymes is about 40 °C. Above this temperature enzymes rapidly become denatured as the weak bonds holding the tertiary structure together become broken. High and low **pH** can also affect enzymes in a similar way (Figure 30). At different pH values the charges on the amino acid residues change, affecting the bonds between them and so disrupting the specific tertiary structure and making the enzyme ineffective. **Heavy metals** can poison enzymes by reacting irreversibly with –SH groups, replacing the hydrogen atom with a heavy metal atom or ion so that the tertiary structure is permanently altered.

Nucleic acids

Nucleic acids are natural polymers with relative molecular masses of up to several million, made up of nucleotides. All cells in the human body, with the exception of red blood cells, contain **DNA** (deoxyribosenucleic acid). A nucleotide of DNA contains deoxyribose (a pentose sugar; Figure 31), which is covalently bonded to a phosphate group and one of four nitrogen-containing bases: adenine, guanine, cytosine or thymine (Figure 32). Nucleotides of ribosenucleic acid, **RNA**, contain a different sugar, ribose, but also

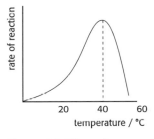

Figure 29 Effect of temperature on enzyme activity.

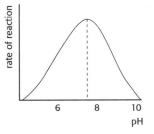

Figure 30 Effect of pH on enzyme activity.

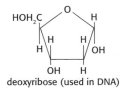

deoxyribose (used in DNA)

ribose (used in RNA)

Figure 31 The different sugars in DNA and RNA. The difference is that deoxyribose lacks an oxygen atom on the second carbon atom.

contain a phosphate group and four nitrogen-containing bases. Three of the bases are the same as those in DNA, but the fourth, uracil, replaces thymine.

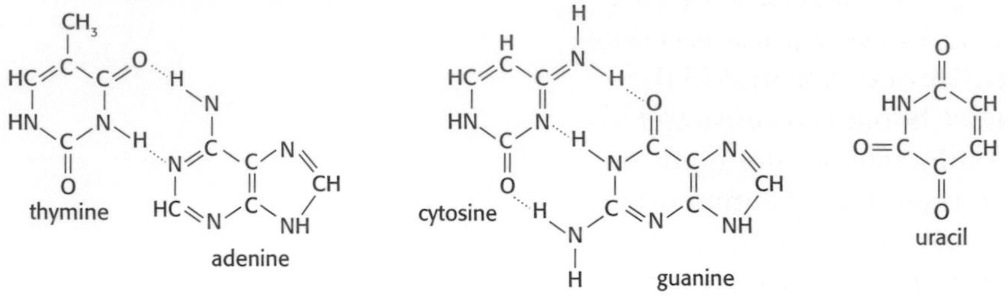

Figure 32 Hydrogen bonding between the nitrogen-containing bases in DNA and RNA.

The nucleotides are condensed together by a phosphodiester bond between the third carbon atom in the sugar and the neighbouring phosphate group to form a polynucleotide (Figure 33). These form the backbone of the nucleic acid, with a repeating pattern of –sugar–phosphate–sugar–phosphate–. The nitrogenous bases are attached to the first carbon atom in the sugar of the backbone.

The **double helix** structure of DNA (Figure 34) was discovered by the American biologist James Watson (1928–) and the British physicist Francis Crick (1916–2004), working together at Cambridge University in the early 1950s. They realized that the backbone was wound into helices, and that the two helices were held together by hydrogen bonds between the bases. The hydrogen bonds are very specific. Cytosine can only hydrogen-bond with guanine and adenine can only hydrogen-bond with thymine (uracil in RNA).

> **From the molecular to the astronomical: a double helix in the stars**
>
> In 2006 astronomers reported the discovery of a highly ordered double helix nebula, possibly made up of spiral galaxies of stars near to the centre of the Milky Way galaxy some 25,000 light-years away from Earth.

The genetic code

When cells divide, the genetic information has to be replicated intact. The genetic information is stored in **chromosomes**, found inside the nucleus. In humans there are 23 pairs of chromosomes. A chromosome is effectively a very long DNA sequence. The DNA in the cell starts to partly unzip as hydrogen bonds between the bases become broken. Sugar base units will be picked up from a pool of nucleotides to form a complementary new strand. Because adenosine can only hydrogen-bond with thymine (A–T) and cytosine can only hydrogen-bond with guanine (C–G), the new strand formed will be identical to the original (Figure 35).

DNA resides mainly in the nucleus, whereas protein synthesis takes places in the cytoplasm part of a cell. The information required to make proteins is transferred from DNA to RNA by a similar enzyme-controlled unzipping process, except that the new strand of RNA contains a different sugar, and uracil in place of thymine.

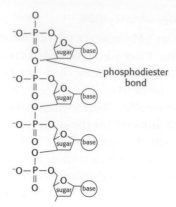

Figure 33 Part of a polynucleotide.

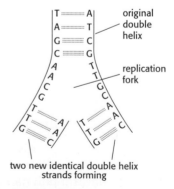

Figure 34 The double helix structure of DNA. Note the hydrogen bonds between the two different strands of polynucleotides.

Figure 35 Replication of DNA.

The coded information held by the DNA lies in the sequence of bases. Each sequence of three bases represents one amino acid, and is known as the **triplet code**. The triplet code allows for up to 64 permutations, known as **codons**. This is more than sufficient to represent the 20 naturally occurring amino acids, and several different codons may represent the same amino acid. In 2000 the **human genome**—the complete sequence of bases in human DNA—was finally determined and published on the Internet.

DNA profiling

Some unsolved crimes that were committed more than 30 years ago are now being reopened, and sometimes solved, thanks to the advances in DNA "fingerprinting". The DNA in traces of blood, semen, hair or saliva is extracted and broken down using restriction enzymes into smaller fragments known as **minisatellites**. Splits occur in the minisatellites in the regions where there are no codons. These regions are unique to each person. The fragments are separated into bands using gel electrophoresis. By labelling with radioactive phosphorus, and then using X-ray film to detect the radiation, a characteristic "fingerprint" of all the fragments is obtained. This is used not only in criminal cases but also to confirm paternity, and to map the evolutionary tree of species.

Respiration

In the body 1 mol of glucose is broken down in a process known as **glycolysis** to 2 mol of pyruvate (2-oxopropanoate). Glycolysis produces energy by storing it in molecules of adenosine triphosphate, ATP. The pyruvate may then break down aerobically or anaerobically. In aerobic decomposition glucose is oxidized by molecular oxygen (which is reduced) to form carbon dioxide and water and release energy. It can be summarized by the following two reactions:

$$C_6H_{12}O_6 + O_2 \rightarrow 2\begin{array}{c}O=C-OH\\|\\O=C-CH_3\end{array} + 2H_2O$$

pyruvate
(pyruvate is the anion of the acid)

then

$$2C_3H_4O_3 + 5O_2 \rightarrow 6CO_2 + 4H_2O$$

overall:

$$C_6H_{12}O_6 + 6O_2 \rightarrow 6CO_2 + 6H_2O$$

Theoretically 1 mol of glucose can produce 36 mol of ATP by aerobic decomposition, but only 2 mol of ATP are formed during anaerobic decomposition (in the absence of oxygen) when the product formed in humans is lactic acid (2-hydroxypropanoic acid), $CH_3CH(OH)COOH$. However, ATP is created more quickly during anaerobic decomposition, so during short bursts of strenuous activity muscle cells may also use anaerobic respiration, which accounts for the build-up of lactic acid in the muscles. In yeast the pyruvate is decomposed not to lactic acid but to ethanol and carbon dioxide. This is the basis of fermentation:

$$C_6H_{12}O_6 \rightarrow 2C_2H_5OH + 2CO_2$$

The ability of transition metals such as iron and copper to form complexes in the form of cytochromes and to show variable oxidation states is important in the electron transfer process that takes place during the enzyme-catalysed oxidation of organic molecules in the body.

Oxidation step:

$$C_6H_{12}O_6 + 6H_2O \rightarrow 6CO_2 + 24H^+ + 24e^-$$
$$Fe^{3+} + e^- \rightarrow Fe^{2+} \text{ (or } Cu^{2+} + e^- \rightarrow Cu^+)$$

Reduction step:

$$O_2 + 4H^+ + 4e^- \rightarrow 2H_2O$$
$$Fe^{2+} \rightarrow Fe^{3+} + e^- \text{ (or } Cu^+ \rightarrow Cu^{2+} + e^-)$$

Iron, in the form of haem, is also responsible for carrying oxygen in the blood during respiration. Haemoglobin contains four large polypeptide groups and four iron(II) ions, each surrounded by a flat porphyrin ligand known as a **haem** group. Haem is a prosthetic group, a group essential for the protein to be able to carry out its function. At high oxygen concentrations the oxygen forms a coordinate bond, with the iron in the haem group functioning as an extra ligand. At low concentrations the reverse process occurs. Carbon monoxide molecules and cyanide ions are poisonous, because they form very stable complex ions with the iron, preventing it from carrying oxygen.

1 Discuss the difference in the catalytic ability of inorganic catalysts, such as nickel, and biological catalysts (enzymes).

2 Find the structure of cytochrome oxidase and the structure of the haem group present in haemoglobin. Describe the similarities and the differences between the two structures.

3 Should the DNA profile of all citizens be held on a central record to aid the police in identifying criminals, or is this a breach of civil rights?

16 Option C: chemistry in industry and technology

Iron, steel and aluminium

Iron and aluminium are both quite reactive metals, and lie above hydrogen in the reactivity series. Neither of them is found uncombined in nature. The common ores of iron are the oxides and sulfides, and they can be reduced to iron metal in a blast furnace using chemical reducing agents such as carbon, carbon monoxide and hydrogen. Aluminium is found mainly as its hydroxide, $Al(OH)_3$ and oxide, Al_2O_3. Aluminium is higher in the reactivity series than iron, and its ores are reduced by electrolysis.

The blast furnace

A modern blast furnace (Figure 1), which uses carbon monoxide, carbon and hydrogen as the reducing agents, is capable of producing 10,000 tonnes of molten iron per day. Most of the molten "pig" iron produced is converted directly into steel, but some is cooled to make cast iron goods such as engine cylinder blocks.

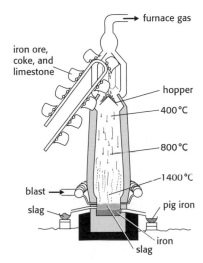

Figure 1 The blast furnace.

Raw materials

The solid charge is fed through the hopper via a conveyor belt. It consists of:

- iron ore: mainly haematite (Fe_2O_3), magnetite (Fe_3O_4) or hydrated oxides (e.g. goethite, FeOH.OH, and limonite, Fe_2O_3.H_2O), and iron oxides obtained by roasting iron sulfides (e.g. iron pyrites, FeS_2) or scrap (recycled) iron
- coke, made by heating coal in the absence of air
- limestone ($CaCO_3$) to dissolve and chemically remove high melting point impurities by forming a "slag"
- preheated air blown in through nozzles known as *tuyères* at the bottom of the furnace. This air is enriched with oxygen and may also include hydrocarbons such as oil or natural gas to replace up to 40% of the coke.

Reactions

The coke burns to form carbon monoxide:

$$2C(s) + O_2(g) \rightarrow 2CO(g)$$

In the reducing conditions, incomplete combustion of the added hydrocarbons occurs. For example:

$$CH_4(g) + \tfrac{1}{2}O_2 \rightarrow CO(g) + 2H_2(g)$$

The reducing gases pass up the furnace, where they reduce the iron oxides in a series of stages, depending on the temperature and composition of the gas. Examples of overall reactions taking place include:

$$Fe_2O_3(s) + 3CO(g) \rightarrow 2Fe(l) + 3CO_2(g)$$

$$Fe_3O_4(s) + 4H_2(g) \rightarrow 3Fe(l) + 4H_2O(g)$$
$$FeO(s) + CO(g) \rightarrow Fe(l) + CO_2(g)$$

In addition, coke itself can directly reduce iron ore in the hottest regions. For example:

$$Fe_2O_3(s) + 3C(s) \rightarrow 2Fe(l) + 3CO(g)$$

The partially oxidized gases (furnace gas) that emerge from the top of the furnace are used as a fuel to preheat the air blasted in through the *tuyères*.

At high temperatures the limestone decomposes.

$$CaCO_3(s) \rightarrow CaO(s) + CO_2(g)$$

The carbon dioxide reacts with coke to produce carbon monoxide, and the coke can also react with water from hydrocarbons to give more carbon monoxide and hydrogen:

$$CO_2(g) + C(s) \rightarrow 2CO(g) \text{ and } H_2O(g) + C(s) \rightarrow H_2(g) + CO(g)$$

The calcium oxide reacts with high melting point impurities to form a complex aluminosilicate "slag" that also contains most of the silicon impurities. For example:

$$CaO(s) + SiO_2(s) \rightarrow CaSiO_3(l)$$

Products

At the very high temperatures at the bottom of the furnace the molten iron and liquid slag separate into two layers, with the less dense slag on top. Both are tapped off as more raw materials are added to the furnace in a continuous process. The molten iron (known as **pig iron**) contains phosphorus and sulfur, together with small amounts of other elements such as manganese and silicon, and about 4–5% carbon. The slag is used for road-making, or is treated to make by-products such as cement and thermal insulation.

Steel

Molten iron from the blast furnace is added to a vessel known as an **oxygen converter** (Figure 2). Preheated oxygen is injected at high pressure into the vessel, and the impurities are oxidized. For example:

$$C + O_2 \rightarrow 2CO_2$$
$$4P + 5O_2 \rightarrow P_4O_{10}$$
$$Si + O_2 \rightarrow SiO_2$$

The silicon and phosphorus oxides then combine with lime added to the converter to form a slag of calcium phosphate, $Ca_3(PO_4)_2$, and calcium silicate, $CaSiO_3$.

As the reactions are highly exothermic, the temperature is controlled by adding scrap steel. The dissolved oxygen in the steel must be removed by adding controlled amounts of aluminium or silicon before the steel is suitable for casting or rolling. During this process other elements such as chromium and nickel are also added to form the precise alloy required.

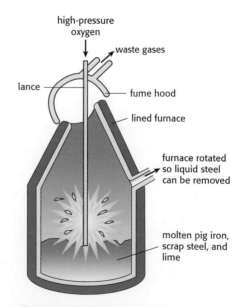

Figure 2 The basic oxygen converter.

Properties and uses of steel alloys

An alloy is a homogeneous mixture of metals, or a mixture of metals and a non-metal. Transition metals often form alloys with each other, because their atoms have similar atomic radii and the crystal structure is not seriously disrupted. Steel is an alloy of iron, carbon and other metallic and non-metallic elements. It has a wide range of uses, and by adjusting its composition it can be tailor-made with specific properties. For example, chromium increases the resistance of steel to corrosion. Stainless steel used for kitchen knives, sinks etc. contains about 18% chromium and 8% nickel. Toughened steel for use in drill bits, which need to retain a sharp cutting edge at high temperatures, contains up to 20% molybdenum.

The mechanical properties of steel can also be affected by heat treatment. This is the controlled heating and cooling of the steel, which changes the nature of the crystals in the steel but does not alter its shape. Steel can be made less brittle by **tempering**. This involves heating it to about 400–600 °C and letting it cool slowly. To make the steel more ductile a process known as **annealing** is used. This involves heating the steel to a higher temperature (about 1040 °C) followed by slow cooling. If hard steel is required, then after annealing the steel must be **quenched** by rapid cooling.

Production of aluminium

The uses of aluminium, the environmental impact of aluminium production and the need for recycling have already been discussed fully in Chapter 9. The worldwide electrolytic production of aluminium currently amounts to more than 20 million tonnes. The main ore of aluminium, bauxite, is mainly in the form of the hydroxide, $Al(OH)_3$, and the principal impurities are iron(III) oxide and titanium oxide. The impurities are removed by heating powdered bauxite with concentrated sodium hydroxide solution. The aluminium hydroxide reacts and dissolves because it is amphoteric.

$$Al(OH)_3(s) + NaOH(aq) \rightarrow NaAlO_2(aq) + 2H_2O(l)$$

The aluminate solution is filtered, leaving the impurities behind. Seeding with pure aluminium hydroxide then reverses the reaction. The pure recrystallized aluminium hydroxide is then heated to produce aluminium oxide (alumina):

$$2Al(OH)_3(s) \rightarrow Al_2O_3(s) + 3H_2O(l)$$

In a separate process hydrogen fluoride is added to the aluminate solution, followed by sodium carbonate to precipitate sodium hexafluoroaluminate(III), Na_3AlF_6, (known as **cryolite**):

$$NaAlO_2(aq) + 6HF(g) + Na_2CO_3(aq) \rightarrow$$
$$Na_3AlF_6(s) + 3H_2O(l) + CO_2(g)$$

The electrolysis of molten alumina takes places in an open-topped steel container lined with graphite. Alumina has a melting point of 2045 °C, so it is dissolved in cryolite to produces a solution with a melting point of about 950 °C so that much less electrical energy is required. The aluminium is produced on the graphite lining, which acts as the negative electrode (cathode). Molten aluminium is more dense than cryolite, so it collects at the bottom of the cell, where it can be tapped off periodically.

$$Al^{3+}(l) + 3e^- \rightarrow Al(l)$$

The positive electrode is made of blocks of graphite. As the oxide ions are oxidized, some of the oxygen formed reacts with the graphite blocks to form carbon dioxide, so that they have to be renewed regularly.

$$2O^{2-}(l) \rightarrow O_2(g) + 4e^-$$
$$C(s) + O_2(g) \rightarrow CO_2(g)$$

A single modern cell can produce up to 2 tonnes of aluminium per day.

Figure 3 Rows of electrolytic cells in an aluminium plant in Bahrain (photo by Geoff Neuss)

The oil industry

It is probable that oil production is currently at, or very near its peak. In Chapter 5 we discussed the problems of global warming caused by the combustion of fossil fuels, and in Chapter 10 we looked at the cost of crude oil, and discussed whether it is wise to use it as an energy source rather than as a chemical feedstock. We have seen that it is an important feedstock, not only for alkanes but also for alkenes, which can be obtained from cracking. There are several types of cracking, each giving different products:

Figure 4 A group of IB teachers visiting an aluminium plant in Bahrain (photo by Geoff Neuss)

- **Steam cracking**: The feedstock is preheated, vaporized and mixed with steam and then converted at 1250–1400 °C to give high yields of low molecular mass alkenes.
- **Catalytic cracking**: The use of a silica/alumina catalyst enables the cracking to take place at the relatively lower temperature of about 1000 °C.
- **Hydrocracking**: The feedstock is mixed with hydrogen at a pressure of about 80 atm and cracked over a platinum or silica/alumina catalyst. This gives a high yield of branched alkanes, cyclic alkanes and aromatic compounds for use in "green" unleaded gasoline (petrol).

Addition polymers

The main use of the alkenes obtained from the cracking of crude oil fractions is to make addition polymers such as poly(ethene) and poly(phenylethene), as described in Chapter 10. Modern society relies heavily on many different types of addition polymer. Their properties depend not only upon which functional groups are attached to the double bond in the monomers, but also on the amount of branching, and the way in which the side groups are arranged in the polymers. The properties can be further modified by using plasticizers, and by the injection of volatile hydrocarbons during their production.

Branching

Depending on the reaction conditions ethene can form high-density poly(ethene), HDPE, and low-density poly(ethene), LDPE. In high-density poly(ethene) there is little branching. This gives long chains that can fit together closely, making the polymer stronger, denser and more rigid than low-density poly(ethene). The presence of side chains in low-density poly(ethene) results in a more resilient and flexible structure, making it ideal for the production of film products such as food wrappings.

Orientation of side groups

In poly(propene) the methyl groups can all have the same orientation along the polymer chain: such polymers are said to be **isotactic**. Because of their regular structure, isotactic polymers are more crystalline and tough. They can be moulded into objects such as car fenders, and drawn into fibres for clothes and carpets. In **atactic** poly(propene) the chains are more loosely held, so the polymer is soft and flexible, making it suitable for sealants and roofing materials. See Figure 5.

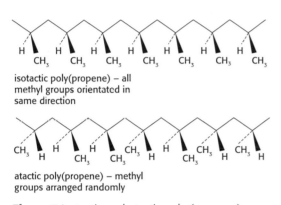

isotactic poly(propene) – all methyl groups orientated in same direction

atactic poly(propene) – methyl groups arranged randomly

Figure 5 Isotactic and atactic poly (propene).

Plasticizers

Plasticizers are small molecules that can fit between the long polymer chains. They act as lubricants and weaken the attraction between the chains, making the plastic more flexible. By varying the amount of plasticizer added, poly(chloroethene), PVC, can form a complete range of polymers whose properties vary from rigid to fully pliable.

Volatile hydrocarbons

If pentane is added during the formation of poly(phenylethene), also known as polystyrene, and the product is heated in steam, the pentane vaporizes, producing expanded polystyrene. This light material is a good thermal insulator, and is also used as packaging because it has good shock-absorbing properties.

Advantages and disadvantages of polymer use

The examples above illustrate how polymers can be tailor-made to perform a variety of functions, based on properties such as strength, density, thermal and electrical insulation, flexibility and lack of reactivity. There are, however, some disadvantages.

● **Depletion of natural resources**. The majority of polymers are carbon based. Currently crude oil is the major source of carbon

Figure 6 This sculpture, entitled *Embankment*, was made out of expanded polystyrene by the British artist Rachel Whiteread (1963–) and exhibited at the Tate Modern in London in 2005–2006.

compounds, although in the past it was coal. Both are fossil fuels and in limited supply.

- **Disposal**. Because of their lack of reactivity, plastics are not easily disposed of. Some, particularly PVC and poly(propene), can be recycled, and others (nylon, for example) are weakened and eventually decomposed by ultraviolet light. Plastics can be burned, but if the temperature is not high enough extremely poisonous dioxins can be produced, along with toxic gases such as hydrogen cyanide and hydrogen chloride.
- **Biodegradability**. Most plastics do not occur naturally and are not degraded by microorganisms. By incorporating natural polymers such as starch into plastics they can be made more biodegradable. However, in the anaerobic conditions present in landfills biodegradation is very slow or will not occur at all.

Catalysts

Catalysts are used in a wide variety of industrial processes. In Chapter 7 we have already discussed the use of iron in the Haber process, and vanadium(V) oxide in the contact process. Many catalysts are either transition metals or their compounds, as they are able to show variable oxidation states, although many other catalysts also exist, such as aluminium oxide, Al_2O_3, and sulfuric acid. As we have seen in Chapter 6, catalysts work by providing an alternative reaction pathway with a lower activation energy than the uncatalysed pathway. Catalysts may be **homogeneous** (the catalyst is in the same phase as the reactants and products) or **heterogeneous**, where the catalyst is in a different phase from the reactants and/or products. The mechanisms for both types of catalysis, which involve either adsorption onto the catalytic surface or the formation of an intermediate compound, have been discussed fully in the AHL section on the catalytic properties of the d-block elements in Chapter 3.

In industry catalysts are chosen to maximize both yields and profits. The advantage of heterogeneous catalysts is that they are easy to remove from the products. They also tend to be less specific than homogeneous catalysts, and so can be used for other reactions. However, because it is only the surface area that is active, they depend very much on particle size, and the smaller particles may aggregate together during the reaction, so making the catalyst less effective. Homogeneous catalysts tend to be more specific for particular reactions, and are often harder to separate from the products. However, they can be more efficient, because all the catalyst, not just the surface, is exposed to the reactants. The choice of catalyst will thus depend upon selectivity: Does the catalyst give a high yield of the desired product? What is its efficiency, its environmental impact, and its ability to work under a range of conditions of temperature and pressure? Many catalysts can be rendered useless if they combine irreversibly with another substance so that the active sites become blocked. Such substances are known as **poisons**. These include sulfur (which is why it must be removed from crude oil before refining takes place), carbon monoxide, and cyanide ions.

Rechargeable batteries and fuel cells

A battery is a general term for a voltaic or electrochemical cell in which chemical energy is converted into electrical energy. The electrons transferred in the spontaneous redox reaction taking place in the voltaic cell produce the electricity. Batteries are a useful way to store and transport relatively small amounts of energy. Some batteries (primary cells) can be used only once, whereas secondary cells can be recharged.

Lead–acid battery

The lead–acid battery (Figure 7) is used in automobiles, and is an example of a secondary cell. Usually it consists of six cells connected in series, producing a total voltage of 12 V. The electrolyte is a solution of sulfuric acid. The negative electrodes are made of lead and the positive electrodes are made of lead(IV) oxide.

Oxidation (− electrode):
$$Pb + SO_4^{2-} \rightarrow PbSO_4 + 2e^-$$
Reduction (+ electrode):
$$PbO_2 + 4H^+ + SO_4^{2-} + 2e^- \rightarrow PbSO_4 + 2H_2O$$

The overall reaction taking place is thus

$$Pb + PbO_2 + 4H^+ + 2SO_4^{2-} \rightarrow 2PbSO_4 + 2H_2O$$

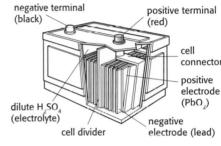

Figure 7 The lead–acid battery.

The reverse reaction takes place during charging. This can be done using a battery charger or through the alternator as the automobile is being driven. As sulfuric acid is used up during discharging, the density of the electrolyte can be measured using a hydrometer to give an indication of the state of the battery. The disadvantages of lead–acid batteries are that they are heavy, and that both lead and sulfuric acid are potentially polluting.

Nickel–cadmium (NiCd) and lithium-ion batteries

Rechargeable nickel–cadmium batteries are used in electronics and toys. They have a cell potential of 1.2 V. The positive electrode is made of nickel hydroxide, which is separated from the negative electrode made of cadmium hydroxide. The electrolyte is potassium hydroxide. During discharge the following reaction occurs:

$$2NiO(OH) + Cd + 2H_2O \rightarrow 2Ni(OH)_2 + Cd(OH)_2$$

This process is reversed during charging. One of the disadvantages of NiCd batteries is that cadmium is an extremely toxic heavy metal, so the batteries need to be disposed of responsibly.

Laptops, cell phones and other handheld devices often use lithium-ion batteries. These contain lithium atoms complexed to other ions, e.g. Li_xCoO_2, and it is these ions rather than lithium itself that undergo the redox reactions. They are much lighter than NiCd batteries, and produce a higher voltage, 3.6 V, but they do not have such a long lifespan.

Fuel cells

A fuel cell (Figure 8) utilizes the reaction between oxygen and hydrogen to produce water. Unlike combustion, the energy is given out not as heat but as electricity. As reactants are used up more are added, so a fuel cell can give a continuous supply of electricity. The electrolyte is aqueous sodium hydroxide. It is contained within the cell using porous electrodes, which allow the passage of water, hydrogen and oxygen.

Oxidation ($-$ electrode):
$$H_2 + 2OH^- \rightarrow 2H_2O + 2e^-$$
Reduction ($+$ electrode):
$$O_2 + 2H_2O + 4e^- \rightarrow 4OH^-$$

The advantages of a fuel cell are that they do not need recharging nor do they pollute, as water is the only product. However, they are very expensive to produce.

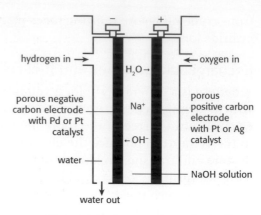

Figure 8 Hydrogen–oxygen fuel cell.

Liquid crystals (1)

Liquid crystals are a phase or state of matter that lies between the solid and liquid states. The molecules in a liquid crystal tend to retain their orientation as in a solid, but can also move to different positions as in a liquid. The physical properties of liquid crystals (such as electrical conductivity, optical activity and elasticity) are dependent on the orientation of the molecules relative to some fixed axis in the material. Examples of substances that can behave as liquid crystals under certain conditions include DNA, soap solution, graphite and cellulose, together with some more specialized substances such as biphenyl nitriles.

Liquid crystals are sensitive to temperature and concentration, and show liquid crystal properties only under certain conditions. Essentially they can be divided into two types. **Thermotropic** liquid crystals are pure substances that show liquid crystal behaviour over a range of temperature between the solid and liquid states, whereas **lyotropic** liquid crystals are solutions that show the liquid crystal phase at certain concentrations. Examples of a thermotropic liquid crystal are biphenyl nitriles used in liquid crystal displays (LCDs) (Figure 9), and examples of lyotropic liquid crystals are soap solutions. At low dilution the polar soap molecules behave randomly, but at higher concentrations they group together into larger units called **micelles**, which in the liquid crystal phase are ordered in their orientation.

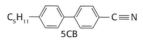

Figure 9 Liquid crystal displays (LCDs) use molecules such as biphenyl nitriles. One such example is 4-pentyl-4'-cyanobiphenyl, known as 5CB.

Within the liquid crystal phase the rod-shaped molecules, which are typically about 2.5×10^{-9} m long, exist in groups or domains. The molecules can flow, and are randomly distributed as in a liquid, but within each domain they all point in the same direction. This is known as the **nematic phase**. As the temperature increases, the orientation becomes increasingly more disrupted, until eventually the directional order is lost and the normal liquid phase is formed (Figure 10).

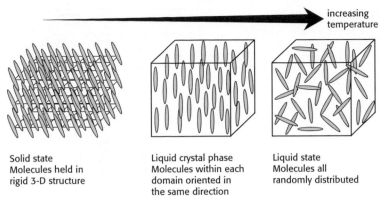

increasing temperature

Solid state
Molecules held in
rigid 3-D structure

Liquid crystal phase
Molecules within each
domain oriented in
the same direction

Liquid state
Molecules all
randomly distributed

Figure 10 The effect of temperature on liquid crystals.

The orientation of the liquid crystal molecules can be controlled by an applied electric field because of the polarity of the molecules. The orientation of the molecules affects the ability of the liquid crystal molecules to transmit light. In liquid crystal displays used in digital watches, calculators and laptops a small voltage is applied across a thin film of the material. This controls the areas of the display that are light and dark, and hence gives the characteristic reading of the time or letters. The great advantage of LCDs over other types of electronic display is that they use extremely small electric currents. The disadvantage is that they work only within a certain temperature range, which explains why a digital watch or laptop screen may give a strange display in very hot or cold temperatures. Hence for use in an LCD a liquid crystal should:

- be a chemically stable compound
- contain polar molecules
- remain stable in the liquid crystal phase over a suitable range of temperature
- be able to orientate quickly (rapid switching speed).

Figure 11 Using liquid crystals, the Light Rail Transit train in Singapore is able to fog out the windows when the train passes by residential flats, to protect the privacy of the flats' occupants.

Nanotechnology

In Chapter 4 we discussed the existence of fullerenes and nanotubes, and the way in which they can be "seen" using scanning tunnelling microscopy. Nanotechnology has been defined as the research and technology of compounds within the range of 1 to 200 hundred nanometres (1.0×10^{-9} m to 1.0×10^{-7} m): that is, on the atomic scale. It creates and uses structures that have novel properties based on their small size.

Nanotechnology covers many separate scientific disciplines. At one extreme it can be seen as an extension of the existing sciences to the nano scale; at the other extreme it can be seen as a completely new discipline of science and technology. There are two main approaches. The **bottom-up** approach involves building materials and devices from individual atoms, molecules or components. The **top-down** approach involves constructing nano-objects from larger entities. Sometimes physical techniques are used that allow atoms to be manipulated and positioned to specific requirements. For example, the image of IBM shown in Chapter 2 was produced using a process known as **dip pen nanolithography**, in which atoms can be deposited in specific positions using an atomic force microscope. It is also possible to use chemical reactions, such as in **DNA nanotechnology**, where the specific base-pairing due to hydrogen bonding can be utilized to build desired molecules and structures.

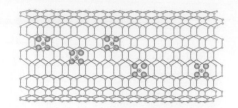

Nanotubes

Much of the excitement in nanochemistry has been in the manufacture and characterization of nanotubes. These are essentially made from just carbon atoms arranged into hexagonal rings with pentagonal rings to close the structure at the ends. Either single or multi-walled tubes made from concentric nanotubes have been formed. An interesting aspect of these tubes is that they have a greatly increased ratio of surface area to volume, which opens up the possibility of their acting as extremely efficient and highly selective catalysts—particularly as a wide variety of different materials, including elements, metal oxides and even small proteins, have been inserted inside the tubes (Figure 12). They also have huge tensile strength: it is estimated that bundles of nanotubes should be in the region of 50 times stronger than steel. Because of their small size the quantum effect is appreciable, which results in an altering of their electronic and optical properties.

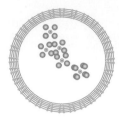

Figure 12 Computer simulations of methane molecules inside a nanotube. (By Tali Mutat of the Computational Physics Group at the Israel Institute of Technology)

Research is currently under way to use carbon nanotubes to create transistors and other electronic devices much smaller than can be created using silicon chips. By inserting silver halides into single-walled nanotubes and then decomposing them to silver, metallic nanowires of pure silver with a diameter of 2.0×10^{-8} m have been made, which are the thinnest electrical "wires" in existence. In 2004, scientists at Los Alamos National Laboratory, USA, succeeded in growing single-walled carbon nanotubes 4 cm long by decomposing ethanol over an iron catalyst.

Implications of nanotechnology

Nanotechnology has the potential to solve many problems, such as increasing food production, preventing, monitoring and curing disease, and improving information and communication technology, although most of these benefits still probably lie somewhere far in the future. The 2004 United Nations Task Force on Science, Technology and Innovation noted that some of the advantages of nanotechnology include: production using little manual labour, land, or maintenance; high productivity; low cost; and modest requirements for materials. As a consequence, many developing countries are investing large sums of money in research and development. However, little is known about the potential risks associated with developing this technology. The hazards associated with small airborne particles are not properly known, or covered by current toxicity regulations. The human immune system may be defenceless against new nanoscale products. There may be social problems, too, as poorer societies may suffer as established technologies become redundant and demands for commodities change rapidly.

1 Can chemical systems have a "memory"? Research the evidence that nickel–cadmium batteries have a "memory", which they retain even after recharging, whereas lithium-ion batteries have no memory effect.

2 Should consumers be free to use crude oil as an energy source, and aluminium to make luxury goods and unnecessary packaging material, or should governments legislate to curtail their use and thus cut down on the emission of carbon dioxide to reduce the increase in global warming?

3 Research into nanotechnology has many implications. Who should decide the direction in which the research is pursued? Should it be industry, whose main aim is to make money, or public bodies such as governments, charities or universities, whose finances are limited? If funds are limited, who should decide on the priorities? How can people best be made aware of the benefits and risks associated with nanotechnology?

HL

Condensation polymers

In Chapter 10 we discussed the difference between addition polymerization and condensation polymerization. Addition polymerization generally occurs with monomers containing a carbon-to-carbon double bond. Condensation polymerization occurs when the monomers contain two reactive functional groups, as for example in the formation of polyamides and polyesters such as polyethylene terephthalate, PET. Other important condensation polymers include phenol-methanal plastics (Bakelite), and polyurethanes.

Phenol-methanal plastics

These are prepared by adding acid or alkali to a mixture of phenol and methanal. The methanal is first substituted in the 2- or 4-position in the phenol, and then the product undergoes a condensation reaction with another molecule of phenol with the elimination of water (Figure 13). Further polymerization then takes place to build up a long chain (Figure 14), followed by covalent cross-linking to form a three-dimensional structure (Figure 15).

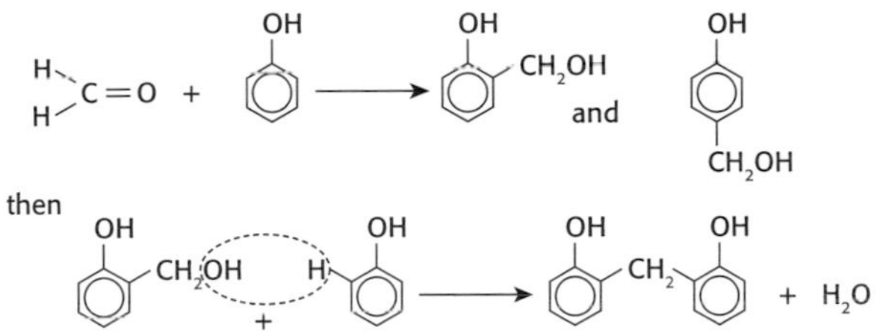

Figure 13 The formation of a phenol-methanal plastic.

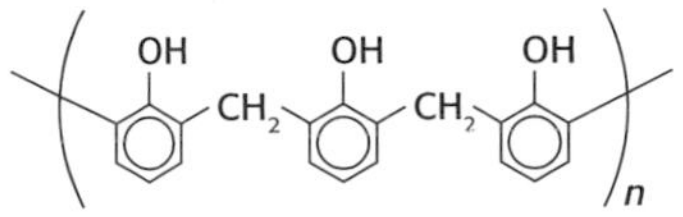

Figure 14 The linear chain in phenol-methanal plastic.

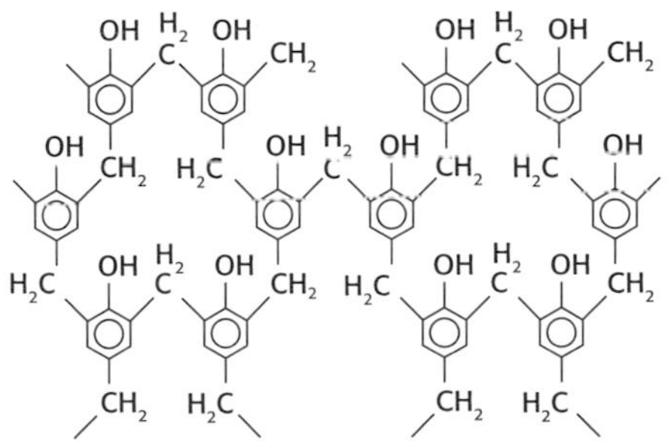

Figure 15 Part of the three-dimensional structure of a phenol-methanal plastic.

Polyurethanes

Polyurethanes are formed from the reaction of polyhydric alcohols (such as diols or triols) with compounds containing more than one isocyanate functional group, –NCO. In general:

In one sense polyurethanes are addition polymers, as no small molecules such as water are released in the reaction, but unlike other addition polymers there are no urethane monomers that can add to each other. They are usually made *in situ* and moulded into the shape required.

Polyethylene terephthalate, PET

This polymer is also known as Rayon or Dacron, and has been discussed in Chapter 10. It is an example of a polyester, and is made by condensing benzene-1,4-dicarboxylic acid with ethane-1,2-diol.

Cross-linking in polymers

When polymers are able to form giant three-dimensional structures, as for example in phenol-methanal plastics, it makes the polymer extremely strong, because a large amount of energy is required to break it down. It also makes them insoluble and generally unreactive. Another example of this is Kevlar, the material from which lightweight bullet-proof vests, composites for motor-cycle helmets and armour are made (Figure 16).

Kevlar is a polyamide made by condensing 1,4,-diaminobenzene (*para*-phenylenediamine) with benzene-1,4-dicarbonyl chloride (terephthaloyl dichloride) (Figure 17). It forms a strong three-dimensional structure due to hydrogen bonding between the long rigid chains (Figure 18).

Figure 16 A bullet-proof vest made out of Kevlar.

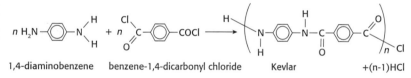

1,4-diaminobenzene   benzene-1,4-dicarbonyl chloride   Kevlar   +(n-1)HCl

Figure 17 Formation of Kevlar by condensation polymerization.

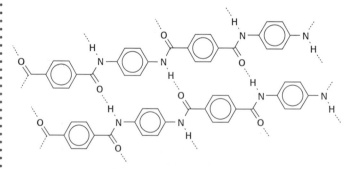

Figure 18 Three-dimensional structure of Kevlar.

Modification of condensation polymers

Like addition polymers, condensation polymers can be modified by various means during their manufacture. For example, air can be blown into polyurethane to make polyurethane foams for use as cushions and thermal insulation. The fibres of polyesters can be blended with other manufactured or natural fibres for making clothes such as shirts or skirts that are more comfortable, and which can hold dyes fast.

Conducting polymers

When ethyne is polymerized using a Ziegler–Natta catalyst (see below), the *cis-* form of poly(ethyne) is found to be an electrical conductor because of the delocalization of the pi electrons (Figure 19).

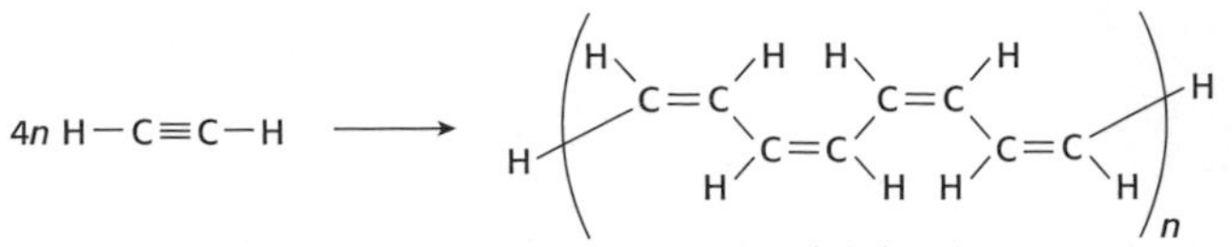

cis-poly(ethyne)

Figure 19 The polymerization of ethyne to give a conducting polymer.

The conductivity can be dramatically increased by doping with chemicals such as iodine, which can remove electrons, or alkali metals, which can add electrons.

Mechanisms in the organic chemicals industry

Manufacture of low-density polythene (LDPE)

The manufacture of low-density polythene is carried out at very high pressures (1000–3000 atm) and a temperature of about 750 °C. An initiator such as an organic peroxide or a trace of oxygen is added. Under these conditions the weak oxygen-to-oxygen single bond breaks homolytically to form free radicals:

$$R–O–O–R \rightarrow 2\ RO^{\bullet}$$

These free radicals then react with an ethene molecule to propagate new radicals:

$$RO^{\bullet} + H_2C{=}CH_2 \rightarrow R–O–CH_2–CH_2{}^{\bullet}$$
$$R–O–CH_2–CH_2{}^{\bullet} + H_2C{=}CH_2 \rightarrow R–O–CH_2–CH_2–CH_2–CH_2{}^{\bullet}\ \text{etc.}$$

Termination takes place when two radicals combine to form a molecule. The average polymer molecule contains between about 4,000 and 40,000 carbon atoms with many short branches. The branches affect both the degree of crystallinity and the density of the material. LDPE generally has a density of about 0.92 g cm^{-3} and is used mostly for packaging.

Manufacture of high-density polythene (HDPE)

High-density polythene is manufactured by polymerizing ethene at a lower temperature (about 600 °C) and pressure (1–50 atm) using a Ziegler–Natta catalyst (Figure 20). The catalyst is a suspension of titanium(III) or titanium(IV) chloride together with an electron-deficient alkyl-aluminium compound such as triethylaluminium, $Al(C_2H_5)_3$. The mechanism is very complex, and still not thoroughly understood. Essentially it involves the insertion of the monomer between the catalyst and the growing polymer chain by an ionic mechanism. The resulting polymer consists mainly of linear chains, with very little branching. This gives it a higher density (0.96 g cm^{-3})

Karl Ziegler

Giulio Natta

Figure 20 Karl Ziegler (1898–1973), working at the Max Planck Institute, Germany, and Giulio Natta (1903–1979), working at the Milan Polytechnic Institute, Italy, shared the Nobel Prize in Chemistry in 1963 for their discovery of an effective catalyst for the polymerization of alkenes.

and a more rigid structure, because the chains can fit together more closely. It is used to make containers and pipes.

Liquid crystals (2)

Liquid crystal molecules

Molecules of liquid crystals need to be polar, rigid, and rod-shaped. The biphenyl nitriles used in liquid crystal displays, such as 5CB (whose structure has been given earlier), fulfil these criteria well. The nitrile functional group gives the molecules sufficient polarity to ensure that the intermolecular forces are strong enough to align them in a common direction. The biphenyl groups help to make the molecule rigid and rod-shaped. The long-chain alkyl group ensures that the molecules cannot pack too closely together, which helps to maintain the liquid crystal phase.

Other similar substances used as liquid crystals with the same properties and approximately the same molecular length are 4-azoxyanisole (PAA) and 4-methoxybenzylidene-4'-butylaniline (MBBA) (Figure 21).

Mechanism of action of liquid crystals

Each pixel in a LCD contains a liquid crystal film sandwiched between two glass plates. The plates have many very fine scratches at right angles to each other, and have the property of polarizing light. The liquid crystal molecules in contact with the glass line up with the scratches and form a twisted arrangement between the plates due to intermolecular forces. This is known as **twisted nematic geometry**. The property utilized by the liquid crystals is their ability to interact with plane-polarized light which is rotated through 90° by the molecules as it passes through the film. When the two polarizers are aligned with the scratches, light will pass through the film and the pixel will appear bright. When a potential difference is applied across the film, the polar molecules will align with the film, thus losing their twisted structure and their ability to interact with

4-azoxyanisole (PAA)

4-methoxybenzylidene-4'-butylaniline (MBBA)

Figure 21 Examples of liquid crystals.

the light. Plane-polarized light will now no longer be rotated, so that the pixel appears dark. See Figure 22.

Liquid crystal properties of Kevlar

The structure of the synthetic polyamide Kevlar has been described above. It consists of rigid, rod-shaped polar molecules with strong intramolecular hydrogen bonding between the chains. It can act as a lyotropic liquid crystal in a solution of concentrated sulfuric acid. It is lyotropic because the alignment of these molecules depends upon the concentration of the solution. In strong acid solution the oxygen and nitrogen atoms in the amide linkage become protonated, and this breaks the hydrogen bonding between the chains.

Silicon and photovoltaic cells

Electrical conductivity of silicon

Metals conduct electricity when a voltage is applied, because they contain delocalized electrons. Non-metals (apart from graphite) are poor conductors at room temperature because the electrons are localized and normally held in fixed positions. Silicon is a **semiconductor**. A crystal of silicon contains a lattice of silicon atoms bonded to each other by shared pairs of electrons. These electrons are in fixed positions, so silicon is a poor conductor under normal conditions. However, the energy required to excite an electron and free it from its bonding position is equivalent to the energy of light with a wavelength of 1.1×10^{-6} metres. Visible light has a shorter wavelength, in the range of $4-7 \times 10^{-7}$ metres. This is higher in energy, and so sunlight is able to excite an electron in silicon. The electron is then free to move through the crystal lattice when a voltage is applied, making it an electrical conductor (Figure 23). This is the basis of the photoelectric effect, and is the theory behind solar-powered batteries. In practice the process is not very efficient, and the cost of purifying the silicon is high. However, solar cells are not directly polluting and do not use up valuable fossil fuel reserves.

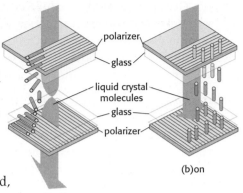
(a)off (b)on
Figure 22 The operation of the twisted nematic liquid crystal display.

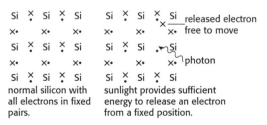

normal silicon with all electrons in fixed pairs.

sunlight provides sufficient energy to release an electron from a fixed position.

Figure 23 The effect of light on silicon.

Doping of silicon to obtain n-type and p-type semiconductors

One method of improving the efficiency of the photoelectric effect is by doping. This process involves adding very small amount of atoms of other elements, usually from Group 3 (Al, Ga or In) or from Group 5 (P or As).

When a Group 5 element is added, the extra electron can move easily throughout the crystal lattice when a voltage is applied, making the doped material a better conductor compared with pure silicon. Such doping produces an n-type semiconductor because the conductivity is due to the movement of negative electrons.

When a Group 3 element is added, the element now has one less electron than silicon. This produces a "hole" in the lattice. When a free electron moves into this hole it produces a new hole where the electron was formerly located. The hole can be regarded as a positive carrier, so the semiconductor is known as a p-type. See Figure 24.

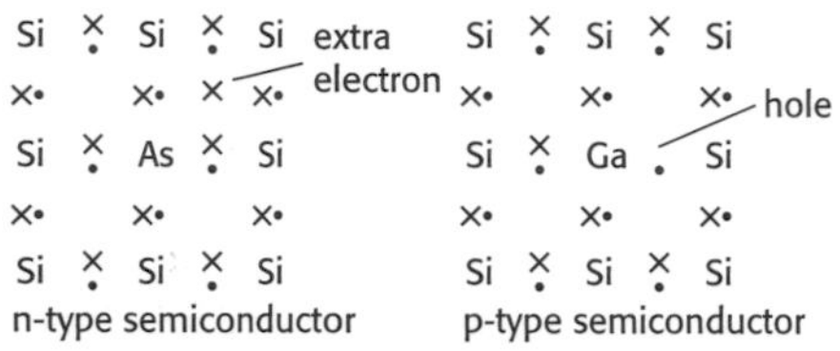

Figure 24 The doping of silicon to form n- and p-type semiconductors.

At a junction between the two different types of semiconductor electrons can flow from an n-type to a p-type, because they are moving from a negatively charged area to a positively charged region. However, they cannot flow the other way. The junction thus allows the current to flow in one direction only. Such a junction is known as a **rectifier**, and can be used to convert alternating current into direct current.

The chlor-alkali industry

Chlorine is a powerful oxidizing agent, with a standard electrode potential of +1.36 V. Apart from fluorine, very few chemical oxidizing agents are powerful enough to oxidize chloride ions to chlorine molecules, so the production depends on using electrons themselves. Chlorine gas is formed during the electrolysis of molten sodium chloride in the industrial production of sodium metal. However, the main source of chlorine is the electrolysis of concentrated aqueous sodium chloride solution (brine), because this requires less energy. Sodium chloride is a cheap raw material, which is readily available, and the process also produces sodium hydroxide and hydrogen, both of which are important industrial products. Hydrogen is used as the feedstock for the production of ammonia and sodium hydroxide in the production of soap, polymers and the manufacture of other substances. The chlorine produced has many important industrial uses, among them chlorinated organic solvents, water purification, pesticides, feedstock for inorganic chemicals (such as hydrochloric acid) and bleaching paper. There are two main methods by which the electrolysis is now achieved – the diaphragm cell and the use of ion exchange membranes. In the past the mercury cell was used.

The mercury cell

The negative electrode (cathode) is made of flowing mercury. Although sodium is above hydrogen in the electrochemical series, sodium ions are preferentially discharged because an alloy (known as an **amalgam**) is formed with the mercury.

$$Na^+(aq) + e^- + Hg(l) \rightarrow Na/Hg(l)$$

The mercury amalgam then flows out of the electrolysis cell into a separate chamber, where it reacts with water to produce hydrogen and sodium hydroxide solution. The resulting mercury is recycled back into the electrolytic cell.

$$Na/Hg(l) + H_2O(l) \rightarrow Na^+(aq) + OH^-(aq) + \tfrac{1}{2}H_2(g) + Hg(l)$$

The cell itself (Figure 25) is made of PVC-lined steel, and the positive electrode (anode) where the chlorine is formed is made of graphite or titanium.

$$2Cl^-(aq) \rightarrow Cl_2(g) + 2e^-$$

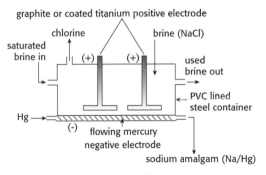

Figure 25 The mercury cell.

The diaphragm cell

In the diaphragm cell (Figure 26) the positive electrode is made of titanium, but the negative electrode is made of steel. Hydrogen is formed at the negative electrode and chlorine at the positive electrode.

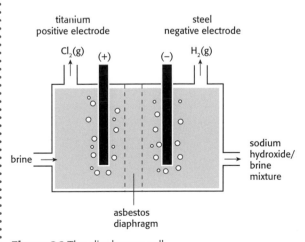

Figure 26 The diaphragm cell.

Negative electrode:
$$2H_2O(l) + 2e^- \rightarrow H_2(g) + 2OH^-(aq)$$
Positive electrode:
$$2Cl^-(aq) \rightarrow Cl_2(g) + 2e^-$$

The diaphragm, made of asbestos, allows the sodium chloride solution to flow between the electrodes but separates the chlorine and hydrogen gas and helps to prevent the hydroxide ions, OH^-(aq), from flowing towards the positive electrode. The sodium hydroxide solution formed accumulates in the cathode compartment and is piped off. The resulting solution contains about 10% sodium hydroxide and 15% unused sodium chloride by mass. It is concentrated by evaporation and the sodium chloride crystallizes out leaving a 50% solution of sodium hydroxide. A more modern version of the diaphragm cell (known as an ion exchange membrane cell) uses a partially permeable ion exchange membrane rather than asbestos. The membrane is made of a fluorinated polymer and is permeable to positive ions but not negative ions.

Environmental impact of the chlor-alkali industry

In many parts of the world the mercury cell has been replaced by the diaphragm or ion exchange membrane cell. The ion exchange membrane cell is much cheaper to run, thanks to the development of modern polymers. The main reason why the use of the mercury cell has discontinued is environmental. In theory all the mercury is recycled, but in practice some leaks into the environment, and can build up in the food chain to toxic levels.

Concern is mounting over the use of chlorinated organic compounds. Several have been shown to be carcinogenic, and the carbon-to-chlorine bond can break homolytically in the presence of ultraviolet light at higher altitudes to form chlorine radicals, which can contribute to ozone depletion.

1 Using polyethylene terephthalate, PET, polyurethanes and phenol-methanal plastic as examples, discuss the advantages and disadvantages of polymer use. Include in your discussion factors such as strength, density, insulation, lack of reactivity, use of natural resources, disposal and biodegradability.

2 Suggest a reason why it is sensible to store bullet-proof clothing made of Kevlar well away from any possible contact with acids.

17 Option D: medicines and drugs

Pharmaceutical products

For centuries man has used natural materials to provide relief from pain, heal injuries and cure disease. Many of these folk remedies have been shown to be very effective, and the active ingredients have been isolated and identified. Morphine was extracted from the poppy *Papaver somniferum* early in the nineteenth century, and later salicylic acid, the precursor of aspirin, was isolated from the bark of the willow tree, *Salix alba*.

Generally a drug or medicine is any chemical (natural or synthetic) that does one or more of the following:

- alters the physiological state (including consciousness, activity level or coordination)
- alters incoming sensory sensations
- alters mood or emotions.

Drugs and medicines are normally (but not always) taken to improve health. They accomplish this by assisting the body in its natural healing process. The mechanism of drug action is still not fully understood, and there is some evidence that the body can be "fooled" into healing itself naturally through the "placebo" effect.

Opium in the nineteenth century

Kubla Khan

> A damsel with a dulcimer
> In a vision once I saw:
> It was an Abyssinian maid,
> And on her dulcimer she play'd,
> Singing of Mount Abora.
> Could I revive within me
> Her symphony and song,
> To such a deep delight 'twould win me,
> That with music loud and long,
> I would build that dome in air,
> That sunny dome! Those caves of ice!
> And all who heard should see them there,
> And all should cry, Beware! Beware!
> His flashing eyes, his floating hair!
> Weave a circle round him thrice,
> And close your eyes with holy dread:
> For he on honey-dew hath fed,
> And drunk the milk of Paradise.

The closing lines of "Kubla Khan" written in 1816.

$$Mg(OH)_2 + 2HCl(aq) \rightarrow MgCl_2(aq) + 2H_2O(l)$$
$$Al(OH)_3(s) + 3HCl(aq) \rightarrow AlCl_3(aq) + 3H_2O(l)$$

Antacids are commonly combined with alginates and anti-foaming agents. **Alginates** form a protective neutralizing coating over the walls of the stomach and oesophagus. This prevents heartburn, which is caused when the stomach acid rises up the oesophagus. **Anti-foaming agents** are used to help prevent flatulence. The most usual anti-foaming agent is dimethicone.

Analgesics

Modes of action of mild and strong analgesics

Analgesics relieve pain. Mild analgesics such as aspirin and paracetamol (acetaminophen) achieve this by intercepting the pain stimulus at the source of the injury. Aspirin, for example, prevents the synthesis of a specific enzyme, prostaglandin synthase. This enzyme is involved in the synthesis of prostaglandins, which produce fever and swelling, and the transmission of pain from the site of an injury to the brain. Strong analgesics, such as morphine, which are normally available only on prescription, work by interacting temporarily with receptor sites in the brain so that pain signals within the brain and spinal cord are blocked.

Aspirin

In the second half of the nineteenth century 2-hydroxybenzoic acid (salicylic acid) was in wide use as an analgesic (pain killer) and antipyretic (fever depressant) (Figure 1). However, salicylic acid has the undesirable side effect of irritating and damaging the mouth, oesophagus and stomach membranes. In 1899 the Bayer Company of Germany introduced the ethanoate ester of salicylic acid, naming it "Aspirin". Since that time mild analgesics containing aspirin have appeared under many different brand names.

2-hydroxybenzoic acid
(salicylic acid)

2-acetoxybenzoic acid
(acetylsalicylic acid or "Aspirin")

Figure 1 Structures of salicylic acid and Aspirin.

Aspirin is less irritating to the stomach membranes, and it is hydrolysed in the body to salicylic acid and ethanoate ions only when the aspirin reaches the more alkaline conditions of the small intestine.

Because of its anti-inflammatory properties aspirin can also be taken for relieving the symptoms of arthritis and rheumatism. Aspirin also has an ability to prevent blood clotting, and is sometime taken to prevent strokes or the reoccurrence of heart attacks. The most common side effect of aspirin is that it causes bleeding in the lining of the stomach. A few people are allergic to aspirin, with just one or two tablets leading to bronchial asthma. The taking of aspirin by children under 12 has been linked to Reye's disease—a potentially fatal liver and brain disorder with the symptoms of vomiting, lethargy, irritability and confusion. Exceeding the safe dose of aspirin can be fatal, because the salicylic acid leads to acidosis caused by a lowering of the pH of the blood.

Methyl salicylate (oil of wintergreen)

Salicylic acid contains a phenol and a carboxylic acid group. It is the phenolic group that reacts with ethanoic acid to form the ester aspirin. However, another ester, methyl salicylate, can be made by reacting the carboxylic acid group of salicylic acid with methanol (Figure 2).

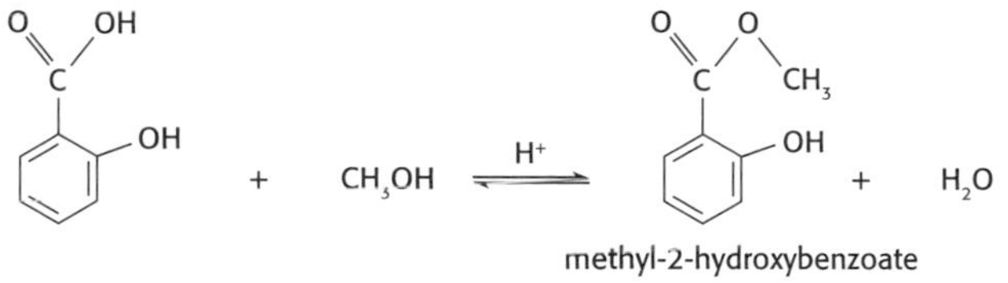

Figure 2 Methyl salicylate, or "oil of wintergreen".

Methyl salicylate has a characteristic smell, and occurs naturally in some plants, such as sweet birch, *Betula lenta*. It is poisonous if taken internally, but is used in liniments to rub into the skin to relieve the inflammation that causes muscle pain.

Paracetamol (acetaminophen)

Paracetamol (Figure 3) (known as acetaminophen in the USA) is often preferred to aspirin as a mild pain reliever, particularly for young children, because its side effects are less problematical than those of aspirin, although in rare cases it can cause kidney damage and blood disorders. Serious problems can arise, however, if an overdose is taken. Even if the overdose does not result in death it can cause brain damage and permanent damage to the liver and kidneys.

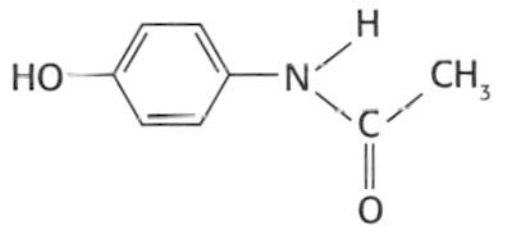

Figure 3 Paracetamol (acetaminophen).

Opiates

Strong analgesics are given to relieve the severe pain caused by injury, surgery, heart attack or chronic diseases such as cancer. The most important naturally occurring strong analgesics are morphine and codeine found in the opium poppy. Morphine, codeine and semi-synthetic opiates (obtained by simple structural modifications to morphine), such as heroin, all possess the same basic structure. In codeine one of the hydroxyl groups has been substituted for an ether group, and heroin is the diester formed from the condensation of morphine with ethanoic anhydride. See Figure 4.

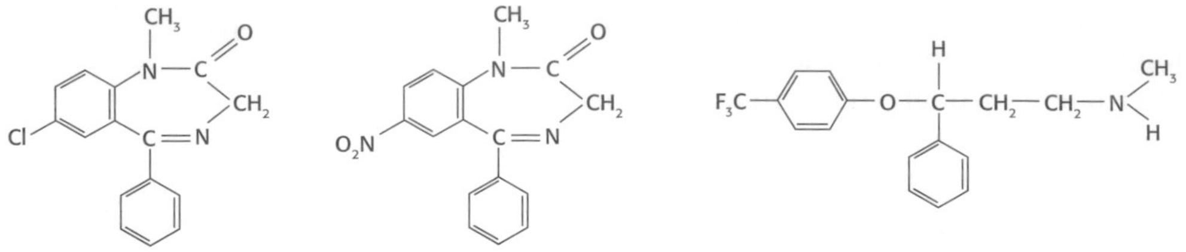

diazepam
(Valium)

nitrazepam
(Mogadon)

fluoxetine
(Prozac)

Figure 6 Some common depressants. Both Valium and Mogadon contain an amide group. The only difference in their structures is the substitution of the chloro- group in Valium for the nitro- group in Mogadon. Fluoxetine contains an asymmetric carbon atom, and so exists in two different enantiomeric forms.

Stimulants

Stimulants are drugs that increase a person's state of mental alertness. All stimulants tend to increase concentration. but other physiological effects vary according to the particular stimulant.

Amphetamines

Amphetamine is chemically related to adrenaline (cpinephrine), the "fight or flight" hormone. It is a sympathomimetic drug: that is, one that mimics the effect of stimulation on the sympathetic nervous system. This is the part of the nervous system that deals with subconscious nerve responses such as speeding up the heart and increasing sweat production. Amphetamines were initially used to treat narcolepsy (an uncontrollable desire for sleep), and were issued to airmen in World War II to combat fatigue (Figure 7). In the 1950s and 1960s they were used as anti-depressants and slimming pills. Regular use can lead to both tolerance and dependence. Short-term effects include increase in heart rate and the rate of breathing, dilation of the pupils, decrease in appetite followed by fatigue and possible depression as the effects wear off. Long-term effects include weight loss, constipation and emotional instability.

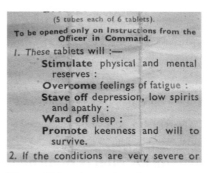

(5 tubes each of 6 tablets).
To be opened only on Instructions from the Officer in Command.
1. These tablets will :—
   Stimulate physical and mental reserves :
   Overcome feelings of fatigue :
   Stave off depression, low spirits and apathy :
   Ward off sleep :
   Promote keenness and will to survive.
2. If the conditions are very severe or

Figure 7 Amphetamines known as "Wakey-Wakey" pills were given to WWII airmen to prevent sleepiness.

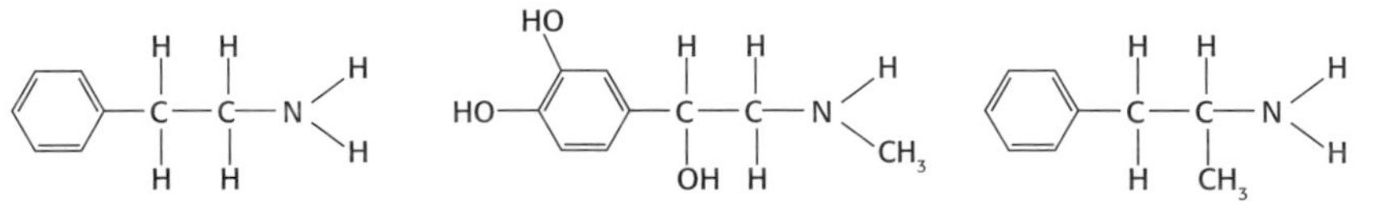

Both amphetamine and adrenaline are derived from the phenylethylamine structure (Figure 8).

phenylethylamine

adrenaline (epinephrine)

amphetamine

Figure 8 The similarities of the structures of adrenaline and amphetamine to phenylethylamine.

> ### Recreational drugs: "speed" and "ecstasy"
>
> Methamphetamine (speed) and ecstasy are also derived from phenylethylamine (Figure 9). Amphetamine is a primary amine whereas methamphetamine and MDMA are secondary amines.

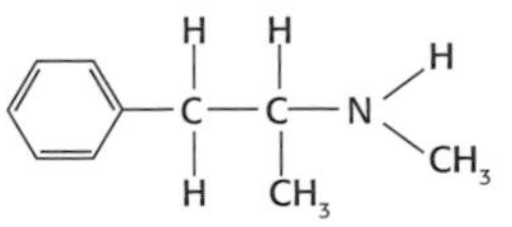

methamphetamine (speed) 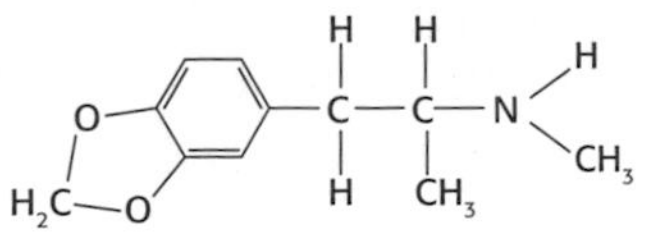   methylenedioxymethamphetamine (MDMA)
                                        ("E" or ecstasy)

Figure 9 Recreational drugs.

Ecstasy is an example of a "designer drug". It has no medical use, and is frequently abused. It produces mental relaxation, increased sensitivity to stimuli, and sometimes hallucinations. It reduces the amount of serotonin in the brain, and can lead to emotional disorders. It can produce severe and sometimes fatal effects, even after a single dose.

Caffeine and nicotine

Like morphine, codeine and cocaine, caffeine and nicotine are alkaloids. Alkaloids are nitrogen-containing compounds of plant origin containing heterocyclic rings (rings containing other atoms as well as carbon) and a tertiary amine group. Caffeine, unlike nicotine, also contains amide groups (Figure 10).

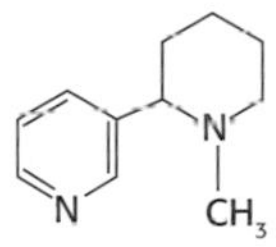

Nicotine is the addictive component of tobacco. Like amphetamines, nicotine is sympathomimetic. It increases concentration, and other short-term effects include the relief of tension, increased heart rate and blood pressure, and a reduction in urine output. The long-term effects include an increased risk of heart disease and coronary thrombosis. Its stimulatory effects may also lead to excess production of stomach acid, leading to an increased risk of peptic ulcers. Because of the way it is usually taken, the other well-known risks of smoking include chronic lung diseases, adverse effects on pregnancy, and cancers of the lung, mouth and throat.

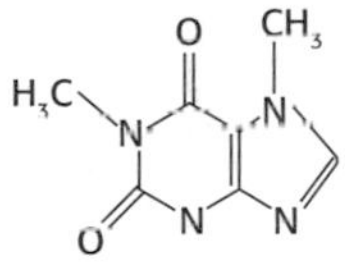

Caffeine is the most widely used stimulant in the world. It is present in coffee, tea, chocolate and cola drinks, and is also found in some painkillers and other medicines. Caffeine is a diuretic (causes frequent urination), and respiratory stimulant. It increases alertness, concentration and restlessness, but when consumed in large amounts it can cause anxiety, irritability and sleeplessness. It is included in many common painkillers, because it speeds up their effects.

Antibacterials

Most bacteria consist of a single cell with a protective wall made up of complex structures consisting of sugars covalently bonded to peptides and associated with lipids. Each bacterial cell contains a single chromosome consisting of a strand of deoxyribonucleic acid, DNA. Some bacteria are aerobic and are more likely to infect surface areas such as the skin or the respiratory tract. Others are anaerobic and multiply in oxygen-free or low-oxygen surroundings such as the bowel. Not all bacteria cause disease, and some are beneficial.

Figure 10 The structures of nicotine and caffeine.

Antibacterials are chemicals that prevent the growth and multiplication of bacteria. The first effective antibacterial, the dye **trypan red**, was developed by a German chemist Paul Ehrlich (1854–1915) to cure sleeping sickness. In 1908 Ehrlich also developed an arsenic-containing compound, arsphenamine (manufactured under the trade name of Salvarsan), which was effective against syphilis. In 1935 the first "sulfa drug" prontosil, which was effective against streptococcal bacteria, was developed. However the real breakthrough in antibacterials came with the discovery and subsequent isolation of penicillin.

The structure of Salvarsan (arsphenamine): a knowledge issue?

Most textbooks show the structure of Salvarsan (the first drug effective against syphilis) as a dimer (Figure 11). However, in 2005 a team led by Professor Brian Nicholson at the University of Waikato, Hamilton, New Zealand, showed that it is in fact a mixture of a trimer and a pentamer (Figure 12).

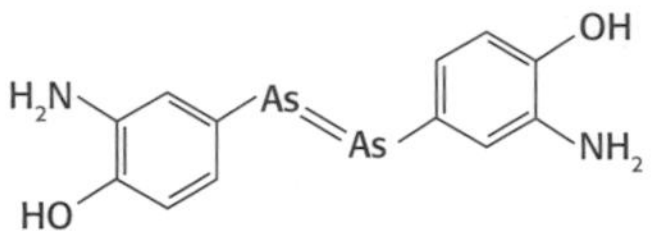

Figure 11 Original "structure" of Salvarsan.

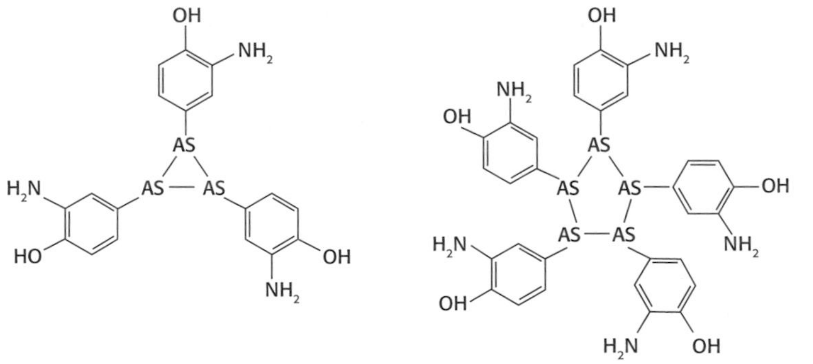

Figure 12 True "structure" of Salvarsan – a mixture of a trimer and a pentamer

---

Historical development of penicillin

The use of moulds to prevent infections has existed in folklore for a long time. In 1923 a Costa Rican, Clodomiro Picado Twight (1887–1944), working in Paris recorded the action of the fungal genus *Penicillium sp.* on the growth of bacteria, and the work was published in 1927. However, the discovery of penicillin is usually attributed to the Scotsman Alexander Fleming (1881–1955).

Fleming, a bacteriologist, was working with cultures of *Staphylococcus aureus*, a bacterium that causes boils and other types of infection. In 1928 he left an open petri dish containing one of the cultures in the laboratory while he went away on holiday. Upon his return he noticed that a mould had developed and had inhibited growth of the bacterium. He deduced that the mould (*Penicillium notatum*) produced a compound (which he called penicillin) that inhibited the growth of bacteria.

Although Fleming published his results in 1929, he did not pursue his discovery. It was an Australian, Howard Florey (1898–1968), and the German-born Ernst Chain (1906–1979) who overcame the problems associated with isolating and purifying penicillin. In 1941 they used penicillin on a policeman who was dying of septicaemia (blood poisoning). They recorded a dramatic improvement in his condition, but tragically their meagre supply ran out before the policeman was cured, and he relapsed and died.

The search was on to produce penicillin in bulk. It was solved in America by growing strains of the penicillin mould in large tanks containing corn-steep liquor. In the 1950s the structure of penicillin was determined, and this enabled chemists to synthesize different types of penicillin and other antibiotics (antibacterials originating from moulds) in the laboratory without recourse to moulds. Fleming, Florey and Chain received the Nobel Prize in Physiology or Medicine in 1945.

> ### Clodomiro Picado Twight (1887–1944)
> Clodomiro Picado Twight was born in Nicaragua, but moved to Costa Rica at the age of two. He worked at the Pasteur Institute in Paris, and in 1923 his notebooks record his work on the effect of penicillin mould on the growth of *Staphylococcus* and *Streptococcus* bacteria. This work was published in 1927 but received little recognition.

1 The discovery of penicillin is often quoted as an example of serendipity—the accidental discovery of something useful while looking for something else. Another example is the discovery of the anti-cancer drug *cis*-platin. Use the Internet to research these and other examples of the role that serendipity plays in scientific discovery. How important are serendipity, open-mindedness and creativity in the scientific method?
2 Fleming was working at St Mary's Hospital in Paddington, London, when he made his discovery of penicillin. The chairman of the hospital board was the press baron Lord Beaverbrook, who used his newspapers to make Fleming famous. Is it ethical for the media to be controlled by a few individuals, or should all media be controlled by the state?

The mode of action of penicillin

Penicillins work by interfering with the enzymes that bacteria need to make normal cell walls. As the bacterium cell swells, the osmotic pressure causes the weakened cell wall to disintegrate, and the bacterium dies. Over time some bacteria have become resistant to penicillin, because they contain an enzyme called penicillinase, which is able to deactivate penicillin.

The original penicillin was benzylpenicillin (penicillin G) (Figure 13). This had to be injected, because it is broken down by the acid in the stomach. All penicillins contain the same basic structure, which includes a four-membered ring containing a nitrogen atom and known as a **beta-lactam ring**. By changing the side chain, penicillins can be made that are resistant to stomach acid and to the enzyme penicillinase.

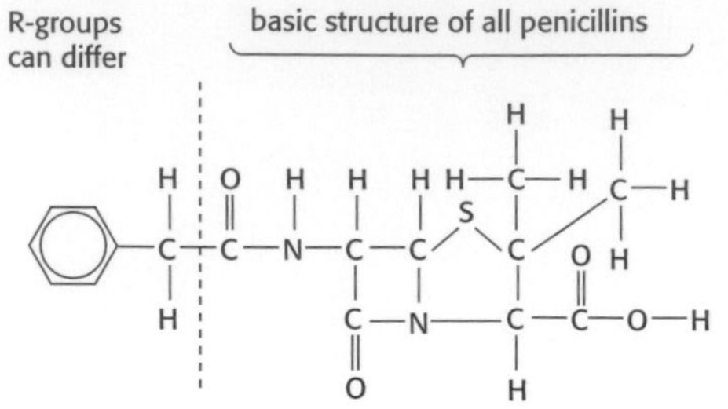

R-groups
can differ

basic structure of all penicillins

Figure 13 The structure of benzylpenicillin (penicillin G)

Because of the indiscriminate use of penicillin to cure minor ailments, and its use in animal feedstock, some bacteria are now extremely resistant to penicillin. Examples include the methicillin-resistant *Staphylococcus aureus* (MRSA), which has been christened the "super bug" because it can make beta-lactamase, which can degrade penicillin, and strains of *Mycobacterium tuberculosis*, which cause tuberculosis, TB. To treat these infections a strict adherence to a treatment regime often involving a "cocktail" of different antibiotics is required to prevent the risk of further resistance developing.

Antivirals

Viruses and bacteria

Viruses are much smaller than bacteria. There are many different types of virus, and they vary in their shape and structure. All viruses, however, have a central core of deoxyribosenucleic acid, DNA, or ribonucleic acid, RNA, surrounded by a coat (capsid) of regularly packed protein units called **capsomeres**, each containing many protein molecules. Unlike bacteria they have no nucleus or cytoplasm, and are therefore not cells. They do not feed, respire, excrete or grow, and they can reproduce only inside the cells of living organisms, using the materials provided by the host cell.

Although viruses can survive outside the host they can replicate only by penetrating the living host cell and injecting their DNA or RNA into the cell's cytoplasm. The virus then "takes over" the biochemical machinery inside the cell. This causes the cell to die or become seriously damaged, giving rise to the symptoms of the viral infection. The cell is made to produce new DNA or RNA, and forms large numbers of new viruses. These are then released and move on to infect other healthy cells.

Treatment of viral diseases

Common viral infections include the common cold, influenza, and childhood diseases such as mumps and chickenpox. The body's own immune system is usually strong enough to overcome infections such as these, so that drugs are given more to remove the associated pain, reduce the fever or counteract secondary infections. One difficulty in treating viral infections is the speed with which the virus multiplies. By the time the symptoms have appeared, the viruses are so numerous that antiviral drugs will have little effect.

During the past few years some drugs have been developed to fight specific viral infections. They can work in different ways. Some work by altering the cell's genetic material so that the virus cannot use it to multiply. An example of this is acyclovir (Figure 14), which is applied topically to treat cold sores caused by the herpes virus. Its structure is similar to that of deoxyguanosine, one of the building blocks of DNA. It "tricks" the viral enzymes into using it as a "building block" for the synthesis of viral DNA, and thus prevents the virus from multiplying. However, it is difficult to eliminate the virus completely, so the infection may flare up again at a later date. Ganciclovir, a drug with a similar chemical structure to acyclovir, is used to treat severe cases of genital herpes when they have become resistant to acyclovir.

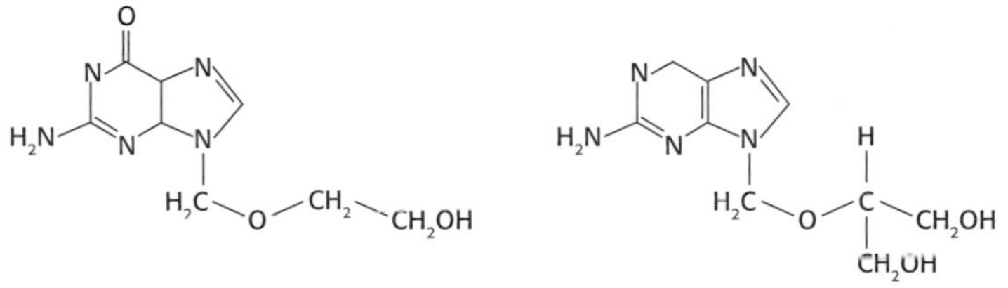

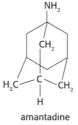

acyclovir   ganciclovir

Figure 14 Acyclovir and ganciclovir have similar structures.

Other antiviral drugs work by preventing the new viruses formed from leaving the cell and multiplying. One such drug is amantadine (Figure 15), which is active against the influenza virus. One of the enzymes used by all influenza viruses to stick to the host cell wall as it leaves is called neuraminidase, and the drug works by inhibiting the active site on this enzyme.

amantadine

Figure 15 The chemical structure of amantadine, $C_{10}H_{17}N$, used to treat influenza and shingles.

The HIV/AIDS problem

One of the problems with developing antiviral drugs is that the viruses themselves are regularly mutating. This is particularly true with the human immunodeficiency virus (HIV), which can lead to AIDS (acquired immune deficiency syndrome), leaving the individual prone to secondary infections and certain types of tumour. AIDS is one of the most destructive epidemics in recent history, and it is estimated that it has led to the death of more than 25 million people since it was first recorded in 1981. In some countries now, particularly in Africa, more than one third of the population is HIV positive.

AIDS is caused by a retrovirus: that is, it contains RNA rather than DNA. The virus invades certain types of white blood cell known as **T helper cells** (specifically the CD4⁺ T cells), which normally activate other cells in the immune system, with the result that the body is unable to fight infection (Figure 16). Once the virus invades a healthy cell, its first task is to make viral DNA from the RNA template using an enzyme called **reverse transcriptase**. This is the opposite process to that which takes place in normal cells, in which

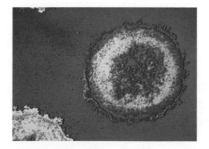

Figure 16 A T cell covered with new HIV particles ready to spread and infect other cells.

RNA is made from a DNA template using RNA polymerase as the enzyme.

There are various ways in which a drug may be developed to counteract the HIV virus. One drug that has met with some success is AZT (zidovudine), which has a remarkably similar structure to acyclovir (Figure 17).

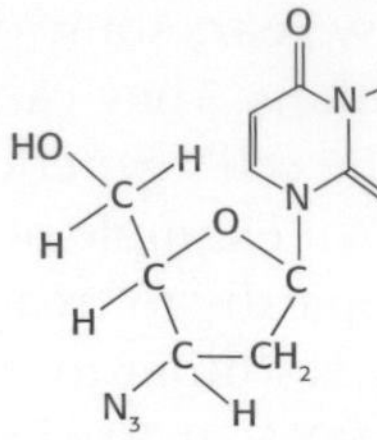

Figure 17 Structure of AZT (zidovudine).

AZT combines with the enzyme that the HIV virus uses to build DNA from RNA, and clogs up its active site. It is therefore a reverse-transcriptase inhibitor. As it is only retroviruses that use this enzyme, AZT does not affect normal cells. Although there is still no known cure or vaccine, treating HIV positive individuals with a cocktail of antiretroviral drugs can prolong the period until AIDS develops, and prolong the average survival time of sufferers with AIDS. At present, from a medical and scientific viewpoint the best way to combat AIDS is still to educate the population to use protective barriers to stop the exchange of bodily fluids, although this clashes with some religious beliefs.

Factors affecting the mechanism of drug action

Stereochemistry

The effectiveness of particular drugs depends on their ability to interact with specific molecules in the body, or in the organisms causing a disease. The precise stereochemistry of the drug molecules may therefore be extremely important.

When a drug molecule contains a carbon-to-carbon double bond, or a ring where there is restricted rotation, then it may exhibit geometric isomerism. The two different geometric isomers may have very biological different effects. One compound that exemplifies this extremely well is diamminedichloroplatinum(II). This compound, which has the formula $Pt(NH_3)_2Cl_2$, is square planar and has a *cis*- and *trans*- isomer (Figure 18).

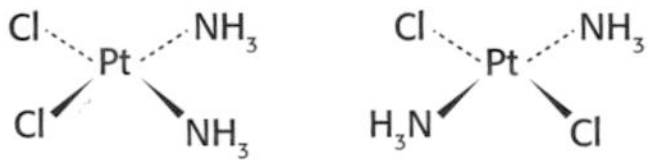

Figure 18 Geometric isomers of diamminedichloroplatinum(II).

The *cis*-isomer is highly effective in the treatment of testicular and ovarian cancers, as well as other forms of cancer. *Trans*-platin is not an effective anti-cancer drug. Both *cis*-platin and *trans*-platin have no overall charge, and so can diffuse through the cancer cell membrane. Inside the cell they can exchange a chloride ion for a molecule of water to form $[Pt(NH_3)_2Cl(H_2O)]^+$. This complex ion enters the cell nucleus. However, because of the stereochemistry only the *cis*-isomer is then able to bind to two guanine molecules present in the DNA by exchanging another chloride ion to form $[Pt(NH_3)_2(DNA)]^{2+}$. This alters the cancer cell's DNA so that when the cell tries to replicate it cannot be copied correctly and the cell dies.

Drug molecules that possess an asymmetric (chiral) carbon atom will exist in two different enantiomeric forms. Examples include the analgesic ibuprofen, amphetamine, and thalidomide (Figure 19).

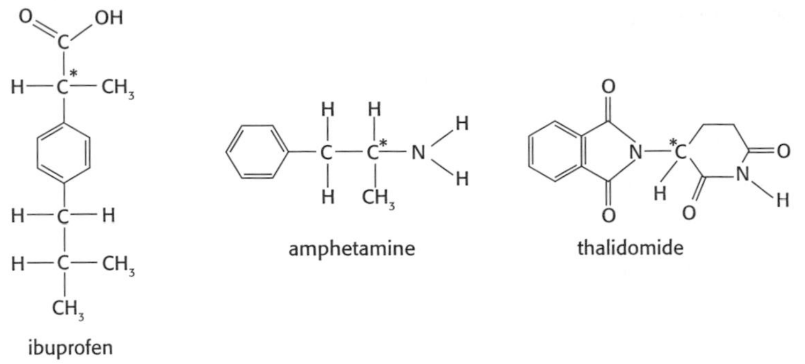

ibuprofen

amphetamine

thalidomide

Figure 19 Chiral molecules. The asymmetric carbon atom is indicated with an asterisk, *.

The two enantiomeric forms may have very different effects in the body. One enantiomer of thalidomide alleviates the symptoms of morning sickness in pregnant women; the other causes severe deformities in the limbs of the fetus.

Ring strain

The effective action of a drug may also depend upon the weakness of carbon-to-carbon or carbon-to-nitrogen bonds in a strained ring. All penicillins contain a four-membered beta-lactam ring. The ring contains two carbon atoms that are sp^3 hybridized, a nitrogen atom that is sp^3 hybridized, and a carbon atom that is sp^2 hybridized. The group containing the beta-lactam ring is very similar to a combination of the two amino acids cysteine and valine (Figure 20).

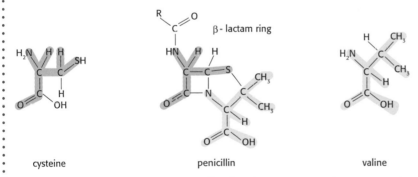

cysteine

penicillin

valine

Figure 20 The β-lactam ring in penicillins breaks to form two fragments with structures similar to cysteine and valine.

Because of the restrictions of the ring, the normal bond angles of 109.5° and 120° are not able to be obtained, and so the ring becomes strained. This makes the amide group highly reactive. The ring opens so that the penicillin can become covalently bonded to the enzyme that synthesizes the cell walls of the bacterium, thus blocking its action.

Polarity

Why is heroin so much more potent than morphine, and why does it produce a much greater feeling of euphoria? This can be explained by the difference in the polarity of the two substances. Morphine molecules contain two polar hydroxyl groups; in heroin molecules these are replaced by ethanoate groups (Figure 21). These two ester groups make it much less polar than morphine and so much more soluble in lipids, which are non-polar. Heroin is thus able to rapidly penetrate the lipid-based blood/brain barrier and reach the brain in higher concentrations than morphine.

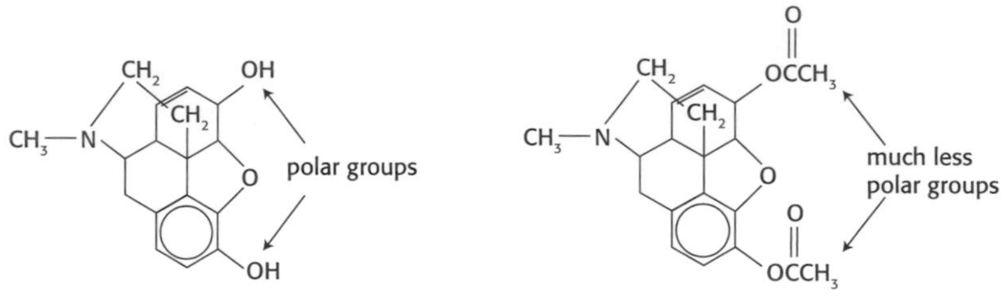

Figure 21 The polarity of morphine compared to heroin.

Drug design

"Libraries" of compounds and combinatorial chemistry

New drugs essentially come from three different traditional sources:

- A natural substance such as a plant or fungus known to have healing properties may be analysed and the active component isolated and purified. Morphine and salicylic acid are two such examples.
- Chemists then synthesize analogues of these substances by changing the side chain or functional groups, and then testing the new molecules for their pharmacological properties. Aspirin and heroin are examples of this.
- The third source is essentially serendipity. A substance is accidentally found to have therapeutic properties, and then analogues of this substance are made. Examples include *cis*-platin and penicillin.

The details of all the molecules isolated or synthesized, and the results of their pharmacological studies, are then kept in a "library" of compounds. This used to be a very time-consuming exercise, because relatively large amounts of materials had to be isolated or synthesized and then purified in order to determine their structure and properties.

The advent of very sensitive analytical techniques such as mass spectrometry to identify extremely small amounts of substances has led to the development of **combinatorial libraries**. In combinatorial chemistry very large numbers of related compounds can be prepared quickly. One way is to use computer-controlled syringes to carry out repetitive chemical techniques, but much use is now made of solid-phase organic chemistry. The starting material for the reaction is covalently bonded to very small beads (100 μm in diameter) of polystyrene-based resin. A process called **mix and split** is then used (Figure 22). Imagine the process for just three amino acids. After the first coupling, all the resin beads are then split into individual portions for the next step, so that when reacted again all the nine (3^2) possible combinations of dipeptides are formed. After another step all 27 (3^3) possible combinations of tripeptides have been formed. If the process is scaled up for all the 20 naturally occurring amino acids, then each cycle produces 400 dipeptides, 8,000 tripeptides, 160,000 and so on. By using a large excess of the second and subsequent amino acids the reaction can be made to give a high yield.

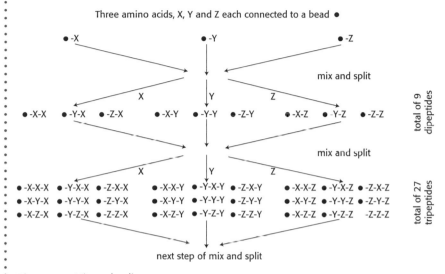

Figure 22 Mix and split.

The final products can be purified easily by filtering off the beads from the reaction mixture and washing. Preliminary screening for drug activity can then take place either *in vitro* or *in vivo* by measuring the ability of a compound to affect enzymes and bind to receptor cells. This process was done first for amino acids but has been extended to cover many other types of active molecules such as those containing the benzodiazepine group in depressants to form very large combinatorial libraries. Once a particular substance is identified as potentially useful it can be made on a larger scale. In fact it is becoming increasingly unnecessary to actually make the initial compounds. By knowing both the exact three-dimensional shape of the active site on the enzyme or receptor and the shapes of the active groups within a potential drug the use of virtual computer modelling software can produce a virtual library.

Changing the polarity

The structure of the SSRI drug fluoxetine was given earlier. In fact prescriptions normally contain the drug in its ionic form, fluoxetine hydrochloride (Figure 23).

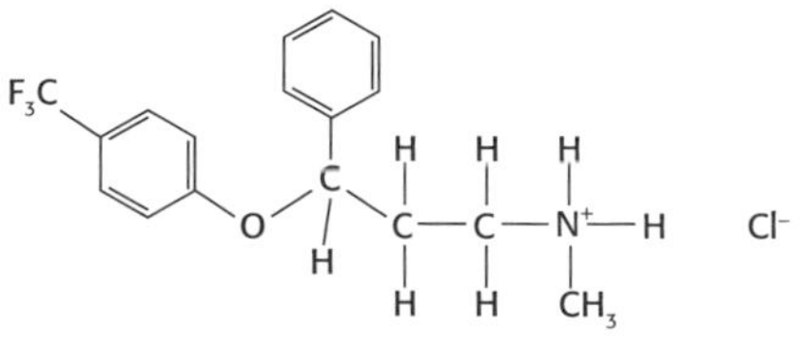

Many other drugs that contain an amine group are similarly administered as their hydrochloride salt, because this increases their solubility in polar environments and hence their transport around the body. Drugs that contain a carboxylic acid group can similarly be made polar by converting them into their anion and administering them as their sodium or calcium salt. This is the case for soluble aspirin. Once the aspirin anion reaches the strongly acidic environment of the stomach it reverts back to the non-ionized form.

Figure 23 Fluoxetine hydrochloride (Prozac).

Cannabis (also known as marijuana) is a mild hallucinogen made from the flowering tops, stems, leaves and seeds of the hemp plant *Cannabis sativa*. Hashish, made from the resin of the plant, is about five times stronger than marijuana. The active ingredient in marijuana is tetrahydrocannabinol (THC) (Figure 28). Unlike the previous three hallucinogens THC (also known as marijuana) does not contain either the indole ring or an amine group.

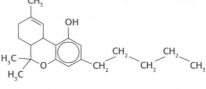

Figure 28 Tetrahydrocannabinol (THC).

The short-term effects include a feeling of relaxation, and enhanced auditory and visual perception. Loss of the sense of time, confusion and emotional distress can also result. Hallucinations may occur in rare cases. It can have a synergistic effect and increase the risk of sedation with depressants. The long-term effects include apathy and lethargy and a lowering of fertility, together with all the risks associated with tobacco smoking.

❷

1 List the arguments for and against the legalization of cannabis. Do you think it should be legalized in your country?
2 Explain why salts of drugs containing amines are converted to their non-polar form in the blood, whereas salts of drugs containing a carboxylic acid group are converted into their non-polar form in the stomach.
3 Which of the following substances consist of chiral molecules?
   (a) tetrahydrocannabinol;   (c) caffeine;
   (b) ganciclovir;   (d) lysergic acid diethylamide, LSD.

Option E: environmental chemistry

Air pollution

An air pollutant can be described as a substance that is not normally present in air, or a substance that is normally present, but in excess amounts. The main **primary air pollutants** are carbon monoxide, oxides of nitrogen, sulfur dioxide, particulates and volatile organic compounds (Table 1). **Secondary air pollutants** are compounds formed when primary pollutants react in the air.

Methods of removal

Thermal exhaust reactor

Exhaust from the car engine is combined with more air, and reacts because of the heat of the exhaust gases. Carbon monoxide is converted into carbon dioxide, and unburned hydrocarbons are also combusted.

$$2CO(g) + O_2(g) \rightarrow 2CO_2(g)$$

$$C_xH_y(g) + \left(x + \frac{y}{4}\right)O_2(g) \rightarrow xCO_2(g) + \frac{y}{2}H_2O(g)$$

Lean-burn engines

By adjusting the carburettor the ratio of air to fuel can be altered. The higher the ratio, the less carbon monoxide emitted, as more

Table 1 Primary air pollutants

Primary pollutant	Natural source	Man-made source	Effect on health	Methods of reduction
Carbon monoxide (CO)	Incomplete oxidation of methane $$CH_4 + 1\frac{1}{2}O_2 \rightarrow CO + 2H_2O$$	Incomplete combustion of fossil fuels e.g. $$C_8H_{18} + 8\frac{1}{2}O_2 \rightarrow 8CO + 9H_2O$$	Prevents haemoglobin from carrying oxygen by forming carboxyhaemoglobin	Use of lean-burn engine, thermal exhaust reactor or catalytic converter
Oxides of nitrogen (NO_x) e.g. N_2O, NO and NO_2	Electrical storms and biological processes	At high temperatures inside internal combustion engine $$N_2 + O_2 \rightarrow 2NO$$	Respiratory irritant leading to respiratory tract infections plus problems caused by acid deposition	Use of lean-burn engine, recirculation of exhaust gases or catalytic converter
Sulfur dioxide (SO_2) (can be oxidized in the air to also form SO_3)	Oxidation of H_2S produced by volcanoes, and decay of organic matter	Combustion of sulfur-containing coal and smelting of sulfide ores $$S + O_2 \rightarrow SO_2$$	Respiratory irritant leading to respiratory tract infections plus problems caused by acid deposition	Removal of sulfur from fossil fuels before combustion. Alkaline scrubbing. Fluidized bed combustion
Particulates	Soot, ash, dust asbestos, sand, smoke, pollen, bacterial and fungal spores	Burning of fossil fuels, particularly coal and diesel	Can affect the respiratory system and cause lung diseases such as emphysema, bronchitis and cancer	Sedimentation chambers. Electrostatic precipitation
Volatile organic compounds (VOCs) (C_xH_y or R–H)	Plants e.g. rice. Many plants emit unsaturated hydrocarbons called terpenes	Unburned or partially burned gasoline and other fuels. Solvents.	Some (e.g. benzene) are known to be carcinogenic. Can form toxic secondary pollutants e.g. peroxyacylnitrates (PANs)	Catalytic converters

complete combustion occurs. Unfortunately this produces higher temperatures, so more oxides of nitrogen are produced. At lower ratios less NO_x but more carbon monoxide will be emitted.

Catalytic converter (see Chapter 3)

The hot exhaust gases are passed over a catalyst of platinum, rhodium or palladium. These fully oxidize carbon monoxide and unburned hydrocarbons, and also catalyse the reaction between carbon monoxide and nitrogen oxide.

$$2CO(g) + 2NO(g) \rightarrow 2CO_2(g) + N_2(g)$$

Sulfur dioxide

Sulfur dioxide can be removed from the exhaust of coal-burning plants by "scrubbing" with an alkaline solution of limestone, $CaCO_3$, and lime, CaO. The resulting sludge is used for landfill or as gypsum, $CaSO_4.2H_2O$, to make plasterboard.

$$CaCO_3(s) + SO_2(g) \rightarrow CaSO_3(s) + CO_2(g)$$
$$CaO(s) + SO_2(g) \rightarrow CaSO_3(s)$$
$$2CaSO_3(s) + O_2(g) + 4H_2O(g) \rightarrow 2CaSO_4.2H_2O(s)$$

A more modern method, known as **fluidized bed combustion**, involves burning the coal on a bed of limestone, which removes the sulfur as calcium sulfate(IV), $CaSO_3$, or calcium sulfate(VI), $CaSO_4$, as the coal burns.

Figure 1 A pressurized fluidized bed combustion (PFBC) plant in Japan.

Electrostatic precipitation

Particulates are solid or liquid particles suspended in the air. Larger particles can be allowed to settle under the influence of gravity in sedimentation chambers. For smaller particles an electrostatic precipitation chamber can be used. The charged particulates are attracted to the oppositely charged electrodes, which are shaken periodically so that the aggregated particulates fall to the bottom of the precipitator, where they can be removed.

Acid deposition

Acid deposition is more commonly called "acid rain", and we have discussed this in Chapter 8. Technically, acid rain, which also covers precipitation as snow, hail, fog and dew, is known as **wet acid deposition**. The acids in the atmosphere may also be transformed chemically into gases and salts before falling back to Earth under the influence of gravity. This is known as **dry acid deposition** and occurs, for example, with sulfur dioxide, which can be deposited as a gas or a salt. As we have seen, pure rainwater naturally contains dissolved carbon dioxide in the form of carbonic acid, which gives it a pH of 5.65. Acid rain is therefore described as precipitation with a pH of less than 5.6. It is caused by the presence of oxides of sulfur and nitrogen in the atmosphere.

Environmental effects of acid deposition

Acid rain can have serious effects on vegetation, fish and other aquatic life, buildings and humans (Figure 2).

Increased acidity in the soil leaches important nutrients such as Ca^{2+}, Mg^{2+} and K^+. Reduction in Mg^{2+} can cause reduction in chlorophyll levels and consequent reduction in the ability of plants to photosynthesize, resulting in stunted growth and loss of leaves. It also leaches aluminium from rocks into the soil. The presence of aluminium(III) ions, Al^{3+}, damages roots, and prevents trees from taking up enough water and nutrients to survive.

Aquatic life is highly sensitive to pH. Below pH 6 the number of sensitive fish such as salmon and minnow decline, as do insect larvae and algae. Snails cannot survive if the pH is less than 5.2 and below pH 5.0 many microscopic animal species disappear. Below pH 4.0 lakes are effectively dead.

Stone such as marble that contain calcium carbonate are eroded by acid rain. With sulfuric acid the calcium carbonate reacts to form calcium sulfate, which can be washed away by rainwater, thus exposing more stone to corrosion. Salts can also form within the stone that can cause the stone to crack and disintegrate.

$$CaCO_3(s) + H_2SO_4(aq) \rightarrow CaSO_4(aq) + CO_2(g) + H_2O(l)$$

Breathing in acidic air irritates the mucous membranes and increases the risk of respiratory illnesses such as asthma, bronchitis and emphysema. In acidic water there is more probability of poisonous ions such as Cu^{2+} and Pb^{2+} leaching from pipes, and high levels of aluminium in water may be linked to Alzheimer's disease, a form of senile dementia.

The greenhouse effect

The problems of global warming associated with the release of carbon dioxide from the combustion of fossils fuels and the way in which carbon dioxide acts as a greenhouse gas have been discussed in Chapter 5. In fact, because of its abundance, water is the main greenhouse gas in the atmosphere. However, the bonds in the carbon dioxide molecule absorb radiation of a different wavelength than the bonds in water molecules. Although it only constitutes 0.03% of the atmosphere, carbon dioxide therefore plays a key role in keeping the average global temperature at about 15 °C. The effects that other greenhouse gases have on global warming depend both on their concentration in the atmosphere (abundance) and on their ability to absorb infrared radiation. Apart from water, carbon dioxide contributes about 50% to global warming. Chlorofluorocarbons, CFCs, are thousands of times better at absorbing heat than carbon dioxide. However, because their concentration is so low their effect is limited to about 14%.

Table 2 lists the main greenhouse gases.

Figure 2 The effect of acid rain on trees: dead and dying Red Spruce trees in Maine, USA.

Acid deposition is a transboundary problem: that is, the acidic gases produced in one country or region can deposit elsewhere. Research the area you live in to find out the probable sources of any acid deposition, and suggest ways in which they could be counteracted.

Table 2 Greenhouse gases

Gas	Main source	Heat-trapping effectiveness compared with CO_2	Overall contribution to increased global warming
H_2O	evaporation of oceans	0.1	–
CO_2	combustion of fossil fuels and biomass	1	50%
CH_4	anaerobic decay of organic matter caused by intensive farming	30	18%
N_2O	artificial fertilizers and combustion of biomass	150	6%
O_3	secondary pollutant in photochemical smogs	2,000	12%
CFCs	refrigerants, propellants, foaming agents, solvents	10 000 – 25 000	14%

Sulfur hexafluoride

Sulfur hexafluoride, SF_6, has replaced the toxic PCBs in electrical generators as an insulating material. It is a very potent greenhouse gas, with a heat-trapping effectiveness 22 200 times that of carbon dioxide. However, at present it is present in the atmosphere only in small amounts, so does not contribute hugely to global warming.

Figure 3

Influence of greenhouse gases on global warming

Evidence from ice core samples in Greenland shows that there have been large fluctuations in global temperature in the past. However, most scientists now accept that the current rapid rate of global warming is a direct consequence of the increased emission of greenhouse gases, which have increased considerably since the onset of the Industrial Revolution.

The predicted consequences of global warming are complex and there is not always agreement about them, as it is difficult to prove direct links between increased greenhouse gas emissions and changes in weather patterns such as an increase in floods, storms and hurricanes. The two most likely major effects are:

- changes in agriculture and biodistribution as the climate changes
- rising sea levels due to thermal expansion and the melting of the polar ice caps and glaciers.

There is likely to be an increased risk of diseases such as malaria as mosquitoes spread to warmer climates. The economies of certain countries may suffer as tourism patterns change, and the problems of maintaining a sufficient global supply of fresh drinking water are likely to be exacerbated.

Ozone depletion (1)

Formation and depletion of ozone in the stratosphere

The ozone layer occurs in the stratosphere between about 12 km and 50 km above the surface of the Earth. Stratospheric ozone is in dynamic equilibrium with oxygen, and is continually being formed and decomposed. The strong double bond in oxygen is broken by high-energy ultraviolet radiation($\lambda = 242$ nm) from the Sun to form atoms. These oxygen atoms are called **radicals**, as they possess an unpaired electron and are very reactive. One oxygen radical can then react with an oxygen molecule to form an ozone molecule:

$$O{=}O(g) \xrightarrow{\text{UV (high energy)}} 2O^{\bullet}(g)$$
$$O_3(g) + O^{\bullet}(g) \rightarrow 2O_2(g)$$

The bonds in ozone are weaker, so ultraviolet radiation of lower energy ($\lambda = 330$ nm) will break them. When they are broken the reverse process happens, and the ozone breaks down back to an oxygen molecule and an oxygen radical. The radical can then react with another ozone molecule to form two oxygen molecules:

$$O_3(g) \xrightarrow{\text{UV (lower energy)}} O_2(g) + O^{\bullet}(g)$$
$$O_3(g) + O^{\bullet}(g) \rightarrow 2O_2(g)$$

Overall, the rate of production of ozone is equal to the rate of ozone destruction. This type of equilibrium is known as a **steady state**. Because the formation and depletion of ozone absorbs a wide range of ultraviolet radiation, the ozone layer serves to protect the surface of the Earth from this damaging radiation. This natural steady state has been altered by the ozone-depleting pollutants, which are mainly chlorofluorocarbons, CFCs, and oxides of nitrogen, NO_x. The evidence for this, and the Montreal Protocol, have already been discussed in Chapter 10.

Alternatives to CFCs

Since the implementation of the Montreal Protocol in 1987 the use of CFCs is being phased out. Even so, because of their low reactivity they are expected to remain in the atmosphere for at least the next 80 years. They destroy the ozone layer because the ultraviolet radiation breaks the relatively weak carbon-to-chlorine, C–Cl, bond. Substitutes for CFCs must have similar properties but not contain a bond that can be broken in ultraviolet radiation to form radicals. The properties required are low reactivity, low toxicity and low flammability, as well as no weak C–Cl bonds. They also should not absorb infrared radiation, otherwise they will act as greenhouse gases. The most immediate replacements are HCFCs—hydrochlorofluorocarbons such as CHF_2Cl—because they decompose more readily and do not build up in the stratosphere. Other alternatives actively being investigated for refrigerants are HFCs—hydrofluorocarbons such as CF_3CH_2F and hydrocarbons such as 2-methylpropane C_4H_{10}—but they suffer from being flammable and also contribute to global warming.

Dissolved oxygen in water
Importance of dissolved oxygen in water and biological oxygen demand

At a pressure of one atmosphere and a temperature of 20 °C the maximum solubility of oxygen in water is only about 9 ppm (i.e. 0.009 g dm^{-3}). Although this value is small, it is crucial, because most aquatic plants and animals require oxygen for aerobic respiration. Fish require the highest levels, and bacteria require the

Figure 4 A pile of old refrigerators. Before old refrigerators can be broken up or recycled, the CFCs they contain must first be removed to prevent them from being released directly into the atmosphere.

lowest levels. Fish need at least 3 ppm in order to be able to survive, but to maintain a balanced and diversified aquatic community the oxygen content should not be less than 6 ppm.

When organic matter decomposes aerobically in water it uses up the dissolved oxygen. The **biological (or biochemical) oxygen demand** (**BOD**) is a measure of the dissolved oxygen (in ppm) required to decompose the organic matter in water biologically. It is often measured over a set time period of 5 days. Water that has a high BOD without the means of replenishing oxygen will rapidly fail to sustain aquatic life. A fast-flowing river can recover its purity, because the water becomes oxygenated through the mechanical action of its flow. Lakes have relatively little flow, and reoxygenation is much slower or will not happen at all. Pure water has a BOD of less than 1 ppm; water with a BOD above about 5 ppm is regarded as polluted.

Measurement of BOD

The BOD of a sample of water can be determined by the **Winkler method**. The sample of the water is saturated with oxygen so that the initial concentration of dissolved oxygen is known. A measured volume of the sample is then incubated at a fixed temperature for five days while micro-organisms in the water oxidize the organic material. An excess of a manganese(II) salt is then added to the sample. Under alkaline conditions manganese(II) ions are oxidized to manganese(IV) oxide by the remaining oxygen:

$$2Mn^{2+}(aq) + 4OH^-(aq) + O_2(aq) \rightarrow 2MnO_2(s) + 2H_2O(l)$$

Potassium iodide is then added, which is oxidized by the manganese(IV) oxide in acidic solution to form iodine:

$$MnO_2(s) + 2I^-(aq) + 4H^+(aq) \rightarrow Mn^{2+}(aq) + I_2(aq) + 2H_2O(l)$$

The iodine released is then titrated with standard sodium thiosulfate solution:

$$I_2(aq) + 2S_2O_3^{2-}(aq) \rightarrow S_4O_6^{2-}(aq) + 2I^-(aq)$$

By knowing the number of moles of iodine produced, the amount of oxygen present in the sample of water can be calculated, and hence its concentration determined.

Aerobic and anaerobic decomposition

If sufficient oxygen is present, organic material will decay aerobically, and oxides or oxyanions are produced. Anaerobic decay involves organisms that do not require oxygen. The products are in the reduced form, and are often foul smelling and toxic (Table 3).

Table 3 Aerobic and anaerobic decomposition

Element	Aerobic decay product	Anaerobic decay product
carbon	CO_2	CH_4 (marsh gas)
nitrogen	NO_3^-	NH_3 and amines
hydrogen	H_2O	CH_4, NH_3, H_2S and H_2O
sulfur	SO_4^{2-}	H_2S ("rotten eggs" gas)
phosphorus	PO_4^{3-}	PH_3 (phosphine)

Eutrophication

Nitrates from intensive animal farming and excess use of artificial fertilizers, together with phosphates from artificial fertilizers and detergents, accumulate in lakes. They act as nutrients, and increase the growth of plants and algae. This can also happen in slow-moving areas of sea water. Normally, when plants and algae die they decompose aerobically and form carbon dioxide and water. However, if the growth is excessive and the dissolved oxygen is not sufficient to cope, anaerobic decomposition will occur. The hydrides formed, such as ammonia, hydrogen sulfide and phosphine, not only smell foul but they poison the water. More species will die, resulting in more anaerobic decay, and the lake itself becomes devoid of life—a process known as **eutrophication** (Figure 5).

Thermal pollution

The solubility of oxygen in water is temperature dependent. As the temperature is increased, the solubility drops. At the same time the metabolic rate of fish and other organisms increases, so the demand for oxygen increases. Many industries use water as a coolant, and the careless discharge of heated water into rivers can cause considerable thermal pollution.

Water treatment

Toxins in polluted water

> #### *Silent Spring*
>
> *Silent Spring* was written by the American zoologist Rachel Carson (1907–1964). Published in 1962, it is widely regarded as the start of the environmental movement in the West. It details the effect of pesticides, particularly DDT, on birds, and led to the banning of the use of DDT in many countries. 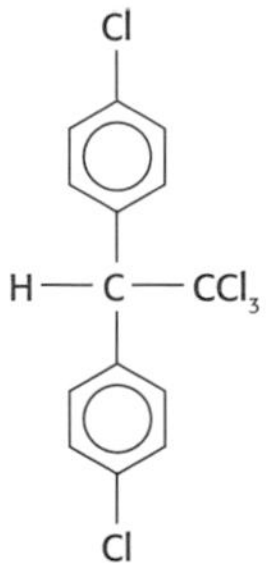
>
> **Figure 6** 1,1,1-trichloro-2,2-bis (4-chlorophenyl)ethane, previously known as *d*ichloro*d*iphenyl*t*richloroethane, DDT.

Pollutants in water can come from a wide variety of sources:

- **Pesticides**: These include insecticides such as DDT, herbicides such as paraquat, and fungicides.
- **Dioxins** (Figures 7 and 8): These are formed when waste materials containing organochlorine compounds are not incinerated at high enough temperatures. They are very persistent in the environment, and extremely toxic as they accumulate in fat and liver cells.

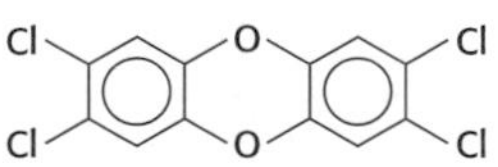

Figure 5 The effects of eutrophication.

Figure 7 Spraying the herbicide Agent Orange, which contained dioxins, during the Vietnam war in the 1960s.

Figure 8 An example of a dioxin: 2,3,7,8-tetrachlorodibenzo-p-dioxin.

- **Polychlorinated biphenyls PCBs** (Figure 9): These contain from one to ten chlorine atoms attached directly to a biphenyl molecule. They are chemically stable and have high electrical resistance, so are used in transformers and capacitors. Like dioxins they persist in the environment and accumulate in fatty tissue. They affect reproductive efficiency, impair learning ability in children, and are thought to be carcinogenic.
- **Nitrates**: Nitrates enter drinking water from intensive animal farming, and from the use of artificial fertilizers. They also enter the water from acid rain. Because all nitrates are soluble they are not easily removed during treatment of sewage, and so tend to accumulate.
- **Heavy metals**: These include cadmium, lead and mercury:
  - **cadmium**: from water effluent in zinc mining areas, rechargeable batteries, metal plating, orange pigment in enamels and some paints
  - **mercury**: from fungicides for seed dressings, batteries, mercury thermometers, industrial electrolysis of brine using a mercury electrode.
  - **lead**: from use of lead tetraethyl in leaded gasoline, old paintwork, lead water pipes, car batteries, fishing weights.

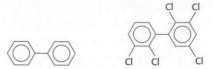

Figure 9 The biphenyl molecule (left), and an example of a polychlorinated biphenyl, PCB (right).

Waste water treatment

Waste water treated at sewage works contains floating matter, suspended matter, colloidal matter, dissolved ions and a range of micro-organisms. The aim of the treatment is to remove this additional material and recycle the fresh water. There are essentially three separate stages.

Primary treatment

Primary treatment effectively removes about 60% of the solid material and about a third of the oxygen-demanding wastes. The incoming sewage is passed through coarse mechanical filters to remove large objects such as sticks, paper, condoms and rags. It is then passed into a grit chamber, where sand and small objects settle. From there it passes into a sedimentation tank, where suspended solids settle out as sludge. A mixture of calcium hydroxide and aluminium sulfate is added to aid this process:

$$Al_2(SO_4)_3(aq) + 3Ca(OH)_2(aq) \rightarrow 2Al(OH)_3(s) + 3CaSO_4(aq)$$

The two chemicals combine to form aluminium hydroxide, which precipitates, carrying with it suspended dirt particles—a process known as **flocculation**. Grease is removed from the surface tank by skimming. The effluent is then discharged into a waterway, or passed on for secondary treatment.

Secondary treatment

Secondary treatment (Figure 10) removes up to 90% of the oxygen-demanding wastes (organic material and dangerous bacteria). The principle is to degrade the waste aerobically using oxygen and bacteria. One method uses trickle filters. The organic material is degraded by bacteria as the waste water trickles through a bed of

Figure 10 A secondary sewage treatment plant at Shek Wu Hui, Hong Kong.

stones. A more effective method is the **activated sludge** process. The sewage is aerated with pure oxygen in a sedimentation tank. The sludge that settles out contains active micro-organisms that digest organic waste, and some of it is recycled. The water that emerges is then discharged into a waterway, where it will be disinfected with chlorine or ozone before being fit for drinking.

Tertiary treatment

Tertiary treatment removes heavy metal ions, phosphates and nitrates. Heavy metal ions and phosphates can be removed by precipitation. Aluminium sulfate or calcium oxide can be used to precipitate phosphates:

$$Al^{3+}(aq) + PO_4^{3-}(aq) \rightarrow AlPO_4(s)$$
$$3Ca^{2+}(aq) + 2PO_4^{3-}(aq) \rightarrow Ca_3(PO_4)_2(s)$$

Heavy metal ions can be precipitated as insoluble hydroxides or basic salts by the addition of calcium hydroxide or sodium carbonate. For example:

$$Cr^{3+}(aq) + 3OH^-(aq) \rightarrow Cr(OH)_3(s).$$

Nitrates are difficult to remove by chemical means as all common nitrates are soluble, so precipitation cannot be used. Anaerobic denitrifying bacteria can reduce them to nitrogen, or the water can be passed through algal ponds, where the algae utilize the nitrate as a nutrient.

> ### NEWater in Singapore
>
> The Public Utilities Board (PUB) in Singapore has a new plant to purify used water to produce NEWater. The water undergoes stringent purification and treatment using advanced dual-membrane (microfiltration and reverse osmosis) and ultraviolet technologies. Figure 10 shows the reverse osmosis membrane in use. This technology even removes nitrate ions from the water.

Figure 11 NEWater in Singapore.

Soil

Soil is a complex mixture of inorganic and organic matter, including living organisms. Soil degradation, which lowers crop production, can be caused by a variety of human-made factors, such as acidification, desertification, contamination, erosion and salinization. In some areas of the world, particularly in countries most dependent upon agriculture for their income, there is considerable concern about the current rate of soil degradation. In Haiti, for example, erosion from timber cutting, overgrazing and other human activities has left up to one-third of the land irreversibly damaged.

- **Salinization** is the result of the continual irrigation of soil. After the water has evaporated the dissolved salts remain behind. The salts themselves can build up to toxic levels, but usually the plants die because they are unable to take up water from the salty soil.

97% of the water in the world is salt water. Research how fresh water can be obtained from sea water by (a) reverse osmosis and (b) multi-stage distillation. Evaluate the two processes in terms of cost and efficiency.

- **Nutrient depletion**: When crops are harvested, they remove the nutrients and minerals they have absorbed from the soil while growing. Because of increasing intensive farming without proper management practices and external inputs, these nutrients are not being replaced.
- **Soil pollution** is caused by the use of pesticides and fertilizers. These chemicals can disrupt the food web, reduce the soil's biodiversity, and ultimately ruin the soil. The pollutants can also run into surface waters and move through the soil to pollute groundwater.

Soil organic matter

It is well known that animal dung and compost are good nutrients for the soil, providing not only minerals but also organic matter. **Soil organic matter (SOM)** is the term generally used to represent the organic constituents of the soil. It includes plant and animal tissues, their partial decomposition products, and the soil biomass.

Identifiable in the SOM from the decomposition of plants are high molecular mass organic materials such as polysaccharides and proteins, and simpler substances such as sugars, amino acids and other small molecules. The end product of this is known as **humus**, which is therefore a mixture of simple and more complex organic chemicals of plant, animal, or microbial origin. Biologically, humus provides a source of energy and a source of the essential nutrient elements phosphorus, nitrogen and sulfur. Physically, humus helps the soil to retain moisture and therefore increases its capacity to withstand drought conditions, and it encourages the formation of good soil structure. Because of its dark colour humus absorbs heat and thus helps to warm the cold soil during spring. Chemically, humus can act rather like clay with its **cation exchange capacity**. It contains active sites that enable it to bind to nutrient cations. This not only makes them more available to plants but also prevents them from being washed away by rain or during irrigation. Toxic cations, such as those from heavy metals, can also bind to the humus, preventing them from entering the wider ecosystem. It also enhances the ability of the soil to maintain a constant pH by acting as an acid–base buffer.

There are many different organic compounds that can pollute the soil. Make a list of the following and research one major source of each type of pollutant.
Agrichemicals, volatile organic compounds (VOCs), solvents, polyaromatic hydrocarbons (PAHs), polychlorinated biphenyls (PCBs), organotin compounds and semi-volatile organic compounds (SVOCs).

Child labour on rubbish tips in Brazil

"Hazardous child labour is a betrayal of every child's rights as a human being and is an offence against our civilization": UNICEF's *The State of the World's Children,* 1997. Even so, many thousands of children still work scavenging on the rubbish tips of major cities such as São Paulo in Brazil.

Figure 12

Waste

Dealing with waste materials is one of the most pressing problems of our times. In the past much waste was just dumped, burned, or placed in landfill sites. Incineration and landfill are still heavily used (see Figure 12), but there is an increasing trend to recycle many materials. Table 4 lists methods of waste disposal.

Table 4 Methods of waste disposal

Method	Advantages	Disadvantages
Landfill	Efficient method to deal with large volumes Filled land can be used for building or other community purposes	Local residents may object to new sites Once filled, needs time to settle, and may require maintenance as methane released
Open dumping	Inexpensive Convenient (to the dumper)	Causes air and ground water pollution Health hazard: encourages rodents and insects Unsightly
Ocean dumping	Source of nutrients. Convenient and inexpensive	Danger to marine animals Pollutes the sea
Incineration	Reduces volume Requires minimal space Produces stable, odour-free residue Can be used as a source of energy	Expensive to build and operate Can cause pollutants, e.g. dioxins, if inefficiently burned Requires energy
Recycling	Provides a sustainable environment	Expensive Difficulty in separating different materials; not possible in all cases

Recycling

The importance of recycling metals such as aluminium has already been discussed in Chapter 9. Recycling is not a new phenomenon. Many materials are now recycled in many countries, and local councils arrange collection times or places for different types of materials. There are some specialist types of recycling, such as used printer cartridges, but many countries now routinely arrange for the recycling of paper, metals, glass and plastics (Table 5).

Table 5 Recycling

Material	Description	Comments
Metals	Mainly aluminium and steel. The metals are sorted, then melted and either reused directly or added to the purification stage of metals formed from their ores.	Particularly important for metals such as aluminium, which require large amounts of energy to produce direct from ore.
Paper	Taken to the plant, sorted into grades. Washed to remove inks etc., made into a slurry to form new types of paper, such as newspaper and toilet rolls.	Energy required to transport. Composting may be as efficient.
Glass	Sorted by colour, washed, crushed, then melted and moulded into new products	Glass is not degraded during the recycling process, so can be recycled many times.
Plastics	Industrial plastic is already sorted, but household waste plastic must first be sorted. Degraded to monomers by pyrolysis, hydrogenation, gasification and thermal cracking, then repolymerized.	Fewer pollutants formed, and energy used to recycle rather than make from crude oil. Better to reuse than recycle.

The American Society of Plastics has developed a code to help consumers identify and sort different types of plastic (Table 6).

Table 6 Plastics recycling codes

Code symbol	Abbreviation	Name	Typical uses
1	PET	polyethylene terephthalate	fizzy drink bottles and oven-ready meal trays
2	HDPE	high-density polyethylene	bottles for milk and washing-up liquids
3	PVC	polyvinyl chloride	food trays, cling film, bottles for squash, mineral water and shampoo
4	LDPE	low-density polyethylene	carrier bags and bin liners
5	PP	polypropylene	margarine tubs, microwaveable meal trays
6	PS	polystyrene	yogurt pots, foam meat or fish trays, hamburger boxes and egg cartons, vending cups, plastic cutlery, protective packaging for electronic goods and toys
7	OTHER	any other plastics that do not fall into any of the above categories	e.g. melamine, which is often used in plastic plates and cups

Radioactive waste

Essentially nuclear waste can be divided into high-level waste and low-level waste. **Low-level waste** includes items such as rubber gloves, paper towels and protective clothing that have been used in areas where radioactive materials are handled. The level of activity is low, and the half-lives of the radioactive isotopes are generally short. **High-level waste** has high activity, and generally the isotopes have long half-lives, so the waste will remain active for a long period. Most high-level waste comes from spent fuel rods or the reprocessing of spent nuclear fuel.

Storage and disposal of nuclear waste

Low-level waste

Different methods are used to dispose of low-level waste. Although many governments have now banned the practice, some is simply discharged straight into the sea, where it becomes diluted. As the decay produces heat, it is better to store the waste in vast tanks of cooled water called **ponds**, where it can lose much of its activity. Before it is then discharged into the sea it is filtered through an ion exchange resin, which removes strontium and caesium, the two elements responsible for much of the radioactivity. Other methods of disposal include keeping the waste in steel containers inside concrete-lined vaults.

High-level waste

During reprocessing of spent fuel about 96% of the uranium is recovered for reuse. About 1% is plutonium, which is a valuable fuel in fast breeder reactors. The remaining 3% is high-level liquid waste. One method used to treat this is **vitrification**. The liquid

1 How can people be encouraged to recycle more of their waste materials? Should it be through education, through strict laws, through a combination of both, or by other means? Is recycling always beneficial?

2 Because of global warming, and since oil production has probably reached its peak, there is increasing pressure to build nuclear power stations to provide the energy to generate electricity. Discuss the advantages and disadvantages of building new nuclear power stations.

waste is dried in a furnace and then fed into a melting pot together with glass-making material. The molten material is then poured into stainless steel tubes, where it solidifies. Air flows round the containers to keep them cool. Because of the high activity and long half-lives some of the waste will remain active for hundreds if not thousands of years. The problem is how to store it safely for this length of time. Currently the best solution seems to be encasing it in concrete and burying it in deep remote places that are geologically stable, such as in disused mines or granite rock. The concern is that the radioactive material may eventually leach into the water table and then into drinking water.

HL : Ozone depletion (2)

The energy of the ultraviolet radiation absorbed by ozone in the ozone layer can be calculated using bond enthalpies. The bond enthalpy at 298 K for the oxygen-to-oxygen double bond, O=O, is given in the IB data booklet as 496 kJ mol^{-1}. For just one double bond this equates to 8.235×10^{-19} J. The wavelength of light (λ) that corresponds to this enthalpy value (E) can be calculated by combining the expressions $E = hf$ and $c = \lambda f$ to give

$$\lambda = \frac{hc}{E}$$

where h represents Planck's constant, and c represents the velocity of light. So

$$\lambda = \frac{6.626 \times 10^{-34}\ (Js) \times 2.998 \times 10^{8}\ (ms^{-1})}{8.235 \times 10^{-19}\ (J)}$$

$$= 241\ nm$$

This is in the high-energy region of the ultraviolet spectrum. Ozone can be described as a hybrid of two resonance structures (Figure 13). An alternative bonding model is to consider the pi electrons to be delocalized over all three oxygen atoms. In both models the bond order is 1.5—that is, between an O–O single bond ($\Delta H = 146$ kJ mol^{-1}) and an O=O double bond—so ultraviolet light with a longer wavelength (lower energy) is absorbed in breaking the ozone bond.

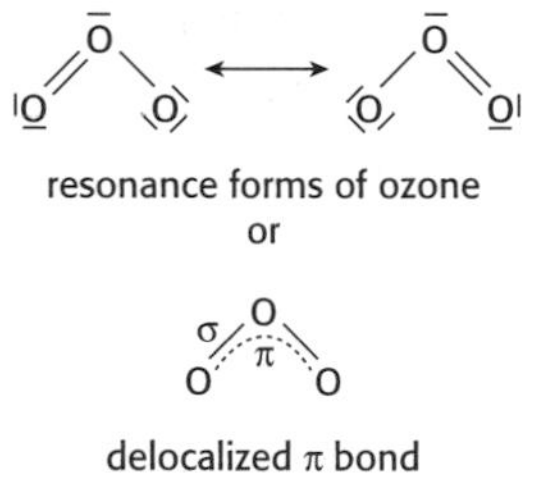

resonance forms of ozone
or

delocalized π bond

Figure 13 Bonding in ozone.

Catalysis of ozone decomposition by CFCs and nitrogen oxides

The free radical mechanism for the breakdown of ozone by CFCs has been discussed in Chapter 10. A complex free radical mechanism also occurs with oxides of nitrogen that are released from jet airplanes. Essentially, nitrogen dioxide reacts with ozone to form nitrogen monoxide and an oxygen free radical:

$$NO_2(g) + O_3(g) \rightarrow NO(g) + O^{\bullet}(g) + O_2(g)$$

Both the oxygen radical and the nitrogen monoxide can react separately to decompose ozone. As more nitrogen dioxide is formed the process is repeated:

$$O^{\bullet}(g) + O_3(g) \rightarrow 2O_2(g)$$
$$NO(g) + O_3(g) \rightarrow NO_2(g) + O_2(g)$$

Reasons for greater ozone depletion in the polar regions

The photographs in Chapter 10 show the "hole" in the ozone layer over Antarctica. During the winter the temperatures get very low in the stratosphere above the poles. At these low temperatures the small amount of water vapour in the air freezes to form crystals of ice. The crystals also contain small amounts of molecules such as hydrogen chloride (HCl) and chlorine nitrate ($ClONO_2$). It is believed that catalytic reactions occur on the surface of the ice crystals to produce species such as hypochlorous acid (HClO) and chlorine (Cl_2). With the advent of spring the sunlight causes these molecules to break down to produce chlorine radicals, which catalyse the destruction of ozone. The largest holes occur during early spring. As the Sun warms the air the ice crystals disperse, warmer winds blow into the region, and the ozone concentration gradually increases again.

Smog

Formation of photochemical smogs

Photochemical smogs tend to occur on windless days in cities such as Beijing and Los Angeles which are "bowl shaped" (that is, surrounded by higher ground in all directions) so that the movement of air is prevented. Smogs are most likely to occur when there is a **temperature inversion**. Normally the temperature decreases with altitude. Warm air rises, taking the pollutants with it, and is replaced by cleaner cooler air. However, sometimes atmospheric conditions cause a layer of still warm air to blanket a layer of cooler air. The trapped pollutants cannot rise, and if these conditions persist the concentration of pollutants in the warm air near the ground can increase to dangerous levels. The primary pollutants are mainly oxides of nitrogen and volatile organic compounds from internal combustion engines. Under the influence of sunlight these are converted into secondary pollutants.

Formation of secondary pollutants in photochemical smogs

During the early morning rush hour there is a build-up of hydrocarbons and nitrogen oxides from car exhausts. As the Sun comes out, the nitrogen dioxide absorbs sunlight and breaks down to form radicals:

$$NO_2(g) \rightarrow NO(g) + O^{\bullet}(g)$$

These oxygen radicals can react with oxygen to form ozone and with water to form hydroxyl radicals. These secondary photochemical oxidants can react with a variety of molecules, including nitrogen oxides to form nitric acid and hydrocarbons to form organic peroxides (ROOR), aldehydes (RCHO) and ketones (RCOR). Peroxides are extremely reactive, and aldehydes and ketones reduce visibility by condensing to form aerosols.

$O^{\bullet}(g) + H_2O(l) \rightarrow 2OH^{\bullet}(g)$	(formation of hydroxyl radicals)
$OH^{\bullet}(g) + NO_2(g) \rightarrow HNO_3(aq)$	(nitric acid formation)
$OH^{\bullet}(g) + RH(g) \rightarrow R^{\bullet}(g) + H_2O(l)$	(radical propagation)
$R^{\bullet}(g) + O_2(g) \rightarrow ROO^{\bullet}(g)$	(peroxide radical formation)

Chain termination can occur when peroxide radicals react with nitrogen dioxide to form **peroxyacylnitrates (PANs)** (Figure 13). These compounds are eye irritants and are toxic to plants.

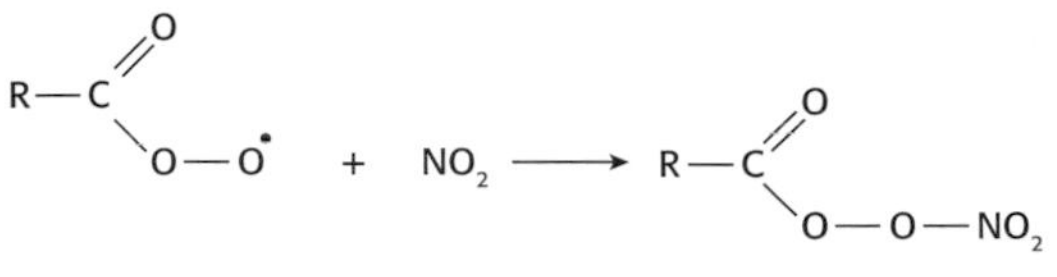

peroxyacylnitrate, PAN

Figure 13 Formation of peroxyacylnitrate, PAN.

Acid deposition (2)

Mechanism of acid deposition

The primary pollutants sulfur dioxide and nitrogen oxides are converted into acids in the air through a radical mechanism involving hydroxyl radicals, OH$^{\bullet}$. These hydroxyl radicals are formed from the reaction of water vapour with either ozone or oxygen radicals formed from the decomposition of ozone:

$$H_2O(g) + O_3(g) \rightarrow 2HO^{\bullet}(g) + O_2(aq)$$

or

$$H_2O(g) + O^{\bullet}(g) \rightarrow 2HO^{\bullet}(g)$$

The hydroxyl radicals then react directly with sulfur dioxide and nitrogen oxides in the presence of water to give the dissolved acids:

$$HO^{\bullet}(g) + NO_2(g) \rightarrow HNO_3(aq)$$
$$HO^{\bullet}(g) + NO(g) \rightarrow HNO_2(aq)$$
$$HO^{\bullet}(g) + SO_2(g) \rightarrow HOSO_2^{\bullet}(g)$$

then

$$HOSO_2^{\bullet}(g) + O_2(g) \rightarrow HO_2^{\bullet}(g) + SO_3(g)$$

followed by

$$SO_3(g) + H_2O(l) \rightarrow H_2SO_4(aq)$$

The role of ammonia in acid deposition

Ammonia can to some extent neutralize the acids in the atmosphere to form ammonium sulfate, $(NH_4)_2SO_4$, and ammonium nitrate, NH_4NO_3. These ammonium salts, which are the product of a weak base and a strong acid, are slightly acidic. As they sink to the ground or are washed out by precipitation the ammonium ion is deposited and enters the soil, where acidification and nitrification can occur:

$$NH_4^+(aq) + 2O_2(g) \rightarrow 2H^+(aq) + NO_3^-(aq) + H_2O(l)$$

Water and soil

Removal of heavy metal ions and phosphates from water by precipitation

We have seen that heavy metal ions and phosphate ions can be removed from water during the tertiary treatment of waste water by precipitation. This assumes that the salts formed are insoluble in water. In fact, even so-called insoluble salts such as silver chloride dissolve to a very small extent in water, so not all the heavy metal will be precipitated out. For a salt formed from a metal M with a non-metal X:

$$MX(s) \rightleftharpoons M^+(aq) + X^-(aq)$$

The equilibrium expression will be

$$K_{sp} = [M^+(aq)] \times [X^-(aq)]$$

where the equilibrium constant K_{sp} is known as the **solubility product**. This is actually an example of heterogeneous equilibrium; MX(s) does not appear in the equilibrium expression as it is a solid. Salts that are very insoluble in aqueous solution will have very small values for K_{sp}. For example, the solubility product of silver chloride is equal to 1.8×10^{-10} at 298 K. Using this value we can calculate the quantity of silver ions that will actually dissolve in one litre of water at this temperature:

$$AgCl(s) \rightleftharpoons Ag^+(aq) + Cl^-(aq)$$
$$K_{sp} = [Ag^+(aq)] \times [Cl^-(aq)]$$

but

$$[Ag^+(aq)] = [Cl^-(aq)]$$

Therefore

$$K_{sp} = [Ag^+(aq)]^2 = 1.80 \times 10^{-10} \text{ at 298 K}$$

and

$$[Ag^+(aq)] = (1.80 \times 10^{-10})^{1/2} = 1.34 \times 10^{-5} \text{ mol dm}^{-3}$$

Hence the concentration of silver ions in the aqueous solution is 1.34×10^{-5} mol dm^{-3}. As the relative atomic mass of silver is 107.87, the mass of silver ions that dissolves in 1 litre of water at 298 K is therefore $107.87 \times 1.34 \times 10^{-5}$ g, which is equal to 1.45×10^{-3} g. Even though this is small, it is still an appreciable amount.

We can reduce this much further by adding more chloride ions to the solution so that now the concentration of the silver ions is not the same as the concentration of the chloride ions. This is known as the **common ion effect**. If the concentration of the chloride ions is made to be 1.00 mol dm^{-3} then since

$$K_{sp} = [Ag^+(aq)] \times [Cl^-(aq)]$$

and

$$[Cl^-(aq)] = 1.00 \text{ mol dm}^{-3}$$

$$K_{sp} = [Ag^+(aq)] = 1.80 \times 10^{-10} \text{ at 298 K}$$

The concentration of silver ions remaining in the solution at 298 K is now only 1.80×10^{-10} mol dm^{-3}, and the mass of silver ions remaining in 1 litre is just 1.94×10^{-8} g.

For the phosphate and heavy metal ions found in waste water it is slightly more complicated as they are not uni-negative or uni-positive ions, but the principle is the same. For example, chromium ions in waste water can be precipitated by adding hydroxide ions:

$$Cr^{3+}(aq) + 3OH^- \rightleftharpoons Cr(OH)_3(s)$$

The solubility product for chromium(III) hydroxide at 298 K is 1.00×10^{-33}.

To calculate how much chromium would remain dissolved in 1 litre of water we can write the equilibrium expression for the reverse reaction:

$$K_{sp} = [Cr^{3+}(aq)] \times [OH^-(aq)]^3$$

Assuming that all the hydroxide ions in the solution come just from the dissolved chromium(III) hydroxide then

$$[OH^-(aq)] = 3[Cr^{3+}(aq)]$$

Therefore

$$K_{sp} = 1.00 \times 10^{-33} = [Cr^{3+}(aq)] \times (3[Cr^{3+}(aq)])^3 = 27[Cr^{3+}(aq)]^4$$

Hence

$$[Cr^{3+}(aq)] = (1.00/27 \times 10^{-33})^{1/4} = (3.70 \times 10^{-35})^{1/4}$$
$$= 5.93 \times 10^{-9} \, mol \, dm^{-3}.$$

This small amount can be reduced even more by using the common ion effect, that is, by adding more hydroxide ions.

Some metal hydroxides, however, re-dissolve in the presence of excess hydroxide ions because of the formation of soluble complexes, such as $[Zn(OH)_4]^{2-}(aq)$. Heavy metal ions such as zinc (Zn^{2+}), mercury (Hg^{2+}) and cadmium (Cd^{2+}) are precipitated as insoluble sulfides by bubbling hydrogen sulfide into the water: for example,

$$Cd^{2+}(aq) + H_2S(g) \rightarrow CdS(s) + 2H^+(aq)$$

Cation exchange capacity

Cation exchange capacity (CEC) has already been referred to regarding soil organic matter (SOM). Both the soil organic matter and the clay particles in the soil have a negative charge, and will attract and bond to positive ions (cations). These ions may be classified as basic ions such as calcium (Ca^{2+}), magnesium (Mg^{2+}), sodium (Na^+) or potassium (K^+), or acidic ions such as hydrogen (H^+) or aluminium ions (Al^{3+}). The quantity of positively charged cations that a soil can hold is described as the CEC. The larger the CEC the more cations the soil can hold. These cations are exchanged with cations such as hydrogen ions on the root hairs of plants and thus provide nutrients to the plant. When soil is analysed, the total concentration of basic cations compared with the total concentration of acidic cations can be obtained. The more acidic the soil is (that is, the lower the pH), the higher the percentage of acidic cations. Soil pH is important, because acid cations such as aluminium ions are harmful to plants. This is the problem caused by acid rain, because it lowers the soil pH, which increases the quantity of aluminium ions. Above pH 5 aluminium ions are not present because they are precipitated out of the soil solution. Soil has a buffering capacity, but even so it is sometimes necessary to add lime to soil to raise the pH and increase the concentration of basic cations held by the clay and soil organic material.

In addition to the nutrient cations required by the plants, the SOM can also bind to organic and inorganic compounds in the soil, which helps to reduce the negative environmental effects of contaminants such as pesticides, heavy metal ions and other pollutants.

1 Ozone absorbs ultraviolet radiation with a wavelength of 330 nm (3.30×10^{-7} m). Use this information to calculate the value for the bond enthalpy of the oxygen-to-oxygen bond in ozone. How does this compare with the bond enthalpy for the oxygen-to-oxygen bond in (a) oxygen gas and (b) organic peroxides?

2 The solubility product for aluminium phosphate at 298 K is 1.40×10^{-21}. Phosphate ions can be removed from waste water by adding aluminium ions. Write the equation for the precipitation reaction that occurs, and determine the concentration of phosphate ions remaining in the water after the precipitation has occurred (assume that no excess aluminium ions are present).

3 Suggest practical solutions to reduce the amount of photochemical smog in cities.

Food groups: requirements for a healthy diet

Food can be defined as any substance, whether processed, partially processed or raw, that is intended for human consumption. It includes beverages, chewing gum and any substance used in the preparation or manufacture of "food", but does not include cosmetics, tobacco, or substances used only as drugs.

The function of nutrients that are obtained from food is to keep the body functioning and healthy. Nutrients provide energy, regulate growth, and replenish chemicals for the maintenance and repair of the body's tissues. Good health requires a balanced diet that includes all the essential nutrients taken from as wide a variety of foods as possible.

Nutrients can be divided into six main groups:

- proteins
- carbohydrates
- lipids
- vitamins
- minerals
- water.

The amount of each required depends on several factors, such as age, weight, gender and occupation. A well-balanced diet consists of about 60% carbohydrate, 20–30% protein, and 10–20% fats. Foods containing these three components will also provide the essential vitamins and the 15 essential minerals, which include calcium, magnesium, sodium, iron and sulfur along with trace elements such as iodine and chromium. In addition, a daily intake of about 2 dm^3 of water is required. Malnutrition can occur when either too little or too much of these essential components is taken.

You should know the chemical composition and structures of lipids (which consist mainly of fats and oils), carbohydrates (compounds with the empirical formula CH_2O, which contain one carbonyl group and at least two hydroxyl groups), and proteins. These have all been discussed in Chapter 15.

Guidelines for healthy eating

- Eat a variety of foods.
- Maintain a healthy weight.
- Eat a diet low in fat, saturated fat and cholesterol.
- Include plenty of fruit and vegetables.
- Use salt and sugar sparingly.
- Moderate the intake of alcohol.

Fats and oils

As described in Chapter 15, fats and oils are triesters (triglycerides) formed from the condensation reaction of propane-1,2,3-triol (glycerol) with long-chain carboxylic acids (fatty acids). You should read this section and familiarize yourself with the difference between saturated and unsaturated fatty acids, and the difference between *cis*-fatty acids and *trans*-fatty acids. Most naturally occurring fats contain a mixture of saturated, mono unsaturated and poly unsaturated fatty acids, and are classified according to the predominant type of unsaturation present. Solid fats tend to be more saturated than oils (liquid fats).

Melting points of fats

The melting point of fatty acids increases with increasing relative molecular mass. It also depends on the amount of unsaturation. Stearic acid (melting point: 69.6 °C) and linoleic acid (melting point: −5.0 °C) contain the same number of carbon atoms and have similar molecular masses (Figure 1). However, linoleic acid contains two carbon-to-carbon double bonds. Generally, the more unsaturated the fatty acid is, the lower is its melting point and the less crystalline it will be. The regular tetrahedral arrangement or carbon "backbone" of saturated acids means that they can pack together closely, so the van der Waals' forces holding molecules together are stronger as the surface area between them is greater. As the bond angle at the C=C double bonds changes from 109.5° to 120° in unsaturated fatty acids it produces a "kink" in the chain. The molecules are unable to pack so closely, and the van der Waals' forces between the become weaker, which results in lower melting points.

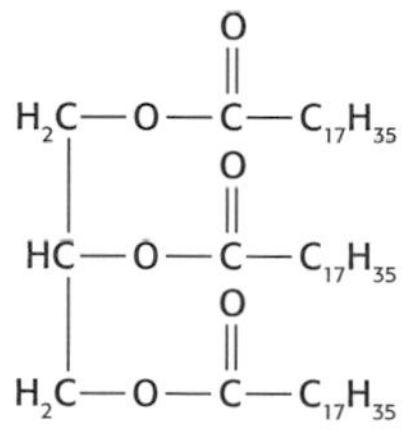

1 Draw the structural formulas and give the IUPAC names for: stearic acid, palmitic acid, oleic acid and linoleic acid. Why do you think these acids are still mainly referred to by their common names?

2 Draw the structural formula of a fat molecule formed from one molecule of propane-1,2,3-triol (glycerol), one molecule of stearic acid, one molecule of palmitic acid, and one molecule of oleic acid. How would you classify this fat molecule?

3 What is meant by the term *omega-three fatty acid*?

A saturated fat which is a solid at room temperature made from glycerol and stearic acid, a saturated fatty acid with a melting point of 69.6 °C

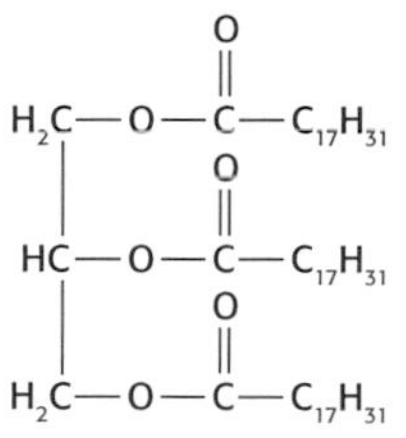

An unsaturated fat (oil) which is liquid at room temperature made from glycerol and linoleic acid, a polysaturated fatty acid with a melting point of −5.0 °C

Figure 1 Saturated and unsaturated fats.

The way in which unsaturated fatty acids can pack also depends on the geometrical isomerism of the double bonds. *Trans*-fatty acids can pack more closely, and so have higher melting points than *cis*-fatty acids, where both hydrogen atoms are on the same side of the double bond. These packing arrangements in fatty acids are similar in fats and oils and explain why unsaturated fats (oils) have lower melting points than saturated fats.

Examples of fats and oils

● Mainly saturated fats: palm, coconut, lard, butter, shortening, tallow

- Mainly mono unsaturated oils: olive, canola, peanut
- Mainly poly unsaturated oils: safflower, sunflower, linoleic, linolenic

Chemical stability of fats and oils

Unsaturated fats are less stable, chemically, than saturated fats. This is because the carbon-to-carbon double bonds in unsaturated fats can react with water and undergo hydrolysis in the presence of heat or enzymes. They can also react with oxygen (auto-oxidation), light (photo-oxidation) and hydrogen (hydrogenation).

The addition of hydrogen in the presence of heat (140–225 °C) and a finely divided nickel (or zinc or copper) catalyst increases the amount of saturation, and the importance of this in the manufacture of margarine has already been discussed in Chapter 10. The advantages of hydrogenating fats and oils include increasing the melting point and hardness and increasing the chemical stability by making them less susceptible to oxidation. However, there are also disadvantages. Oils, naturally, contain only *cis*-fatty acids, but under partial hydrogenation oils containing *trans*-fatty acids can form.

Unlike natural mono- and poly unsaturated oils, which increase HDL cholesterol and are healthier for the heart (see Chapter 15), oils containing *trans*-fatty acids behave more like saturated fats and increase levels of LDL cholesterol, leading to an increased risk of heart disease. Fats containing *trans*-fatty acids are harder to metabolize and excrete, so they accumulate in fatty tissue. They are also a lower-quality energy source.

Shelf life

In many areas of the world almost all packaged food on sale has one or more dates stamped on it. These may be labelled as *best before*, *use by* or *display until*. Essentially these dates are detailing the **shelf life** of the food, which is when the food no longer maintains the expected quality desired by the consumer because of changes in flavour, smell, texture and appearance (such as colour), and because of microbial spoilage. A food that has passed its shelf life may still be safe to consume, but optimal quality is no longer guaranteed.

Chemical factors that cause a decrease in shelf life include:

- **water content**: a change in the water content can cause loss of nutrients, browning and rancidity. Dry foods become more vulnerable to microbial spoilage if they absorb water.
- **change in pH**: can cause changes in flavour, colour, browning and loss of nutrients. light. Can lead to rancidity, loss of vitamins and fading of natural colours.
- **temperature**: a higher temperature increases the rate of other forms of spoilage.
- **exposure to air**: can increase the rate of oxidation, causing browning, changes in flavour and colour, and loss of nutrients.

Rancidity

Rancidity occurs in lipids, and is perceived by our senses as when they have gone "off" owing to a disagreeable smell, taste, texture or

Figure 2 Packaging showing the "display until" and "use by" dates. (photo by Geoff Neuss)

appearance. It is caused either by the hydrolysis of the triesters or by the oxidation of the fatty acid chains.

H₂C—O—C(=O)—R
|
HC—O—C(=O)—R' + 3H₂O ⟶ RCOOH + R'COOH + R"COOH +
|
H₂C—O—C(=O)—R"
fat

H₂C—OH
|
HC—OH
|
H₂C—OH

propane-1,2,3-triol
(glycerol)

Figure 3 Hydrolysis of a fat.

Hydrolytic rancidity is caused by the breaking down of a lipid into its component fatty acids and propane-1,2,3-triol (glycerol) (Figure 3). This process is thus the reverse of esterification. It occurs more rapidly in the presence of enzymes such as lipase, and with heat and moisture. In deep frying the water present in the food and the high temperatures increases the rate of hydrolysis to fatty acids. Examples of fatty acids responsible for rancidity include:

- butanoic acid (C_3H_7COOH), hexanoic acid ($C_5H_{11}COOH$) and octanoic acid ($C_7H_{15}COOH$) in milk
- palmitic acid ($C_{15}H_{31}COOH$), stearic acid ($C_{17}H_{35}COOH$) and oleic acid ($C_{17}H_{33}COOH$), which give chocolate an oily or fatty flavour
- lauric acid ($C_{11}H_{23}COOH$), which gives palm and coconut oil, in cocoa butter substitutes, a soapy flavour
- butanoic acid (C_3H_7COOH) in butter.

Oxidative rancidity is due to the oxidation of the fatty acid chains, typically by the addition of oxygen across the carbon-to-carbon double bond in unsaturated fatty acids. Oily fishes, such as mackerel, contain a high proportion of unsaturated fatty acids and are prone to oxidative rancidity. The process proceeds by a free radical mechanism catalysed by light in the presence of enzymes or metal ions. This photo-oxidation process leads to the formation of hydroperoxides. Hydroperoxides have the formula R–O–O–H and break down to form free radicals owing to the weak oxygen-to-oxygen single bond they contain.

Prolonging shelf life and minimizing the rate of rancidity

Traditional methods of prolonging the shelf life of food include: fermentation, preserving, pickling, salting, drying and smoking.

Preserving fish by smoking

One of the old ways of preserving herring (a fish found in the North Sea) was by smoking them over oak wood smoke to make "kippers".

Figure 4

More modern methods that slow down the rate of deterioration of food include processing, packaging and the use of additives (Table 1).

Table 1 Methods for prolonging the shelf life of food

Processing	Packaging	Use of additives
• Refrigeration: storing dairy products at low temperature slows lipase hydrolysis.	• Using an inert gas to minimize contact with oxygen.	• Sodium sulfite, sodium hydrogensulfite and citric acid: delay the onset of enzymatic browning.
• Reducing light levels by using coloured glass or keeping in a dark place.	• Using hermetic sealing or low-gas permeability packaging film.	• Sodium and potassium nitrite and nitrate for curing meat, fixing colour and inhibiting micro-organisms.
• Keeping moisture levels low during processing by smoking, or by adding salt or sugar.	• Keeping jars full to minimize the amount of air in the headspace above oil.	• Sodium benzoate and benzoic acid as antimicrobial agents in fruit juices, carbonated drinks, pickles and sauerkraut.
	• Sealing in tin cans.	• Sorbic acid, propanoic acid, calcium propanoate and sodium propanoate for delaying mould and bacterial growth in breads and cheeses.
		• Ethanoic acid and benzoic acid for delaying mould and bacterial growth in pickled meats and fish products, and to add flavour.

Antioxidants

Antioxidants delay the onset of oxidation, or slow down the rate at which itoccurs. Although some occur naturally, they are also added to extend the shelf life of food. Naturally occurring antioxidants include:

- **vitamin C** (ascorbic acid): found in citrus fruits, green peppers, broccoli, green leafy vegetables, strawberries, redcurrants and potatoes
- **vitamin E** (tocopherols): found in wheatgerm, nuts, seeds, whole grains, green leafy vegetables, vegetable oils (canola) and soya beans
- **β-carotene**: found in carrots, squash, broccoli, sweet potatoes, tomatoes, kale, cantaloupe, melon, peaches and apricots
- **selenium**: found in fish, shellfish, red meat, eggs, grains, chicken and garlic.

Synthetic antioxidants (Figure 4) include:

- butylated hydroxyanisole (BHA)
- butylated hydroxytoluene (BHT)
- propyl gallate (PG)
- trihydroxybutyrophenone (THBP)
- *tert*-butylhydroquinone (TBHQ).

All of them contain a phenolic group, and many contain a carbon atom bonded directly to three methyl groups on a central carbon atom, which is known as a **tertiary butyl group**. Both the phenolic group and the tertiary butyl group are **free radical scavengers**. They react with and remove the free radicals involved in the oxidation of the food and thus prolong the shelf life.

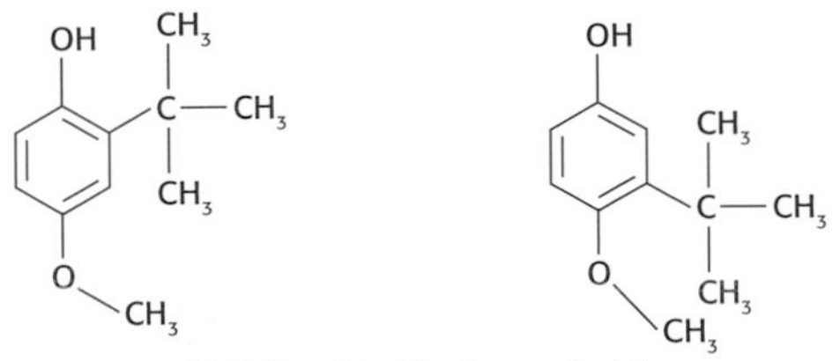

BHA (butylated hydroxyanisole)
BHA is a mixture of 3-*tert*-butyl-4-hydroxyanisole
and 2-*tert*-butyl-4-hydroxyanisole

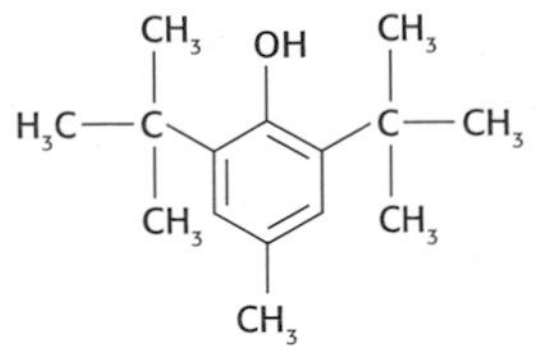

BHT
(butylated hydroxytoluene)

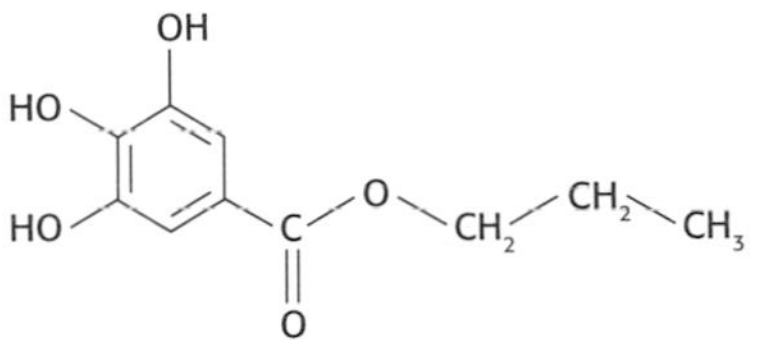

PG (propyl gallate)

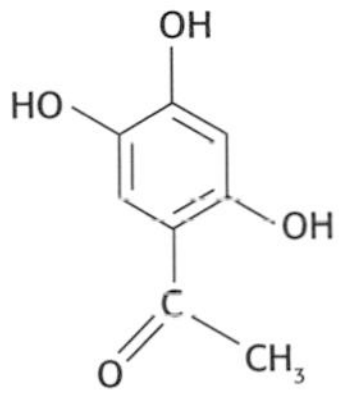

THBP
2,4,5-trihydroxybutyrophenone

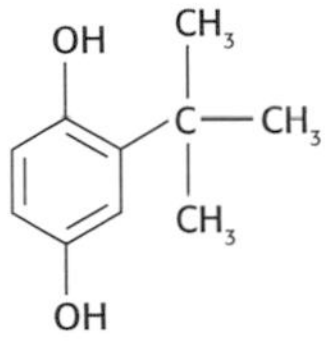

TBHQ
tert-butylhydroquinone

Figure 5 Some synthetic antioxidants

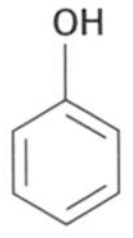

phenol group
(present in phenolic
compounds)

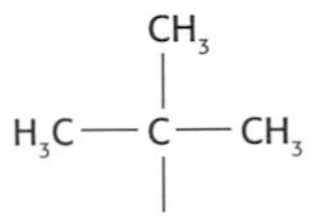

tert-butyl group

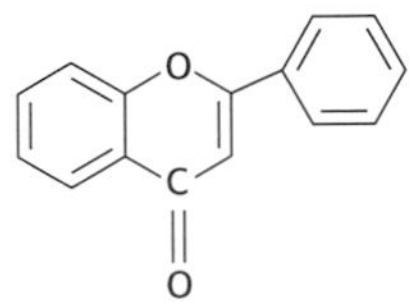

flavone
(present in flavonoids)

Figure 6 Basic structures of some radical scavengers found in antioxidants.

Table 2 The advantages and disadvantages of antioxidants in food

Advantages	Disadvantages
• Naturally occurring vitamins C, E and carotenoids reduce the risk of cancer and heart disease by inhibiting the formation of free radicals. • Vitamin C is vital for the production of hormones and collagen. • β-carotene can be used as an additive in margarine to provide it with a yellow colour and act as a precursor for vitamin A synthesis. • They are believed to enhance the health effects of other foods, and boost overall health and resilience.	• Consumers perceive synthetic antioxidants to be less safe, as they do not occur naturally. • Natural antioxidants are more expensive and less effective than synthetic antioxidants, and can also add unwanted colour and leave an aftertaste to food. • Synthetic antioxidants are classed as food additives, and need to be regulated by policies and legislation to ensure their safe use in food. • Policies regarding the safe use and labelling of food additives can be difficult to implement and monitor, especially in developing countries and across borders.

Antioxidants in traditional food

Many traditional foods found in different cultures contain natural antioxidants. Vitamin C and carotenoids have already been mentioned. Another class of natural antioxidants is the **flavonoids**. These are found in all citrus fruits, green tea, red wine, oregano and dark chocolate (containing at least 70% cocoa). They have been linked to lowering levels of LDL cholesterol and blood sugar levels, which reduce high blood pressure, and to preventing the development of cancerous cells.

Turmeric

Turmeric is a yellow spice used in Indian cooking, and is known for its antioxidant properties. The active ingredient in turmeric is curcumin, whose structure in solution is shown below. Note the presence of two phenolic groups.

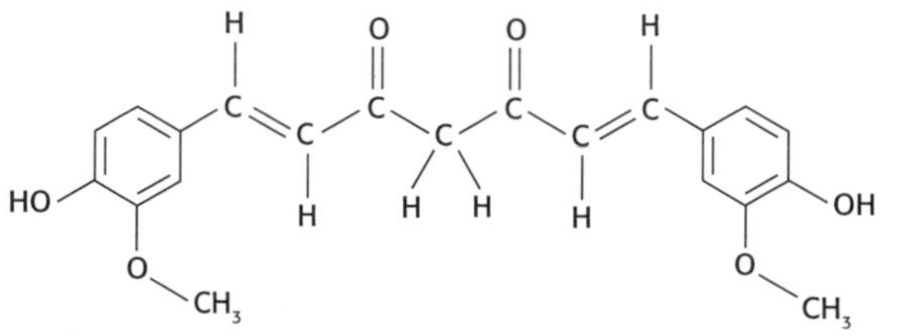

Figure 7 Curcumin.

---

1 Ethoxyquin is an antioxidant that is used in pet foods to prevent fats going rancid. There are some reports that it can cause health problems for pets, but so far no strong evidence has been found to support this. The structure of ethoxyquin is given in Figure 8. Identify the functional groups labelled A and B.

2 In 2003 it was reported that researchers in Scotland and Italy had found that the antioxidant properties of dark chocolate, which may help protect against heart disease and cancer, were cancelled out by taking milk at the same time. They recruited 12 healthy volunteers and showed that those who ate just dark chocolate had increased levels of antioxidants of up to 20% in their blood. However, there was no such effect in those who ate milk chocolate or drank milk with dark chocolate. It was speculated that this showed that proteins in milk bind with the antioxidants in chocolate. Discuss the scientific reliability of this report.

3 Identify foods in your own diet that contain antioxidants.

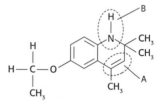

Figure 8 Ethoxyquin.

Colour

Food can be coloured naturally or artificially. A **pigment** is a naturally occurring colour found in the cells of plants and animals. In the context of food chemistry a **dye** is a food-grade, synthetic, water-soluble colour. Both pigments and dyes are coloured because of their ability to absorb light in the visible region of the electromagnetic spectrum and transmit the remaining light of the visible spectrum that has not been absorbed. The light that is transmitted is known as the **complementary colour**. Thus a dye or pigment that absorbs red light will transmit a blue-green colour. This is explained in detail in Chapter 14.

Why poppies are red

Red poppies contain an anthocyanin called cyanidin. The sap of poppies is acidic, and under these conditions the cyanidin absorbs light at 375 nm and at 530 nm. The absorbance spectrum is shown in Figure 9.

    The absorbance at 375 nm lies in the ultraviolet region of the electromagnetic spectrum, and so does not affect the colour. The absorbance at 530 nm lies in the blue-green region of the spectrum, and so the cyanidin in poppies transmits the complementary colour to blue-green, which is red.

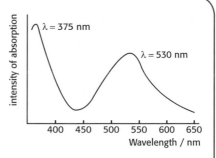

Figure 9 UV-visible spectrum of cyanidin.

Pigments

Naturally occurring pigments include anthocyanins, carotenoids, chlorophyll and haem (Table 3).

Table 3 Naturally occurring pigments

Anthocyanins	Carotenoids	Chlorophyll
• Most widely occurring pigments in plants. • Responsible for the pink, red, purple and blue colours in fruits and vegetables, including cranberries, blueberries, strawberries and raspberries.	• Most widespread pigments in nature (the large majority are produced by algae). • Act as a precursor for vitamin A synthesis: those found in fruit and vegetables contribute 30–100% of the vitamin A requirement in humans. • Colours range from yellow to red-orange. Found in bananas, carrots, tomatoes, watermelon, peppers and saffron. • Red astaxanthin (complexed to a protein) is responsible for the blue or green colour of live lobsters and crabs, and the pink colour of salmon and flamingos.	• Major pigments necessary for photosynthesis found in green plants. **Haem** • Myoglobin is responsible for the purple-red colour of fresh meat.

Factors affecting the colour stability of pigments

Many factors affect the colour stability of pigments. They are essentially factors that change the structure of the molecule, because this will affect the precise wavelength of visible light that the pigments absorb and thus the complementary colours transmitted. These factors include: oxidation, temperature changes, pH changes, and the presence of metal ions.

Anthocyanins

In aqueous solution anthocyanins exist in different forms, which are in equilibrium with each other. Changing the pH and the temperature affects the position of equilibrium and thus the predominant species responsible for the colour. They are most highly coloured at low pH (in acidic solution) and at low temperatures. When exposed to heat the equilibrium moves to the right, and the compounds are less thermodynamically stable. This causes a loss of colour, and browning.

$$\text{quinoid (A)} \rightleftharpoons \text{flavylium (AH}^+\text{)} \rightleftharpoons \text{carbinol} \rightleftharpoons \text{chalcone}$$
$$\quad\text{(blue)} \qquad\qquad \text{(red)} \qquad\qquad \text{(colourless)} \qquad \text{(colourless)}$$

Anthocyanins also form complexes with metal ions such as aluminium ions (Al^{3+}) and iron(III) irons (Fe^{3+}). These ions are present in the metal of "tin" cans to which canned fruit is exposed, which can lead to discoloration.

Carotenoids

Carotenoids contain many alternate carbon-to-carbon single and double bonds, which, when bonded together, account for the fact that they are coloured. The structure and visible spectrum of β-carotene is given in Chapter 14. It absorbs light in the violet-blue region of the visible spectrum, which means that it transmits yellow light. Because of the unsaturation due to the carbon-to-carbon double bonds carotenoids are susceptible to oxidation. This oxidation process can be catalysed by light, metals or hydroperoxides. This changes the type of bonding, and results in the bleaching of colour, loss of vitamin A activity, and the bad smells formed.

Carotenoids are stable up to 50 °C and in a pH range of 2–7, and are therefore not degraded by most forms of food processing. When heated, the naturally occurring *trans-* isomer rearranges to the *cis-* isomer.

Chlorophyll

Chlorophyll contains a group called a porphin, which contains four nitrogen atoms. The porphin ring forms a very stable complex with a magnesium ion (Figure 10). The stability of chlorophyll towards heat depends on the pH. In a basic solution with a pH of 9 it is thermodynamically stable, but in acidic solution with a pH of 3 it is unstable. When heated, the cell membrane of the plant deteriorates, releasing acids, which decrease the pH. At this lower pH the magnesium ion is displaced by two hydrogen ions, resulting in the formation of an olive-brown pheophytin complex. The breakdown of the cell during heating also increases the susceptibility of chlorophyll to decomposition by light.

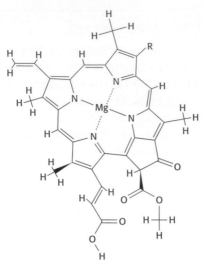

Figure 10 Basic structure of chlorophyll (the various forms of chlorophyll have different groups attached in the R-position).

Robert Woodward

Chlorophyll was first synthesized by Robert (Bob) Woodward (1917–1979), an American chemist who is considered by many to be the greatest synthetic chemist of the last century. His total synthesis of chlorophyll in 1960 required 55 separate steps. He was awarded the Nobel Prize in Chemistry in 1965.

Figure 11

Haem

The haem group also contains a porphin ring, but it is complexed to an iron ion. During oxidation, oxygen binds to purple-red myoglobin (Mb) and red oxymyoglobin (MbO_2) forms. In both Mb and MbO_2 the iron in the haem group is in the form of iron(II), Fe^{2+}. Through auto-oxidation of Mb and MbO_2 the oxidation state of the iron is changed to iron(III), Fe^{3+}. In the Fe^{3+} state it is called metmyoglobin (MMb) and has an undesirable brown-red colour. Interconversion between these three forms occurs readily.

$$MbO_2 \rightleftharpoons Mb \rightleftharpoons MMb$$

oxymyoglobin	myoglobin	metmyoglobin
(red, Fe^{2+})	(purple-red, Fe^{2+})	(brown, Fe^{3+})

In order to minimize the rate of formation of brown metmyoglobin from auto-oxidation, meat can be stored free of oxygen. Packaging films with low gas permeability are used. Air is removed from the package and a storage gas, normally pure carbon dioxide, is injected.

Synthetic colourants (dyes)

If you have a yearning for a particular food it may be worth examining the package. Most countries now require the ingredients of all foods to be stated. For example, a package of raspberry jelly may never have seen a raspberry. Instead, flavourings and colourants are added to make the "food" look and taste like raspberry. Food additives are given numbers so they can be identified. In some parts of the world these are called E numbers; elsewhere an International Numbering System (INS) is used. Unfortunately many artificial dyes have been shown to be carcinogenic. For example, until 1938, 4-dimethylaminoazobenzene (Figure 12) was used to dye butter. This "butter yellow" has now been withdrawn as a food dye because it is carcinogenic.

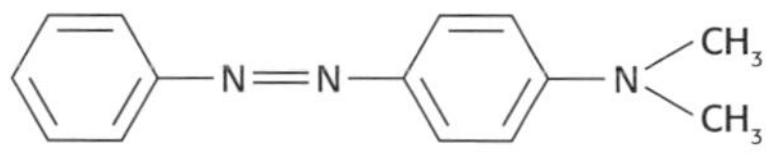

Figure 12 Structure of 4-dimethyl aminoazobenzene.

One of the problems is that the various countries have different regulations. Some dyes that are permitted in some countries are banned in others. With much food crossing international borders this is now of real concern, and there is a need for international legislation on colourant legislation. This topic provides good scope for practical work in international schools, where students can bring different brands of a similar product from different countries. The different dyes in these products can then be identified using chromatography.

Non-enzymatic browning of food

Enzymatic browning such as that caused when an apple is peeled is normally undesirable. However, natural browning through cooking enhances both the flavour and the appearance of food.

Foods high in carbohydrate content, especially sucrose and reducing sugars such as glucose, and lacking nitrogen-containing compounds can be **caramelized**. Although this can be achieved simply by heating, the chemical process is far from simple. Both sucrose and glucose when caramelized form many different products, among them acids, sweet and bitter derivatives, volatile molecules with a caramel aroma, and brown-coloured polymers. Factors that increase the rate of caramelization include acid catalysis at pH values below 3 or base catalysis at pH greater than 9. The temperature is important too. Boiling will not cause caramelization, because a temperature above 120 °C is required, which occurs during the baking and the roasting of foods with a high sugar content. An example of caramelization is the browning on the top of baked egg dishes.

For foods that contain nitrogen the **Maillard reaction** involves the reaction of a carbohydrate, either a free sugar or one bound up in starch, with the amine group on an amino acid, which may also be free or part of a protein chain. Basically it involves a condensation reaction between a reducing sugar and the amine group. The presence of the amino acid lysine results in the most browning colour and cysteine the least. Thus foods that contain lysine, such as milk, brown readily. Because moisture lowers the temperature, in order to make a good stew it is sensible to brown the meat, vegetables and flour well in hot oil to bring out the flavours before adding any liquid. Other examples of Maillard browning include making milk chocolate, and heating sugar and cream to make fudge, toffees and caramels.

Genetically modified foods

Genetic engineering involves the process of selecting a single gene for a single characteristic and transferring that sequence of DNA from one organism to another. Thus a genetically modified (GM) food can be defined as one that is derived or produced from a genetically modified organism. The GM food can be substantially different from or essentially the same as the conventional food in composition, nutrition, taste, smell, texture and functional characteristics. An example of genetically modified food is the FlavrSavr tomato. In normal tomatoes a gene is triggered when they ripen to produce a substance that makes the fruit go soft and eventually rot. In the FlavrSavr tomato the gene has been modified to "switch off" the chemical so that the fruit can mature longer on the vine for fuller taste and have a longer shelf life.

Benefits of GM foods

- Improve the taste, flavour, texture, nutritional value and shelf life of food.
- Anti-cancer substances could be incorporated and exposure to less healthy fats reduced.
- Plants can be made more resistant to disease, herbicides and insect attack. Similarly GM foods can increase resistance to disease in animals, increase productivity and feed efficiency to give higher yields of milk and eggs.
- Environmentally "friendly" bio-herbicides and bio-insecticides can be formed. GM foods can lead to soil, water and energy conservation and improve natural waste management.

Potential concerns of GM foods

- The outcome of alterations is uncertain, as not enough is known about how genes operate.
- They may cause disease, as antibiotic-resistant genes could be passed to harmful micro-organisms.

- Genetically engineered genes may escape to contaminate normal crops, with unknown effects.
- They may alter the balance of delicate ecosystems as food chains become damaged.
- There are possible links to an increase in allergic reactions (particularly for those involved in the food processing).

Texture

Food often appears homogeneous: that is, all the ingredients are in the same phase. However, many ingredients in food are completely immiscible, and so will form separate phases within the food. The reason why the food often appears to be homogeneous is that the sizes of these phases can be very small. A **dispersed system** is a kinetically stable mixture of one phase within another largely immiscible phase.

It follows that there are several types of dispersed system, as the dispersed particles must be in a different phase from the immiscible phase. Usually the continuous phase is a liquid. A solid dispersed in a liquid is known as a **suspension**, a liquid dispersed in a liquid is known as an **emulsion**, and a gas dispersed in a liquid is known as a **foam**. An example of a suspension is molten chocolate, cream is an emulsion, and beer is a foam.

Emulsifiers

There are essentially two types of food emulsion. One is water-in-oil emulsion, such as is found in butter. Water-in-oil emulsions consist essentially of the dispersion of water droplets in a continuous oil phase. The second type is an oil-in-water emulsion, such as milk or salad dressing. This type of emulsion is more common in manufactured foods. Emulsifying agents (emulsifiers) are substances that aid the dispersal of these droplets and stabilize them to prevent them coalescing to form larger globules. Essentially they act as the interface between the solid, liquid or gas phases in the dispersed system, so they are also used in making foams. Good emulsifiers will tend to be soluble in both fats (oils) and water. Common emulsifiers include lecithin (either pure or in egg yolk), milk protein and salts of fatty acids. To physically make an emulsion mechanical energy is also required in addition to adding an emulsifier. This aids dispersion, which explains the need for beating, mixing and whisking in cooking. Stabilizers can also be added to prevent the emulsions and foams from separating out into the separate phases.

HL
Oxidative rancidity

Oxidative rancidity proceeds by a free radical mechanism. In oxidative rancidity the carbon-to-hydrogen bonds are broken homolytically; these are strong bonds with an average bond enthalpy of 412 kJ mol^{-1}. Much energy is required to overcome this high activation energy to initiate radical formation:

$$R-H \rightarrow R^{\bullet} + H^{\bullet}$$

where R represents any unsaturated fatty acid.

Use the Internet to find some of the issues raised over whether or not GM foods should be used. Examine the facts, language, statistics and images used. Who should decide in which direction society should proceed over the GM issue?

Once radicals have been formed they can propagate in the presence of oxygen to form peroxide radicals, which in turn can react with more of the unsaturated fatty acid molecules to form hydroperoxides:

$$R^{\bullet} + O_2 \rightarrow ROO^{\bullet}$$
$$ROO^{\bullet} + RH \rightarrow R^{\bullet} + ROOH$$

It has been proposed that oxygen can react directly with the alkene of the hydrocarbon chain to form hydroperoxides directly, and not necessarily by a free radical mechanism:

$$R\text{-}H + O_2 \rightarrow ROOH$$

Whichever way the hydroperoxide is formed, the weak oxygen-to-oxygen bond it contains then breaks, either photochemically or through catalysis involving a metal or an enzyme, to form two radicals, which can lead to further radical propagation and other products from termination reactions (Figure 13). These other products include aldehydes, ketones and alcohols, all of which can be isolated from oxidizing lipid systems.

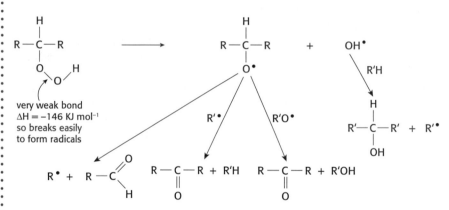

Figure 13 Breakdown of hydroperoxides, leading to both radical and non-radical products.

Other termination steps can also occur when free radicals combine to form non-radical products. These include:

$$R^{\bullet} + R^{\bullet} \rightarrow R\text{-}R$$
$$R^{\bullet} + ROO^{\bullet} \rightarrow ROOR$$
$$ROO^{\bullet} + ROO^{\bullet} \rightarrow ROOR + O_2$$
$$RO^{\bullet} + R \rightarrow ROR$$

Antioxidants

Antioxidants can be subdivided into three types.

The first group work either by interrupting the formation of free radicals in the initiation step of auto-oxidation, or by interrupting the propagation of the free radical chain. Examples of such antioxidants (with the simplified formula AH) include BHA, BHT, TBHQ and vitamin E (tocopherols). These contain groups such as the phenolic group and the *tert*-butyl group, which can scavenge radicals. They do this by forming less stable and less reactive free radicals or non-radical products. For example:

$$R^{\bullet} + AH \rightarrow R\text{-}H + A^{\bullet}$$
$$RO^{\bullet} + AH \rightarrow R\text{-}O\text{-}H + A^{\bullet}$$
$$ROO^{\bullet} + AH \rightarrow R\text{-}O\text{-}O\text{-}H + A^{\bullet}$$
$$R^{\bullet} + A^{\bullet} \rightarrow R\text{-}A$$
$$RO^{\bullet} + A^{\bullet} \rightarrow R\text{-}O\text{-}A$$

The second group work by forming an irreversible complex with metal ions, such as the iron(II) ion, Fe^{2+}, to reduce the concentration of free metal ion in the solution so that the metal ion is unable to catalyse the oxidation reactions. This complexing is known as **chelating**, and effective chelating agents include salts of EDTA (Figure 14), which is the shortened version of its old name, ethylenediaminetetracetic acid, and certain plant extracts, such as rosemary, tea and ground mustard. EDTA has been shown to inhibit the Fe^{2+}-catalysed oxidation of raw beef, but it has not been approved for use in commercial meat products. EDTA can function as a hexadentate ligand (Figure 15). Monodentate ligands such as ammonia or water molecules use one pair of non-bonding electrons to form a coordinate bond with the metal ion. EDTA has six non-bonding pairs of electrons available to form coordinate bonds, so that a single ligand can occupy six different sites on the metal ion.

The third group consists of reducing agents, and includes vitamin C (ascorbic acid) and carotenoids. These are electron donors and they remove or reduce concentrations of oxygen. The half-equation for the oxidation of ascorbic acid is shown in Figure 16.

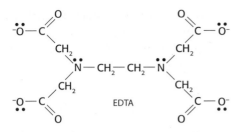

Figure 14 Ethylenediaminetetraacetate ion (EDTA^{4-}).

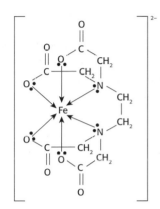

Figure 15 EDTA acting as a hexadentate ligand with Fe^{2+}. The six non-bonding pairs of electrons on a single EDTA ion can coordinate to six sites on the metal ion. Note that as the charge on the ligand is −4 and the charge on the Fe is +2 then the overall charge on the complex is −2.

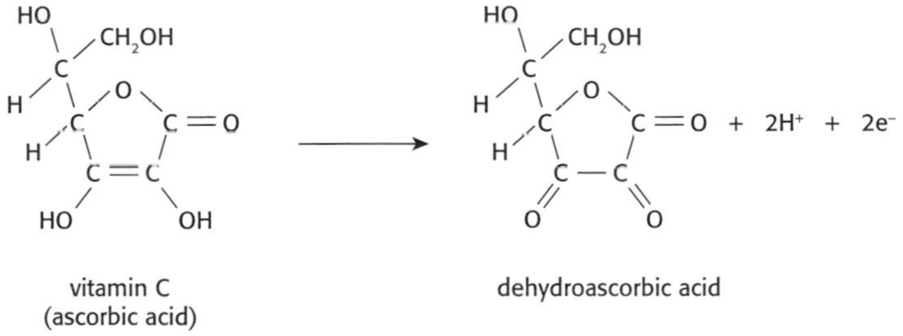

Figure 16 Oxidation of ascorbic acid.

Stereochemistry in food

Conventions for naming enantiomers

Enantiomers occur when a pure compound contains an asymmetric or **chiral** carbon atom. They are non-superimposable mirror images. There are three different conventions for naming the different enantiomers

- **(+) or *d*- and (−) or *l*-isomers.** We have seen in Chapter 10 that enantiomers have the property of rotating the plane of plane-polarized light. The enantiomer or optical isomer that rotates the direction of plane polarized light clockwise, known as **dextrorotatory**, is identified by either by (+) or *d* (from the Latin *dexter* meaning "right"), whereas the one that rotates the plane-polarized light anticlockwise, known as **laevorotatory** (from the Latin *laevus* meaning "left"), is identified either by (−) or by *l*. This form of nomenclature is useful, as it relates directly to a physical property, but it is impossible to state which enantiomer is which simply by looking at their chemical structures.

- **D and L notation**. This notation uses the capital letters D and L, and is based on the absolute configuration of each of the two enantiomers (i.e. on their spatial distribution) rather than on their ability to rotate plane-polarized light in a particular direction. The D and L notation is commonly used when naming the enantiomers of carbohydrates, such as sugars, and amino acids. For naming carbohydrates the basic reference is (+)-2,3-dihydroxypropanal, more commonly known as (+)-glyceraldehyde, which is assigned the letter D. All other carbohydrates are assigned either D or L with reference to this. Figure 16 shows the structure of 2,3-dihydroxypropanal (glyceraldehyde).

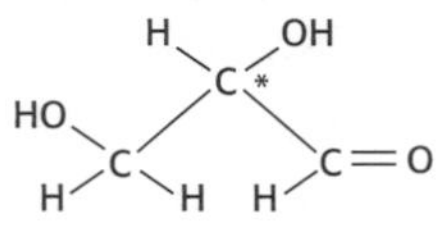

Figure 17 The structure of 2,3-dihydroxypropanal (glyceraldehyde), showing the chiral carbon atom marked by *.

For amino acids the "CORN" rule is applied. This arranges the substituents **CO**OH, **R**, **N**H$_2$ and H around the asymmetric carbon atom with the hydrogen atom pointing *away* from the viewer. If the "CORN" groups are arranged clockwise, then it is the D-enantiomer, and if they are arranged anti-clockwise it is the L-enantiomer.

Glyceraldehyde as a reference for naming enantiomers

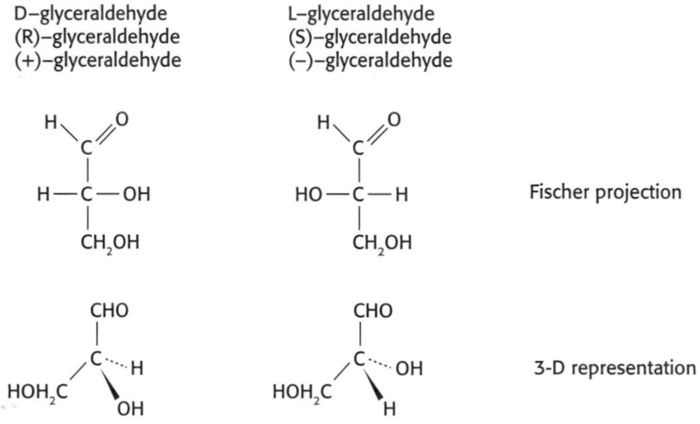

D–glyceraldehyde
(R)–glyceraldehyde
(+)–glyceraldehyde

L–glyceraldehyde
(S)–glyceraldehyde
(–)–glyceraldehyde

Fischer projection

3-D representation

Figure 18 Representations of the enantiomers of glyceraldehyde.

In the case of glyceraldehyde the D and L and R and S forms do match up with the (+) and (−) forms, but this is *not* always the case with other optically active substances. Only the (+) and (−) forms can be found empirically; the others rely on the absolute configuration of the molecules in question, that is, on their spatial arrangement.

The structures can be represented using either the Fischer projection as shown above in Figure 18 or more clearly using dotted and wedge-shaped lines to show the three-dimensional structure in two dimensions.

- **R and S notation**. This notation is mainly used by chemists when dealing with substances other than carbohydrates and amino acids. Each chiral, or asymmetric, carbon centre is labelled R or S according to the Cahn–Ingold–Prelog priority rules. These are based on atomic number. The lowest priority of the four substituents is pointed away from the viewer. If the priority of the remaining three substituents decreases in a clockwise direction it is assigned the R-form; if it decreases in an anti-clockwise direction it is the S-form.

Properties of enantiomers in food

Enantiomers can clearly have different biological effects. The importance of this with drugs was tragically shown with the drug thalidomide in the last century. One enantiomer of the drug relieved morning sickness in pregnant women; the other enantiomer caused severe defects in the limbs of the unborn child (see Chapter 16). Foods, too, can have different effects. The D-form of vitamin C, for example, has no biological activity. Different enantiomeric forms may vary in their tastes, odour and toxicity. Most naturally occurring sugars exist in the D-form, whereas most naturally occurring amino acids are in the L-form. D-amino acids tend to taste sweet, whereas L-forms often have no taste. Both caraway and dill seeds and spearmint contain carvone, but the L-(+) form of carvone in caraway and dill seeds tastes very different from the D-(−) form of carvone in spearmint (Figure 19).

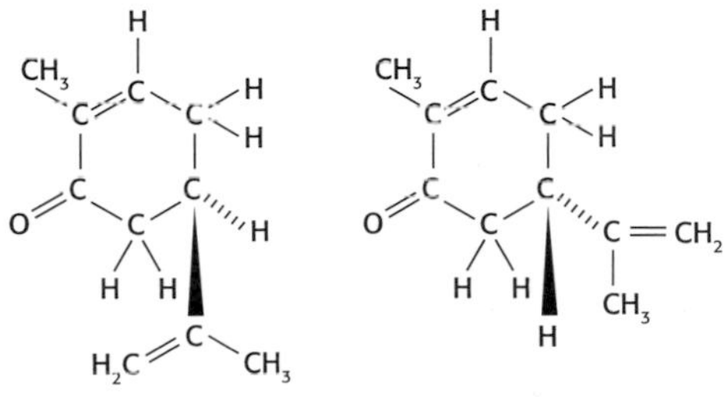

Smells and tastes can appear different to different people, because the olfactory receptors also contain chiral receptor molecules, which can interact differently with the enantiomeric molecules in food. The natural flavour of raspberries is due to R-α-ionone, whereas synthetic raspberry flavourings contain both R- and S-isomers. Other synthetically made foods often contain a racemic mixture of both enantiomers.

Figure 19 The two enantiomeric forms of carvone (2-methyl-5-(prop-1-en-2-yl) cyclohex-2-enone).

Oranges or lemons?

Orange and lemon peel each contain different enantiomers of a compound called limonene, whose structure is shown in Figure 20. The (+) enantiomer has the smell characteristic of oranges and the (−) isomer gives the characteristic smell of lemons.

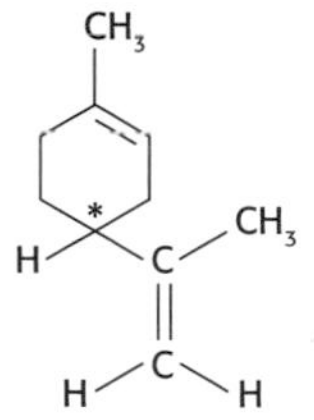

Figure 20 The asterisk * shows the position of the chiral carbon atom.

Chemical structure and colour

Colour is due to the excitation of electrons to a higher energy level, where the energy difference corresponds to the wavelength of visible light. For normal unsaturated compounds the pi electrons in a double bond absorb in the ultraviolet region of the spectrum when they are excited. However, the more delocalization there is in a compound, the lower the energy of the excitation. Thus compounds containing several alternate single and double carbon-to-carbon bonds (known as **conjugation**) tend to absorb in the visible region and transmit light of the complementary colour.

Anthocyanins contain the flavonoid $C_6C_3C_6$ skeleton (shown earlier in the flavonone structure). It is the conjugation of the pi electrons contained in this structure that accounts for the colour of anthocyanins. The more extensive the conjugation, the lower the energy (the longer the wavelength) of the light absorbed. This can be exemplified using cyanidin. The structure of cyanidin has already been given. In acidic solution it forms a positive ion, and there is less conjugation than in alkaline solution, where the pi electrons in the extra double bond between the carbon and oxygen atom are also delocalized (Figure 21).

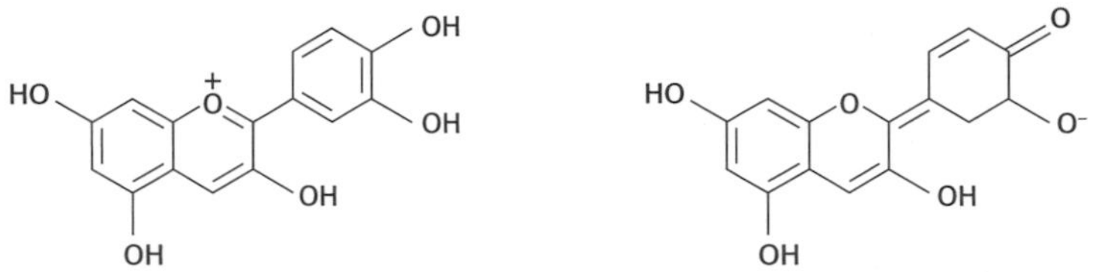

Structure of cyanidin in acidic solution. Less conjugation so absorbs in blue-green region and transmits red light.

Structure of cyanidin in alkaline solution. More conjugation so absorbs in the orange region of the spectrum and transmits blue light.

Figure 21 The effect of pH on the colour of cyanidin.

This difference in colour depending on pH explains why poppies, which have acidic pap, are red whereas cornflowers, which also contain cyanidin but have alkaline sap, are blue. Other anthocyanins differ in the number and types of other groups, such as hydroxyl or methoxy groups, which affect the precise wavelength of the light absorbed and hence the colour transmitted. The addition of other groups also affects anthocyanins in other ways. The basic flavonoid $C_6C_3C_6$ backbone is essentially non-polar. As more polar hydroxyl groups are added the potential for them to form hydrogen bonds with water molecules increases, and many anthocyanins, such as cyanidin with several –OH groups, are appreciably soluble in water for this reason.

Whereas the conjugation in anthocyanins depends on linking unsaturated ring systems together, in **carotenoids** essentially the conjugation is due to a long hydrocarbon chain consisting of alternate single and double carbon-to-carbon bonds. The majority are derived from a (poly)ene chain containing 40 carbon atoms, which may be terminated by cyclic end groups and may be complemented with oxygen-containing functional groups. The hydrocarbon carotenoids are known as **xanthophylls**. Examples include α-carotene, β-carotene and vitamin A. The structure and visible absorption spectrum of α-carotene are given in Chapter 14. α- and β-carotene and vitamin A are all fat soluble and not water soluble. Although vitamin A does contain one polar hydroxyl group, the rest of the molecule is a large non-polar hydrocarbon (Figure 22).

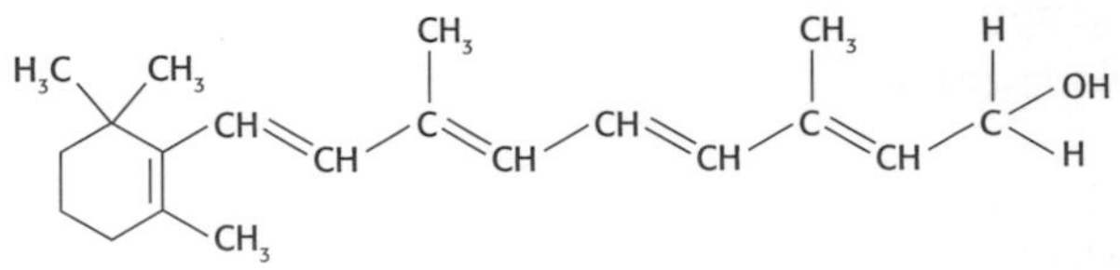

Figure 22 Vitamin A (retinol).

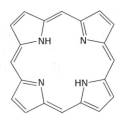

Chlorophyll and **haem** both contain a planar heterocyclic unit with the general name of a porphin (Figure 23). Porphins contain a cyclic system in which all the carbon atoms are sp² hybridized. This results in a planar structure with extensive pi conjugation.

Figure 23 The structure of the free porphin group.

Porphin groups that contain substituents in the 1 to 8 positions are known more specifically as **porphyrins**. Porphins contain four nitrogen atoms. The non-bonding pairs of electrons on the nitrogen atoms enable the porphin to form coordinate bonds with metal ions. The general structure of chlorophyll, which contains magnesium, has been given earlier. In chlorophyll the porphyrin complex with the original double bond between positions 7 and 8 is now saturated, and there is an -R group on the third carbon atom, C_3.

Chlorophyll is found in two closely related forms. In chlorophyll a, the -R group is a methyl group, $-CH_3$, and in chlorophyll b the -R group is an aldehyde group, -CHO. The haem group is present in both myoglobin and haemoglobin. Haemoglobin is the oxygen carrier found in mammalian blood. Myoglobin is the primary pigment in muscle tissue and is a complex of a protein, globin, together with a heme group, which is a porphyrin ring containing a central iron(II) ion (Figure 24).

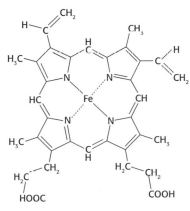

Figure 24 The haem group.

1 The structure of quercetin is given in Figure 25.
   Quercetin is found in tea, red wine, apples, onions, and many other types of fruit and vegetable. It is an antioxidant, an anti-inflammatory, and is effective against skin and prostate cancers. Use the structure to explain your answers to the following questions:
   (a) Identify the class of antioxidants to which quercetin belongs.
   (b) Would you expect quercetin to be coloured?
   (c) Would you expect quercetin to be soluble to some extent in water?
2 The photographs (Figures 25 and 26) show bottles containing chemicals that can be found in a school laboratory.

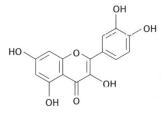

Figure 25 Quercetin.

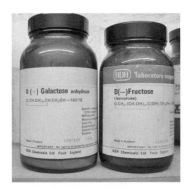

Figure 25 Photograph 1.
(photo by Geoff Neuss)

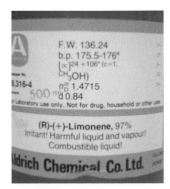

Figure 26 Photograph 2.
(photo by Geoff Neuss)

(a) Photograph 1 shows two different sugars: galactose and fructose. Explain why both of them have a D in front of the name, but one has (+) and one has (−).
(b) Photograph 2 shows the label on a bottle of limonene. Explain the meaning of the suffixes "R" and "(+)". If you took the top off this bottle what would you expect the contents to smell like?

Option G: further organic chemistry

Electrophilic addition reactions

In alkenes each carbon atom has three bonds with bond angles of 120°. This makes the area around the double bond planar. One of the two shared pairs of electrons which form the double bond between the two carbon atoms lies above and below this plane. Reacting species that are attracted to this pair of electrons are known as **electrophiles**. Molecules such as hydrogen bromide are polar because of the higher electronegativity of bromine compared with hydrogen. It is the hydrogen atom containing a small positive charge, δ+, that forms the bond with the carbon atom, releasing a bromide ion in the process. This can be shown using curly arrows to represent the movement of pairs of electrons. During the process a carbocation intermediate is formed. This carbocation then reacts with a bromide ion to form the product (Figure 1).

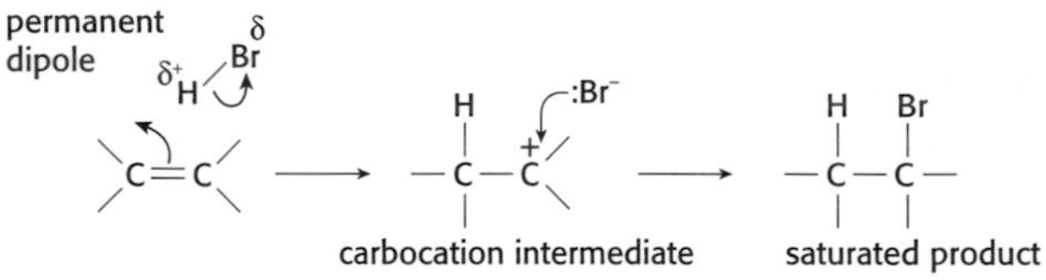

Figure 1 The mechanism of the electrophilic addition of hydrogen bromide to an alkene.

A similar type of electrophilic addition reaction takes place with bromine. The bromine molecule is non-polar, but as it approaches the double bond of the alkene it gains a temporarily induced dipole to form the electrophile (Figure 2).

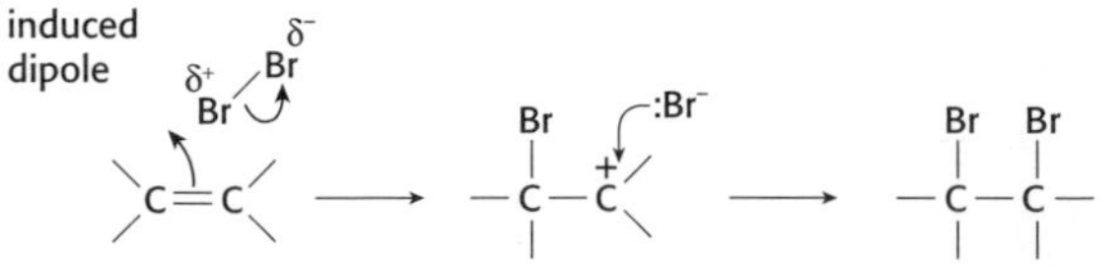

Figure 2 Electrophilic addition of bromine to an alkene.

Markovnikov addition

An interesting situation arises in the case of asymmetrical alkenes. Consider the addition of hydrogen bromide to propene. Two possible carbocations intermediate can be formed, leading to two different products (Figure 3).

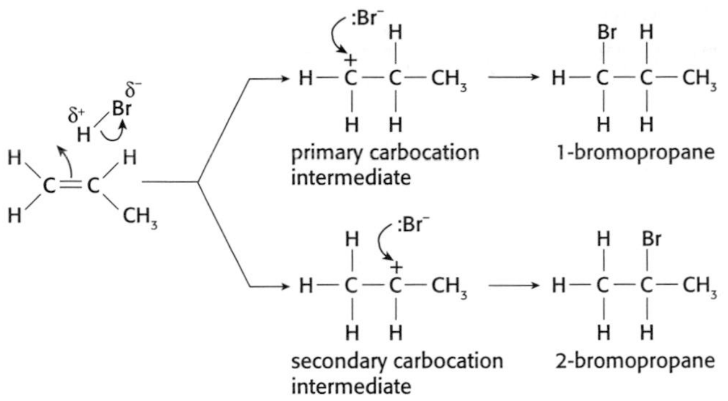

Figure 3 The two possible products from the addition of hydrogen bromide to propene.

A Russian chemist, Vladimir Markovnikov (1838–1904), formulated a rule to predict which isomer is formed. This states that when hydrogen halides add to alkenes the hydrogen atom always adds to the carbon atom that is already bonded to the most hydrogen atoms. This rule only predicts the outcome; it does not explain it. The explanation lies in the relative stabilities of the intermediate carbocations.

As we have seen when considering the S_N1 mechanism for nucleophilic substitution of halogenoalkanes, carbocations can be characterized as primary, secondary or tertiary. Tertiary carbocations are the most energetically stable and primary carbocations the least energetically stable, as the alkyl groups tend to "push" electrons towards the positive carbon ion. This is known as the **positive inductive effect**. In the above example it will be the secondary carbocation that will be formed in preference to the primary, and the major product will be 2-bromopropane not 1-bromopropane.

Nucleophilic addition reactions

Aldehydes and ketones also contain a double bond. However, unlike alkenes, the bond is significantly polar because of the greater electronegativity of the oxygen atom compared with the carbon atom. Nucleophiles are attracted to the slightly positive and electron-deficient carbon atom. One significant nucleophile that reacts in this way is hydrogen cyanide. Hydrogen cyanide is a weak acid and so is slightly dissociated to hydrogen ions and cyanide ions:

$$HCN \rightleftharpoons H^+ + CN^-$$

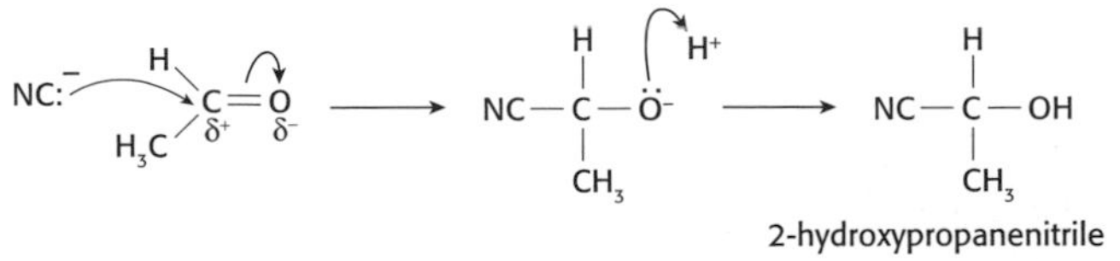

Figure 4 The nucleophilic addition reaction of cyanide ions with ethanol.

It is a non-bonding pair of electrons on the cyanide carbon atom that enables it to act as a Lewis base, that is, a nucleophile (Figure 4). The product of the addition reaction is a carbon atom bonded to a hydroxyl group and a nitrile group: this is known as a

hydroxynitrile or **cyanohydrin**. Nitriles can be hydrolysed in the presence of acid to form carboxylic acids and an ammonium salt:

$$RCN(l) + 2H_2O(l) + HCl(aq) \rightarrow RCOOH(aq) + NH_4Cl(aq)$$

Thus ethanal can be converted into 2-hydroxypropanoic acid (lactic acid) by this route (Figure 5). During this addition reaction the number of carbon atoms in the carbon chain has increased by one, which can be useful in organic synthesis.

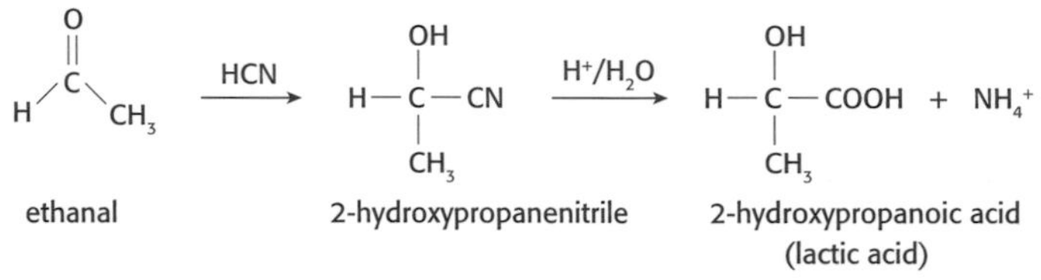

ethanal   2-hydroxypropanenitrile   2-hydroxypropanoic acid (lactic acid)

Figure 5 Conversion of ethanal into lactic acid.

Elimination reactions

Elimination is the removal of a small molecule from a larger molecule. One example has already been covered in the core: the elimination of hydrogen bromide from a bromoalkane using hot alcoholic sodium hydroxide solution to form an alkene. Alkenes can also be formed by the elimination of water from alcohols: this is also known as **dehydration**. One good dehydrating agent is concentrated sulfuric acid, but as this is also an oxidizing agent it is better to use concentrated phosphoric(V) acid, H_3PO_4. The reaction needs to be performed at 180 °C. The function of the acid is to act as a homogeneous catalyst by protonating the oxygen atom in the alcohol. Water is then lost, with the formation of a carbocation intermediate. The carbocation then donates a proton to form the alkene and regenerate the catalyst (Figure 6).

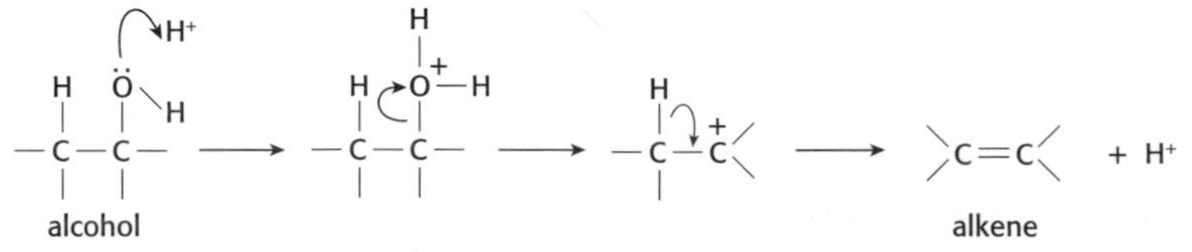

alcohol   alkene

Figure 6 Elimination of water from an alcohol.

As would be expected, tertiary alcohols are more easily dehydrated than secondary and primary alcohols, because the tertiary carbocation intermediate is more energetically stable and so more likely to be formed.

Addition–elimination reactions

A useful method to distinguish between individual aldehydes and ketones is to react them with 2,4-dinitrophenylhydrazine. This acts as a nucleophile because of the non-bonded pair of electrons on the nitrogen atoms on the hydrazine group, and then water is lost, so that overall the mechanism is known as **addition–elimination**, although it can equally well be described as a condensation reaction (Figure 7).

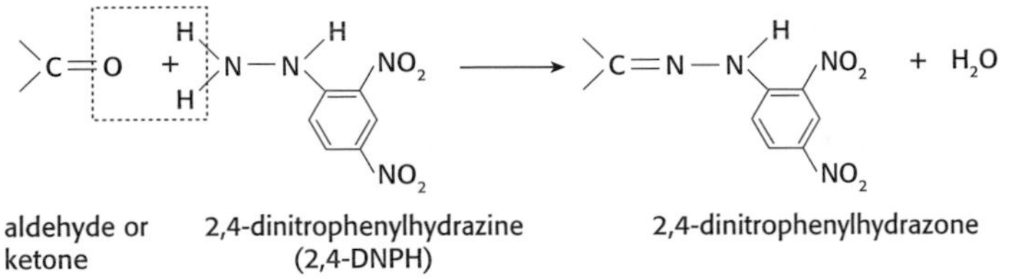

aldehyde or ketone   2,4-dinitrophenylhydrazine (2,4-DNPH)   2,4-dinitrophenylhydrazone

Figure 7 Reaction of aldehydes and ketones with 2,4-DNPH.

The products are orange or red crystalline solids, which can be obtained in a very pure form by recrystallization. Each of the derivatives, known as 2,4-dinitrophenylhydrazones, has a sharp characteristic melting point, from which they can be readily identified by comparing with values in a data book.

Arenes

Structure of benzene

Arenes are compounds that contain the benzene ring. The simplest arene is benzene itself. The Kekulé structure of benzene (cyclohexa-1,3,5-triene) has already been described in the Core. All the chemical and physical evidence supports the fact that the Lewis structure of benzene is a resonance hybrid that lies between the two equivalent resonance forms (Figure 8):

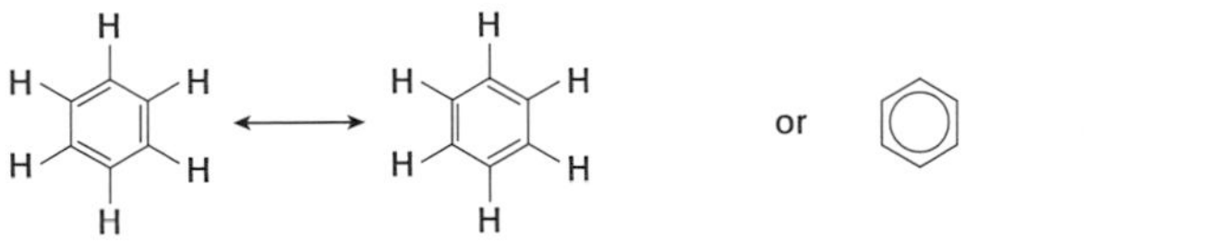

resonance forms of benzene

or   an alternative representation of benzene

Figure 8 Structure of benzene.

- The C–C bond lengths are all the same, and have a value of 0.139 nm, which lies between the values for C–C (0.154 nm) and C=C (0.134 nm).
- Only one isomer exists for 1,2-disubstituted benzene compounds. If there were alternate double bonds two isomers would exist (Figure 9).
- If benzene does simply contain three double bonds then it should undergo electrophilic addition reactions. In fact it will undergo addition reactions only with difficulty, and more commonly undergoes substitution reactions.
- The enthalpy change of hydrogenation of benzene (-210 kJ mol^{-1}) is not equal to three times the enthalpy change of hydrogenation of cyclohexene (-120 kJ mol^{-1}). The difference of 150 kJ mol^{-1} is the extra energy associated with the formation of the resonance hybrid.

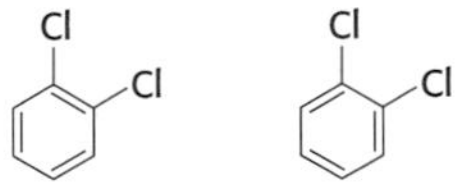

Figure 9 The two isomers of 1,2-dichlorobenzene that would exist if there were alternate double bonds.

Nucleophilic substitution reactions of halogenated arenes

Halogenated arenes can contain the halogen atom substituted either directly on the benzene ring, for example bromobenzene (C_6H_5Br), or onto a side chain, for example bromomethylbenzene ($C_6H_5CH_2Br$) (Figure 10).

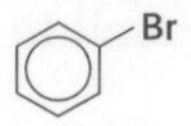

If the halogen atom is on the side chain then the compound behaves like a halogenoalkane, and readily undergoes nucleophilic substitution with aqueous sodium hydroxide solution. For example, bromomethylbenzene is a primary halogenoalkane, and so the substitution proceeds by an S_N2 mechanism to produce phenylmethanol (Figure 11).

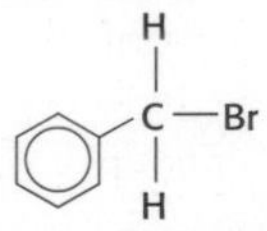

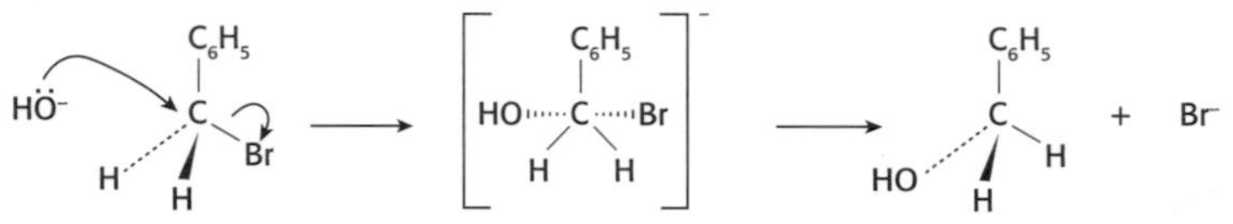

Figure 10 Halogenated arenes.

Figure 11 Nucleophilic substitution of bromomethylbenzene.

However, if the halogen atom is directly attached to the benzene ring then nucleophilic substitution occurs only very slowly or not at all. There are two reasons for this. First, the nucleophile is unable to "attack" the carbon atom bonded to the bromine atom because it is repelled by the dense cloud of electrons surrounding the arene ring. Second, the C–Br bond is stronger in bromobenzene than the C–Br bond in halogenoalkanes. This is because one of the non-bonding pairs of electrons on the bromine atom is able to interact with the electrons in the benzene ring, which makes the bond shorter and stronger, and consequently the bromine a much poorer leaving group.

Grignard reagents

Compounds containing a carbon atom covalently bonded directly to a metal are known as **organometallic compounds**. Examples of organometallic compounds include tetraethyl lead(IV), $Pb(C_2H_5)_4$, which used to be added to petrol to improve its octane rating, and triethylaluminium, $Al(C_2H_5)_3$, which is used as a catalyst for the polymerization of alkenes.

One important group of organometallic compounds that are used in organic synthesis is known as **Grignard reagents**. We have seen that one extra carbon atom can be introduced into a carbon chain by using cyanide ions; Grignard reagents enables the carbon chain to be increased by several carbon atoms. They consist of an alkyl group bonded directly to magnesium. They are usually prepared *in situ* by reacting magnesium metal with a halogenoalkane using dry ethoxyethane (ether) as the solvent.

Victor Grignard

The French chemist Victor Grignard (1871–1935) first wrote about the preparation and reactions of his organometallic reagents for his PhD thesis in 1899. He went on to become professor of organic chemistry at Lyon University, and won the Nobel Prize in Chemistry in 1912 for his invention of Grignard reagents. In one sense he was the counterpart of Fritz Haber, as Grignard too worked on poison gases during the First World War but on the opposite side. Grignard developed phosgene, $COCl_2$, and also worked on the detection of mustard gas, $S(CH_2CH_2Cl)_2$.

Figure 12 Victor Grignard (1871–1935).

Carbon is more electronegative than magnesium so the magnesium to carbon bond is polar, with the negative part of the dipole on the carbon atom. This small negative charge means that the carbon atom can act as a good nucleophile. Grignard reagents can add to the partially positive carbon atom in aldehydes and ketones, and then be hydrolysed in acid solution to form alcohols with a longer carbon chain. For example, with methanal they form a primary alcohol, with other aldehydes they form secondary alcohols, and with ketones they form tertiary alcohols (Figure 13).

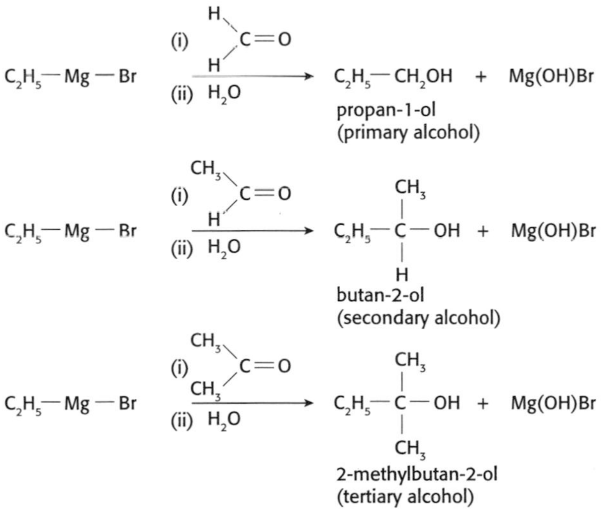

$$C_2H_5-Mg-Br \xrightarrow[\text{(ii) } H_2O]{\text{(i)}} \underset{\substack{\text{propan-1-ol}\\\text{(primary alcohol)}}}{C_2H_5-CH_2OH} + Mg(OH)Br$$

$$C_2H_5-Mg-Br \xrightarrow[\text{(ii) } H_2O]{\text{(i)}} \underset{\substack{\text{butan-2-ol}\\\text{(secondary alcohol)}}}{C_2H_5-C(CH_3)(H)-OH} + Mg(OH)Br$$

$$C_2H_5-Mg-Br \xrightarrow[\text{(ii) } H_2O]{\text{(i)}} \underset{\substack{\text{2-methylbutan-2-ol}\\\text{(tertiary alcohol)}}}{C_2H_5-C(CH_3)(CH_3)-OH} + Mg(OH)Br$$

Figure 13 Reactions of Grignard reagents to form alcohols.

Grignard reagents react in a similar way with carbon dioxide to form carboxylic acids:

$$C_2H_5\text{-}Mg\text{-}Br \xrightarrow[\text{(ii) } H_2O]{\text{(i) } CO_2} \underset{\text{propanoic acid}}{C_2H_5COOH} + Mg(OH)Br$$

During their preparation, and when undergoing reactions, Grignard reagents must be kept completely dry, because they react with water to form an alkane:

$$C_2H_5\text{-}Mg\text{-}Br \xrightarrow{H_2O} \underset{\text{ethane}}{C_2H_6} + Mg(OH)Br$$

Acid–base reactions

Organic acids

Carboxylic acids are weak acids: that is, they are only slightly dissociated in aqueous solution. Their acid strength can be measured in terms of a pK_a value. This is a logarithmic scale. The lower the pK_a value, the more dissociated (stronger) the acid. The difference in the pK_a values for methanoic acid and ethanoic acid provides good evidence for the positive inductive effect of alky groups. Methanoic acid ($pK_a = 3.75$) is almost exactly 10 times stronger than ethanoic acid ($pK_a = 4.76$), and the only difference is that a hydrogen atom has been exchanged for a methyl group. Consider the dissociation of any weak monoprotic acid, HA:

$$HA(aq) \rightleftharpoons H^+(aq) + A^-(aq)$$

where A^- represents the conjugate base of the acid. In general, the greater the charge density of the negative ion then the more the conjugate base will attract a proton, and the weaker the acid will be. Equally, if the negative charge can be delocalized (spread out) so that its charge density is less, then the conjugate base will be weaker and the acid will be stronger.

Carboxylic acids are acidic partly because the negative charge present in the carboxylate anion is spread over the two resonance forms. However, if a methyl group is present, then relative to a hydrogen atom it pushes electrons towards the carboxylate group and so makes the conjugate base stronger and the acid weaker (Figure 14).

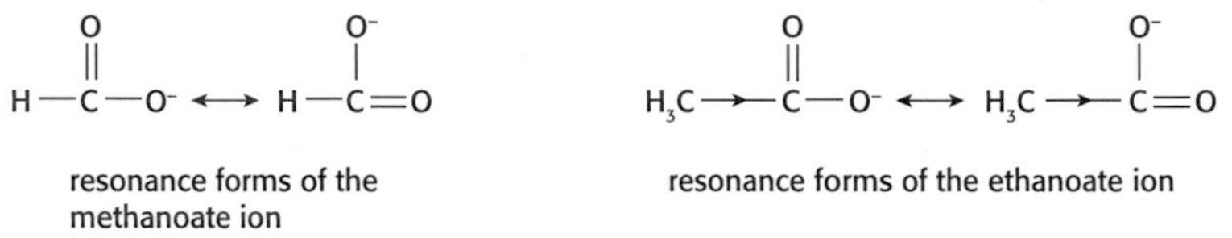

resonance forms of the methanoate ion

resonance forms of the ethanoate ion

Figure 14 The positive inductive effect of the methyl group makes the ethanoate ion a stronger base than the methanoate ion, so that ethanoic acid is a weaker acid than methanoic acid.

Atoms or groups of atoms that draw electrons away from the carboxylate group will make the conjugate base weaker, and so the acid will be stronger. One such atom is the electronegative chlorine atom (Figure 15).

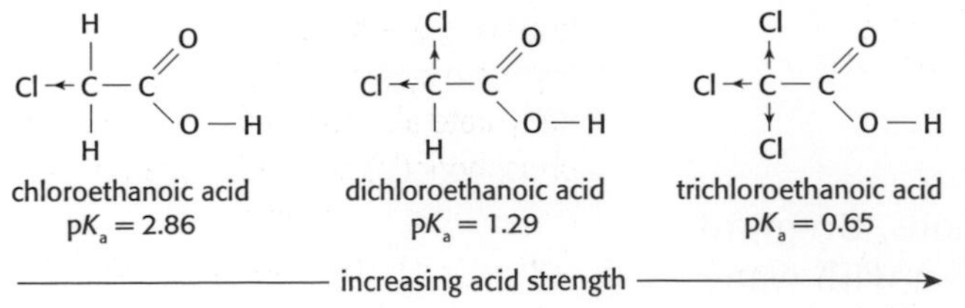

chloroethanoic acid
$pK_a = 2.86$

dichloroethanoic acid
$pK_a = 1.29$

trichloroethanoic acid
$pK_a = 0.65$

———————— increasing acid strength ————————→

Figure 15 The more electron withdrawing groups present, the stronger the carboxylic acid.

Phenols, unlike alcohols, are also acidic, although much less so than carboxylic acids. In ethanol (pK_a ≈ 16) the ethoxide ion formed when it dissociates is an extremely strong base, because the negative charge is concentrated totally on the oxygen atom (Figure 16). However, in phenol (pK_a = 10.0) the negative charge on the oxygen atom in the phenoxide ion can be spread over the entire benzene ring due to the different resonance forms (Figure 17).

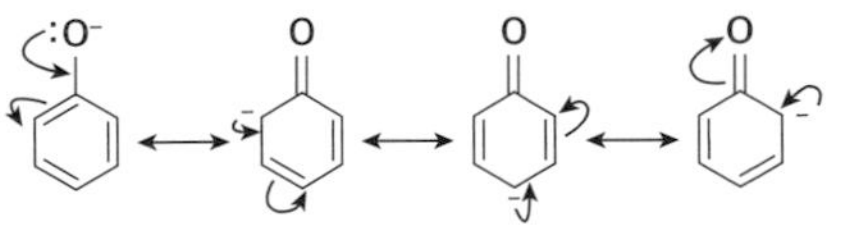

$$C_2H_5OH \rightleftharpoons H^+ + C_2H_5{-}O^-$$
pK_a = 16

ethoxide ion
negative charge localized on oxygen atom

⬡—OH $\rightleftharpoons$ H⁺ + ⬡—O⁻
pK_a = 10.0

phenoxide ion negative charge spread over the benzene ring as well as the oxygen atom

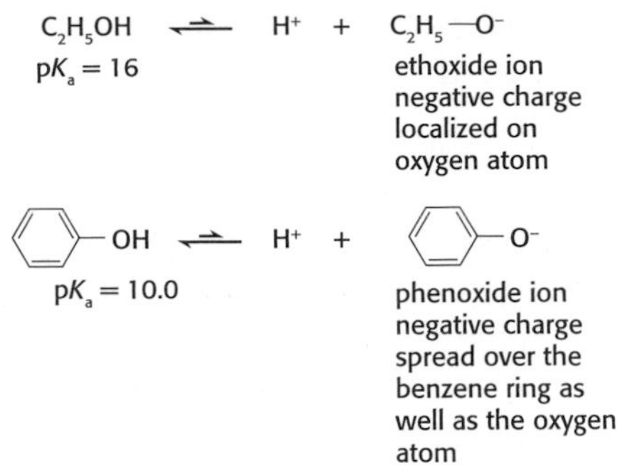

Figure 16 The ethoxide ion is a stronger base than the phenoxide ion.

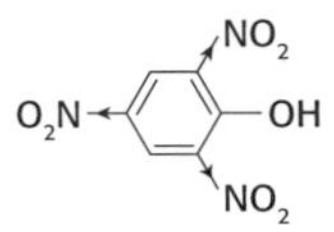

Figure 17 The different resonance forms of the phenoxide ion, showing how the negative charge is delocalised over the benzene ring.

The negative charge can be spread even further if electron-withdrawing groups such as the nitro group, –NO_2, are substituted into the benzene ring, so that 2,4,6-trinitrophenol (pK_a = 0.42) is almost a strong acid, and in fact goes by the common name of picric acid (Figure 18).

Figure 18 2,4,6-trinitrophenol (pK_a = 0.42).

Organic bases

Amines can be thought of as substituted ammonia compounds. Because they still contain the non-bonding pair of electrons on the nitrogen atom, they also act as weak bases. The strength of bases can be illustrated by their pK_b values: the lower the pK_b value, the stronger the base. Due to the positive inductive effect of the alkyl groups, amines are stronger bases than ammonia, because the electron density on the nitrogen atom will be increased so that there will be a stronger attraction between the non-bonding pair of electrons on the nitrogen atom and a proton (Figure 19).

pK_b = 3.27   pK_b = 4.75

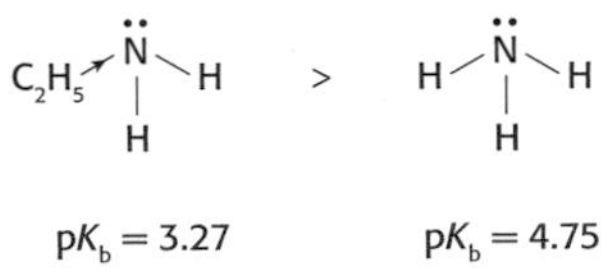

Figure 19 Ethylamine is a stronger base than ammonia.

Secondary amines will be even stronger, due to the increased positive inductive effect of the extra alkyl group. On this basis tertiary amines might be expected to be even stronger, but this is not the case (Figure 20). It is thought that this is because they are less hydrated in aqueous solution, owing to their inability to form hydrogen bonds with water molecules.

ṄH₃	CH₃ṄH₂	(CH₃)₂ṄH	(CH₃)₃Ṅ
ammonia	methylamine	dimethylamine	trimethylamine
pK_b = 4.75	pK_b = 3.36	pK_b = 3.28	pK_b = 4.20

Figure 20 Organic bases.

When amines act as bases, they form ammonium salts. For example, ethylamine will react with hydrochloric acid to form ethylammonium chloride:

$$C_2H_5NH_2 + HCl \rightarrow C_2H_5NH_3{^+}Cl^-$$

In the same way that ammonium salts react with sodium hydroxide to release ammonia, the salts of amines will also react when warmed with aqueous sodium hydroxide to release amines:

$$C_2H_5NH_3{^+}Cl^- + NaOH \rightarrow C_2H_5NH_2 + NaCl + H_2O$$

1 Predict the major product when hydrogen chloride reacts with pent-1-ene, and give the mechanism for its formation.

2 Explain why 1,2-dibromoethane is formed when ethene reacts with bromine in a non-polar solvent, but 2-bromoethanol is formed as the major product when ethene reacts with bromine water.

3 Predict the major product when but-1-ene reacts with iodine monochloride, ICl, and give the mechanism for the reaction.

4 Outline two different methods by which propanoic acid can be made from bromoethane.

5 Explain why it is preferable to dehydrate alcohols with concentrated phosphoric(V) acid rather than with concentrated sulfuric acid.

6 Give the reagents and conditions for converting 2-bromo-2-methylpropane into 2,2-dimethylpropanoic acid in a two-step process. Write the equations for the reactions taking place.

Addition–elimination reactions of acid anhydrides and acid chlorides

We have seen that carboxylic acids react with alcohols to form esters. In this reaction the –OH group of the carboxylic acid is replaced by the –OR group of the alcohol. This is an example of an addition–elimination reaction, because the alcohol acts as a nucleophile and the –OH group of the acid is eliminated as water. However, the –OH group is not a particularly good leaving group. An acid anhydride is a carboxylic acid in which the hydroxyl group has been replaced by the –OCOR group from another molecule of the acid. Similarly, in acyl chlorides the hydroxyl group has been replaced by a chlorine atom. Both –OCOR and –Cl are much better leaving groups than –OH, so acid anhydrides and acyl chlorides are much more reactive with nucleophiles than carboxylic acids (Figure 21).

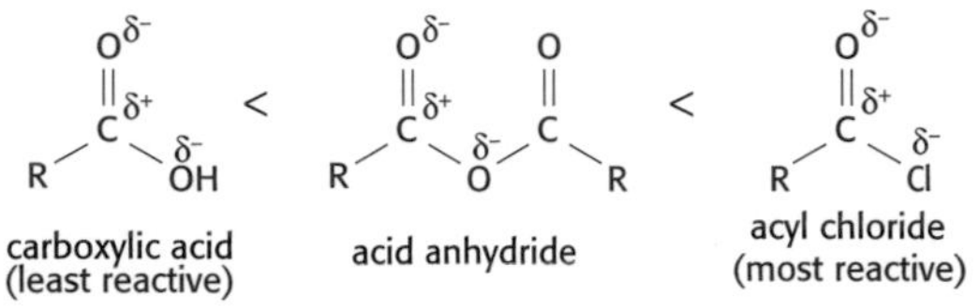

Figure 21 The –OCOR and –Cl groups are more electron-withdrawing than the –OH group. This increases the small positive charge δ^+ on the carbonyl carbon atom, making them more reactive with nucleophiles.

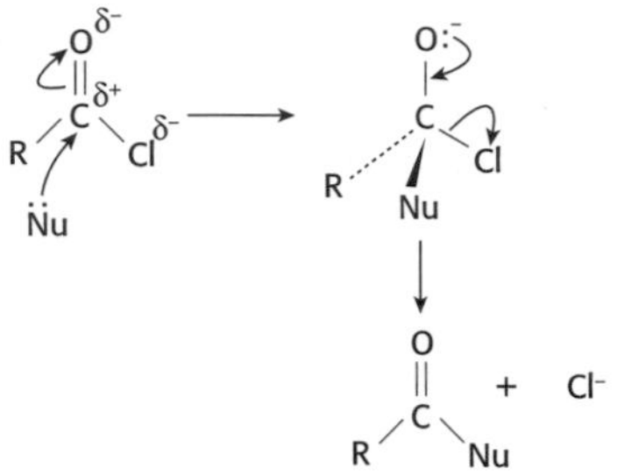

Figure 22 Mechanism of addition–elimination reactions of acyl chlorides. Nu represents a nucleophile, e.g. $H_2\ddot{O}$, $\ddot{N}H_3$, RÖH and substituted amines.

Examples of addition–elimination reactions with acid anhydrides

● With **water**, carboxylic acids are formed. For example:

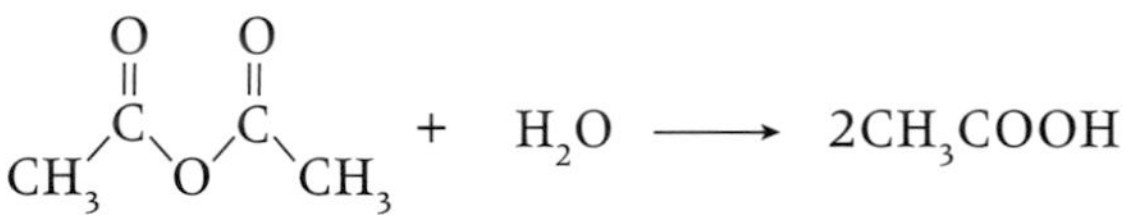

ethanoic anhydride   ethanoic acid

● Acid anhydrides react more readily with **alcohols** than carboxylic acids to form esters. For example:

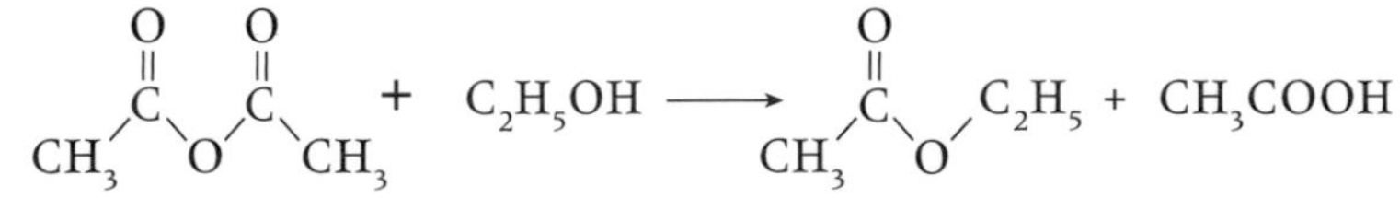

ethanoic anhydride   ethanol   ethyl ethanoate

The laboratory preparation of aspirin is a useful example of the reaction of an alcohol with an acid anhydride. The phenol group in salicylic acid, 2-hydroxybenzoic acid, behaves as an alcohol when it reacts with ethanoic anhydride to produce aspirin, 2-ethanoyloxybenzenecarboxylic acid, and ethanoic acid.

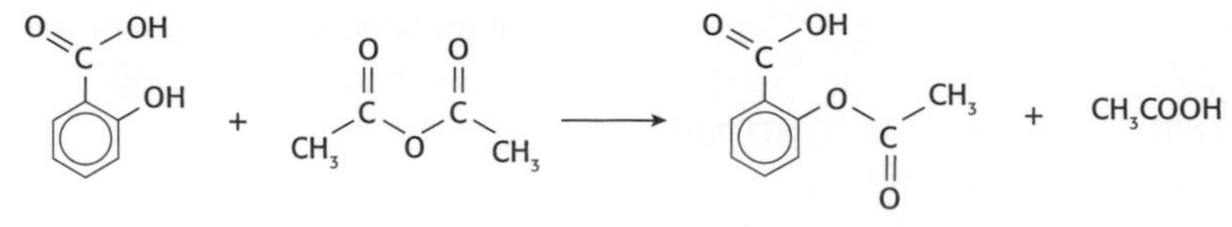

2-hydroxybenzoic acid   ethanoic anhydride   2-ethanoyloxybenzenecarboxylic acid
(salicylic acid)   (aspirin or acetylsalicylate)

- Acid anhydrides react readily with concentrated aqueous **ammonia** to form amides. For example:

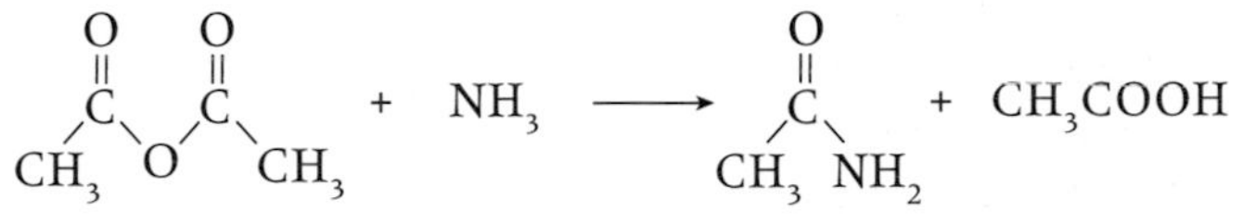

ethanamide

> Secondary amines also contain a non-bonded pair of electrons on the nitrogen atom, and so can also function as nucleophiles. When 4-aminophenol is reacted with ethanoic anhydride in the presence of water at room temperature the product is *N*-(4-hydroxylphenyl)ethanamide. Like aspirin this is also a painkiller and fever reducer, and has the common name of paracetamol in the UK and acetaminophen in the USA.
>
>
> 4-aminophenol   ethanoic anhydride   N-(4-hydroxyphenyl)ethamide (paracetamol or acetaminophen)

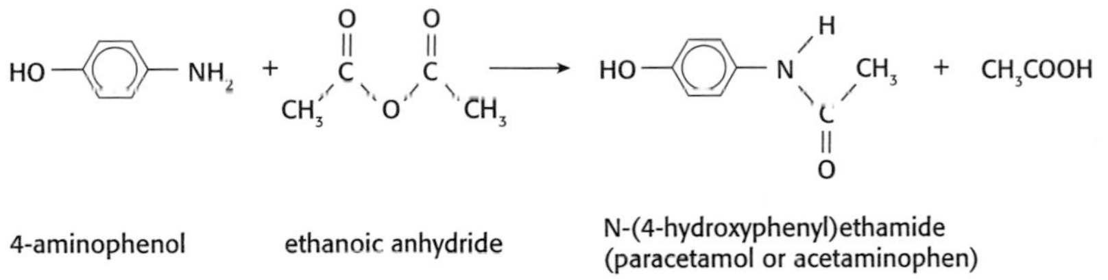

Examples of addition–elimination reactions with acyl chlorides

Acyl chlorides are even more reactive than acid anhydrides, so that when the stopper is taken off a bottle of ethanoyl chloride it fumes in the air as droplets of hydrochloric acid are formed as a result of hydrolysis. However, essentially the reactions are similar, with hydrochloric acid rather than ethanoic acid being the other product (Figure 23).

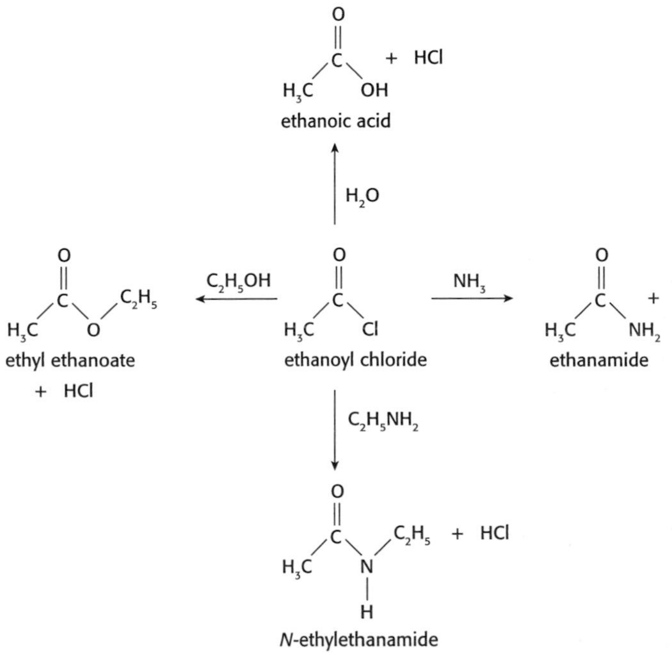

Figure 23 Summary of reactions of ethanoyl chloride.

The reaction of acyl chlorides with secondary amines forms the basis of the preparation of nylons. For example, nylon 6,6 can be made by reacting 1,6-diaminohexane with hexane-1,6-dioyl dichloride.

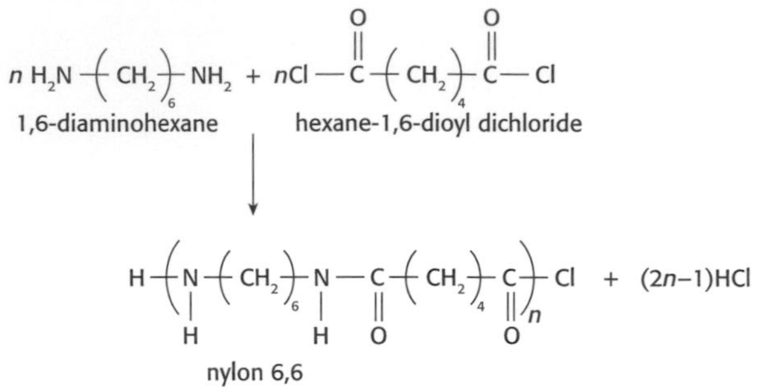

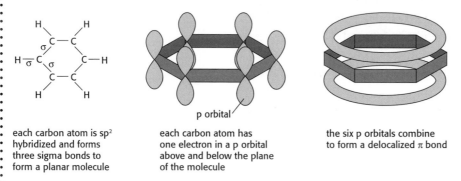

Figure 24 Preparation of nylon 6,6.

Electrophilic substitution reactions

All six carbon atoms in the benzene ring are sp^2 hybridized. The molecular orbital theory of bonding for benzene shows that each carbon atom is joined to two other carbon atoms and a hydrogen atom by three sigma bonds to form a planar molecule. Each carbon atom has one electron remaining in a p orbital, which lies above and below the plane of the molecule. These six unhybridized p orbitals combine to form a delocalized pi bond containing six pi electrons. It is this delocalization that results in the extra thermodynamic stability of the benzene ring by about 150 kJ mol^{-1}. See Figure 25.

each carbon atom is sp^2
hybridized and forms
three sigma bonds to
form a planar molecule

each carbon atom has
one electron in a p orbital
above and below the plane
of the molecule

the six p orbitals combine
to form a delocalized π bond

Figure 25 The delocalization of π electrons in benzene

Electrophiles are attracted to this cloud of delocalized pi electrons, but because of the extra stability substitution rather than addition occurs.

Two examples of electrophilic substitution are nitration and chlorination:

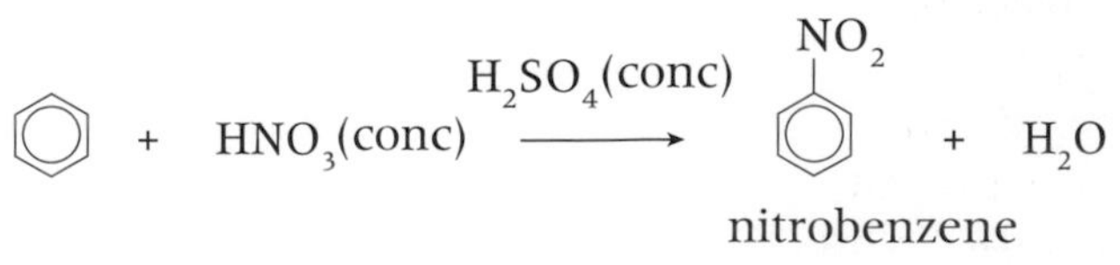

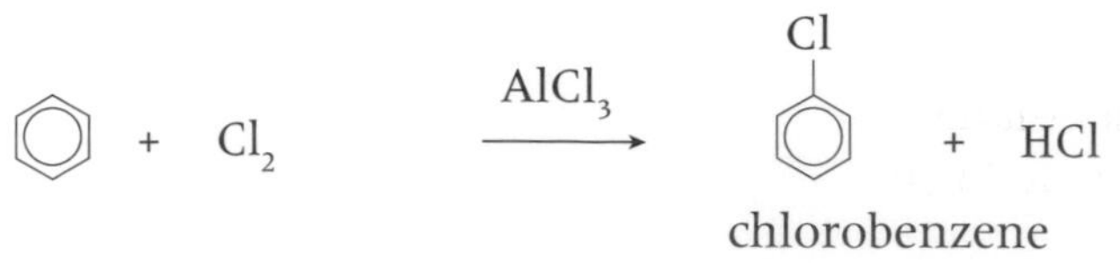

$$\text{benzene} + HNO_3(\text{conc}) \xrightarrow{H_2SO_4(\text{conc})} \text{nitrobenzene (NO}_2) + H_2O$$

$$\text{benzene} + Cl_2 \xrightarrow{AlCl_3} \text{chlorobenzene (Cl)} + HCl$$

In each case extra conditions are necessary to produce the active electrophile. The role of the concentrated sulfuric acid is to protonate the concentrated nitric acid to give an ion, which then breaks down to form water and the nitronium ion, NO_2^+:

$$HNO_3 \quad + \quad H_2SO_4 \longrightarrow \quad H_2NO_3^+ \quad + \quad HSO_4^-$$

$$\downarrow$$

$$NO_2^+ \quad + \quad H_2O$$

A pair of delocalized pi electrons from the benzene ring forms a bond with the nitronium ion, leaving the four remaining electrons delocalized over five carbon atoms. The hydrogen sulfate(VI) ion then removes a hydrogen atom, and the pair of electrons from the carbon to hydrogen bond is transferred back into the benzene ring:

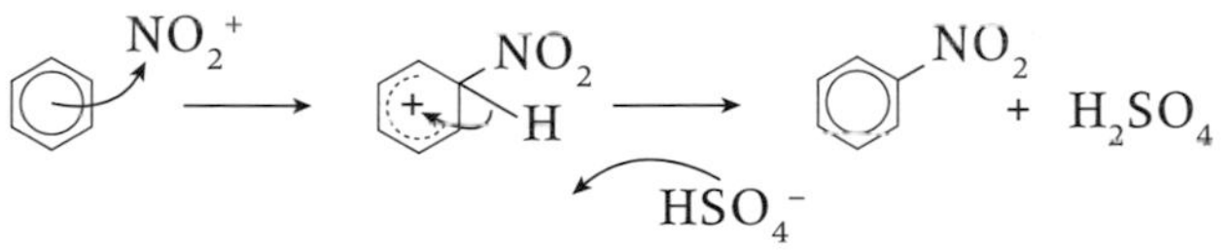

The mechanism for the chlorination of benzene also involves the formation of an active electrophile. Anhydrous aluminium chloride, $AlCl_3$, is electron deficient because there are only six electrons around the central aluminium atom. This means it can act as a Lewis acid and attract a pair of electrons from a molecule of chlorine, so that the chlorine molecule becomes polarized.

Anhydrous aluminium chloride is said to be a **halogen carrier**. Other examples of halogen carriers include anhydrous iron(III) chloride and iron itself in the presence of chlorine as it forms iron(III) chloride *in situ*. Once the electrophile has been formed, the reaction proceeds in a similar manner to nitration:

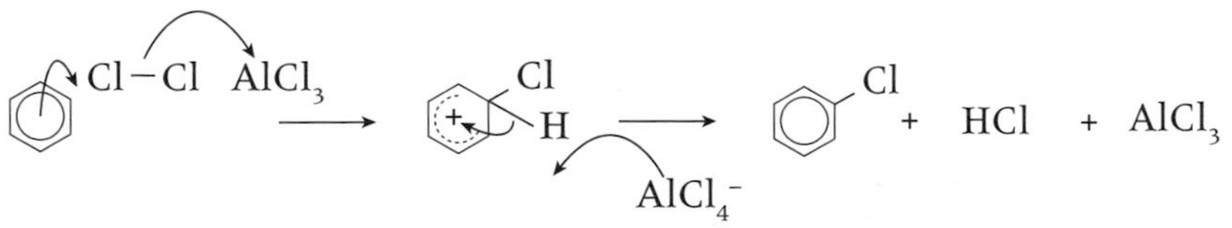

A French chemist, Charles Friedel (1832–1899), collaborating with an American chemist, James Crafts (1839–1917), realized the significance of halogen carriers and reasoned that by reacting them with halogenated organic compounds they could produce electrophiles with positively charged and hence electron-deficient carbon atoms, which could react to substitute alkyl and acyl groups into the benzene ring. These reactions are known as **Friedel–Crafts reactions**, and provide a useful way to synthesize a whole range of organic compounds.

- *Alkylation*: For example, benzene reacts with bromoethane in the presence of anhydrous aluminium tribromide to form ethylbenzene:

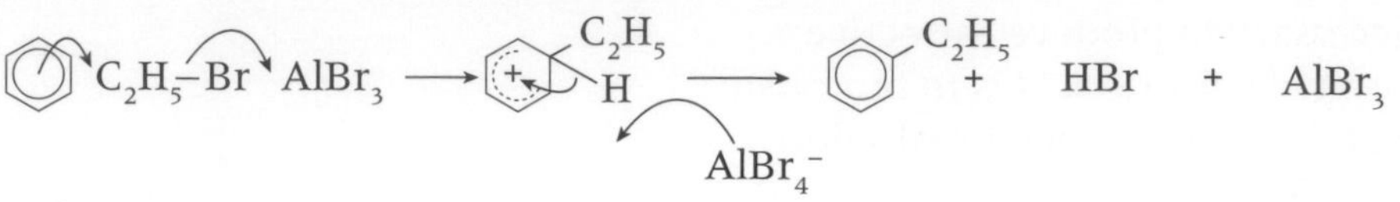

- *Acylation*: For example, benzene reacts with ethanoyl chloride in the presence of anhydrous aluminium chloride to form phenylethanone:

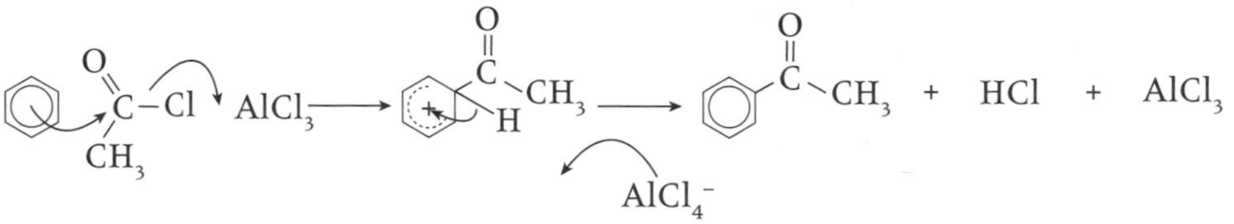

Electrophilic reactions of substituted benzene compounds

When a substituent is attached to the benzene ring it will affect the electron density of the delocalized pi electrons. Essentially the substituents can be grouped into three different classes:

- groups with a positive inductive effect
- groups with a non-bonding pair of electrons
- electron-withdrawing groups.

Groups with a positive inductive effect, e.g. the methyl group in methylbenzene

The electron density of the delocalized pi electrons in the benzene ring will be increased because of the positive inductive effect of the methyl group. This will have two significant effects:

- Methylbenzene is more reactive towards electrophiles than benzene under the same experimental conditions.
- The methyl group in methylbenzene is 2- and 4- directing. Thus when chlorine reacts with methylbenzene in the presence of anhydrous aluminium chloride the two main organic products formed are 2-chloromethylbenzene and 4-chloromethylbenzene in approximately equal amounts. Almost no 3-chloromethylbenzene is formed.

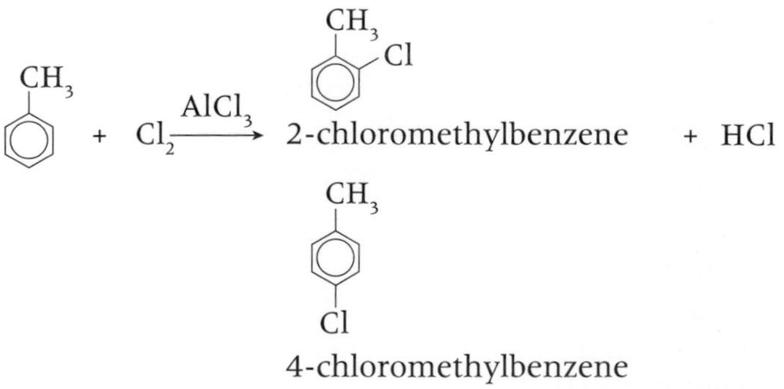

The 2- and 4- directing effect can be explained by considering the stability of the possible intermediates formed. When the new substituent adds to the 2- or 4- position, then in one of the possible resonance forms the positive charge can be accommodated on the carbon atom to which the methyl group is bonded (Figure 26).

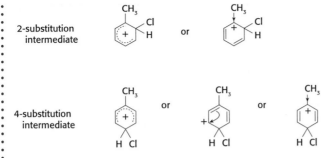

2-substitution
intermediate

4-substitution
intermediate

Figure 26 The positive inductive effect of the methyl group can stabilize the intermediate carbocation when 2- or 4-substitution takes place in methylbenzene

The positive inductive effect of the methyl group will make this intermediate more energetically stable, and so more likely to be formed. In the case of 3- substitution it is not possible to have the positive charge on the carbon atom bonded to the methyl group (Figure 27).

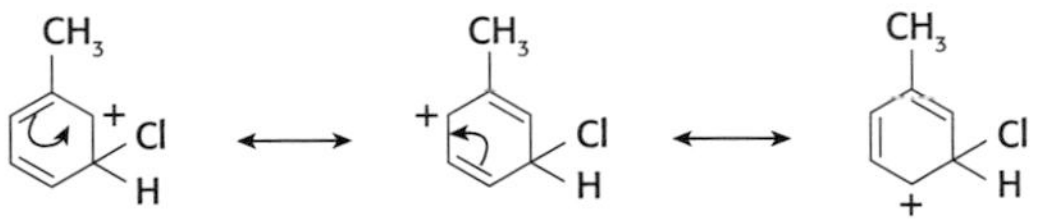

Figure 27 The three resonance forms of the intermediate carbocation formed during the preparation of 3-chloromethylbenzne. The positive charge cannot be located on the carbon atom bonded to the methyl group, and so is unable to be stabilized by the positive inductive effect of the methyl group.

Groups with a non-bonding pair of electrons, e.g. the hydroxyl group in phenol or the amine group in phenylamine

Even though oxygen and nitrogen are electronegative, and so might be expected to draw electrons away from the benzene ring, both phenol and phenylamine are much more reactive than benzene. This is because the non-bonding pair of electrons can contribute to the delocalized ring, so that instead of six pi electrons spread over six atoms there are now eight pi electrons spread over seven atoms, which increases the electron density. This is known as the **mesomeric effect**. Both phenol and phenylamine are 2- and 4-directing, because an additional resonance form is possible.

For example, when phenol is nitrated with a mixture of concentrated nitric and sulfuric acids the four possible resonance forms for the 2- substituted intermediate are

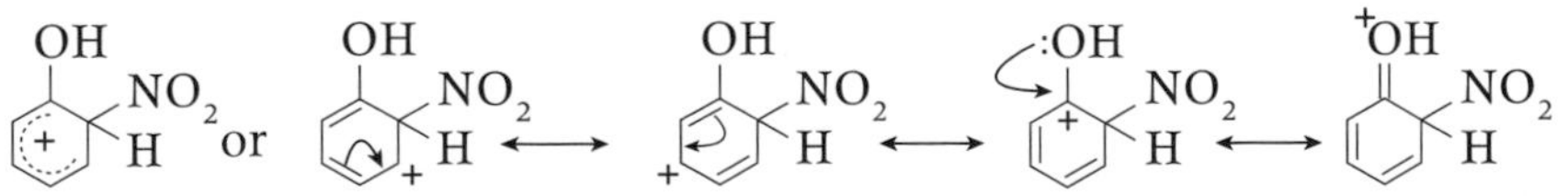

There are similarly four possible resonance forms for the 4-substituted intermediate too but only three for the 3- substituted intermediate.

Chlorination of phenol

Phenol is so reactive that when it reacts with chlorine at room temperature no halogen carrier is necessary, and the substitution rapidly proceeds all the way to produce 2,4,6-**tri**chloro**p**henol. This is known as TCP, and is used as a disinfectant.

OH
+ 3Cl2
OH
Cl
Cl
Cl
+ 3HCl

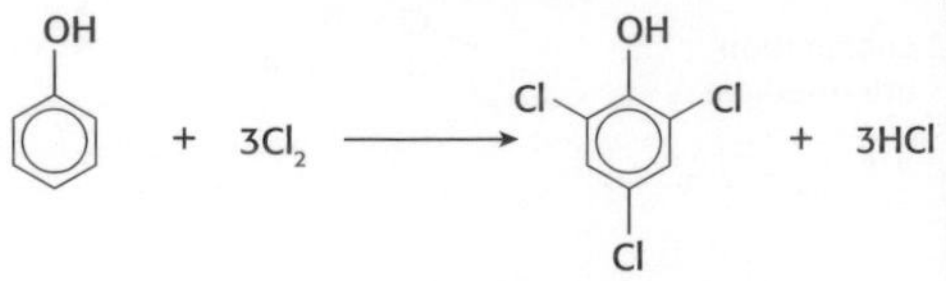

2,4,6-trichlorophenol

Electron-withdrawing groups, e.g. nitrobenzene

In the first two classes, which increase the reactivity of the benzene ring, the substituents are known as **activating groups**. The nitro group, which contains three highly electronegative atoms, in nitrobenzene is an electron-withdrawing group with no non-bonding pair of electrons that can delocalize into the ring. The nitro-group is therefore **deactivating**, and nitrobenzene is much less reactive than benzene. In order to nitrate nitrobenzene the mixture of nitrobenzene with concentrated nitric acid and concentrated sulfuric acid must be refluxed for several hours. The product that is formed is 1,3-dinitrobenzene, because the new substituent will attach at the 3- position.

NO2
+ HNO3
H2SO4
reflux
NO2
NO2
+ H2O

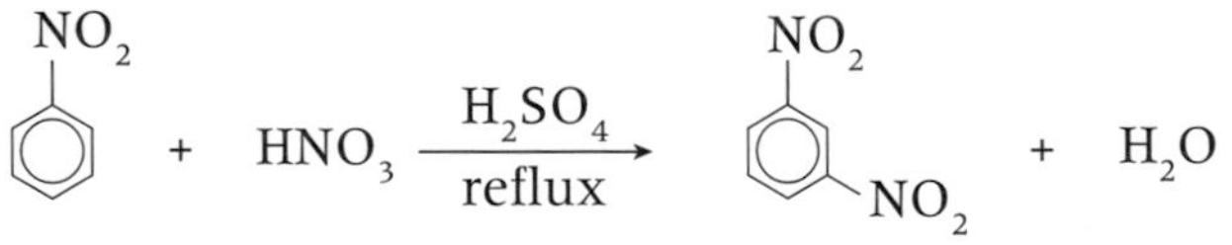

1,3-dinitrobenzene

?

1 Outline a possible reaction pathway to synthesize each of the following organic materials from the named starting material. Each reaction pathway must not involve more than two steps.

(a) Starting with benzene

(i) (a)
CH3
NO2

(ii) (b)
CH3
NO2

(iii) (c)
O
C
C2H5
C2H5

(b) Starting with methylbenzene

(i) (a)
CH2OH

(ii) (b)
CH2Br
Br

(iii)
CH3
CH3
C
N
N–H
NO2
NO2

2 Electrophiles and nucleophiles are useful reagents in organic syntheses. Outline one reaction of halogenoalkanes in which the carbon atom bonded to the halogen atom can be converted into an electrophile, and one reaction in which it can be converted into a nucleophile.

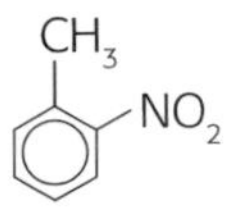

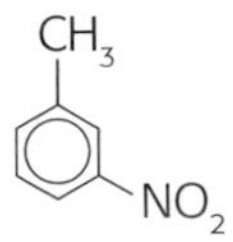

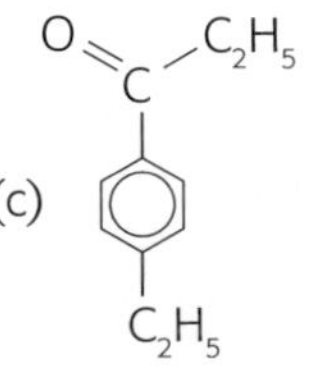

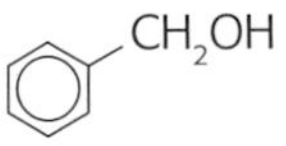

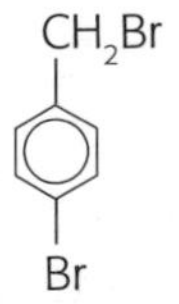

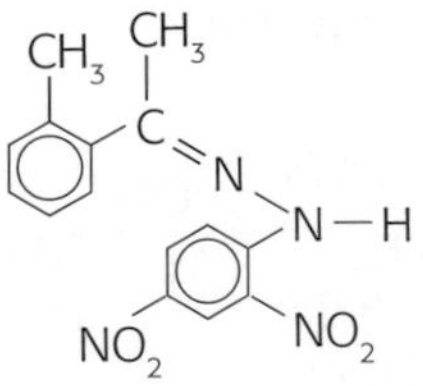

Answers to questions

Chapter 1

page 9
1 a 2 mol or 1.20×10^{24} **b** 0.5 mol or 3.01×10^{23} **c** 3 mol or 1.81×10^{24}

2 a 4.04 g   **b** 8.08 g   **c** 16.1 g   **3 a** 4
b 2.41×10^{24}

page 10
1 1 mol or 6.02×10^{23} **2** 58.33 g   **3** 1.65 mol   **4** 19.3 g

page 12
1 PCl_5 **2** $C_3H_7NO_2$

page 12
a CH_2Br **b** $C_2H_4Br_2$

page 16
1 a 3 mol   **b** 0.203 mol   **c** 2.37×10^{-3} mol
2 a 1.58 g   **b** 1.41 g   **3** 2.10 mol dm^{-3}

page 17
1 High temperature and low pressure
2 atm dm^3 mol^{-1} K^{-1}

page 19
1 2 dm^3 **2** 0.448 dm^3 (448 cm^3)

page 21
1 17.9 cm^3 **2** 62.0%   **3** 94.6%

4 There was nothing else in the aspirin that reacted with the sodium hydroxide and nothing else in the shell that reacted with the hydr chloric acid.

Chapter 2

page 28
1 a Bromine   **b** $^{81}_{35}Br$, 35 electrons, $^{81}_{35}B^-$, 36 electrons   **c** $^{79}_{35}Br$

2 Strontium could become part of the body by replacing calcium in the bones.

3 a 3.16×10^{-26} kg;   **b** (volume = 5.2×10^{-43} m^3) so density = 6.05×10^{16} kg m^{-3} **c** Extremely strong   **d** 1×10^{12}

page 30
1 A_r for Fe = 55.91   **2** 77.5% ^{35}Cl and 22.5% ^{37}Cl

page 34
1 By looking at the light from the stars through a spectrometer and observing lines with characteristic wavelengths for individual elements.
2 Five

page 38
1 a $1s^22s^22p^63s^23p^64s^23d^{10}4p^65s^24d^7$ or $[Kr]5s^24d^7$
b $1s^22s^22p^63s^23p^64s^23d^{10}4p^64d^6$ or $[Kr] 4d^6$

Chapter 4

page 64
1 KOH, $Al(OH)_3$, $CuCl_2$, $(NH_4)_2CO_3$, $MnSO_4$, $Ca_3(PO_4)_2$, $Mg(HCO)_2$ and $Fe_2(SO_4)_3$

2 potassium sulfate, barium hydroxide, copper(II) sulfate, copper(I) sulfate, manganese(III) oxide, manganese(IV) oxide, iron(II) sulfide and iron(III) nitrate

page 70
H_2S: bent, about 105°; $PbCl_2$: tetrahedral, 109.5°; H_2CO: trigonal planar, 120°; NO_3^-: trigonal planar, 120°; PH_3: trigonal pyramidal, about 107°; NO_2^-: bent, about 120° (or a bit less); $POCl_3$: tetrahedral, about 109.5°

page 75
1 Butane (van der Waals) < propanone (dipole–dipole) < propan-1-ol (hydrogen bonding)

2 Each water molecule can form two hydrogen bonds with other water molecules.

3 Hydrogen peroxide, ethylamine, ethanoic acid

page 79
1 Carbon dioxide consists of individual small molecules with weak van der Waals' forces of attraction between them. Silicon dioxide is a giant covalent molecule.

page 81
1 Carbon dioxide consists of individual small molecules with weak van der Waals' forces of attraction between them. Silicon dioxide is a giant covalent molecule.

2 Two of the four bonds around each oxygen atom in water are due to hydrogen bonds and are weaker than the other two, which are covalent. In methane all four bonds around each carbon atom are covalent.

page 87

1 Liquid nitrogen will not be paramagnetic as it possesses no unpaired electrons.

2 $CO < NO < O_2 < F_2$

page 89

1 sp^3, sp, sp; sp^3, sp^2, sp^2, sp^3; sp^3, sp^3, sp^2, sp^2

2 (H_2NNH_2) sp^3, sp^3; (NO_3^-) sp^2; (NO_2^-) sp^2; (HCN) sp; NO_2^+ sp

3 $(H_2C=CHCl)$ 5σ, 1π; $(H_3C–CH_2–CH=CH_2)$ 11 σ, 1π; $(H_3CC≡CCH_3)$ 6s, 2π

page 90

(PF_5) trigonal bipyramid, 120°,180°; (PF_6^-) octahedral, 90°; (PF_4^+) tetrahedral, 109.5°; (BrF_3) trigonal planar, 120°

page 91

2 $H_3C–CH=CH–CH=CH_2$ and vitamin A

page 93–94

1 Li: $1s^22s^1$, so one unpaired electron in outer energy level should be shown, but electrons in first energy level should be shown paired.

O: $1s^22s^22p_x^2 2p_y^1 2p_z^1$, so two unpaired electrons in outer shell should be shown, but electrons in first energy level should be shown paired.

2 The lithium atom should be shown larger than the oxygen atom. Atomic radius of Li = 0.152 nm, atomic radius of O = 0.074 nm.

3 Endothermic, +519 kJ mol^{-1} (for a gaseous lithium atom)

4 It must be turned into a gas, which is an endothermic process with a value of +161 kJ mol^{-1}.

5 Electron affinity. Endothermic. +702 kJ mol^{-1}

6 It is much smaller (0.060 nm compared with 0.152 nm)

7 It is much larger (0.140 nm compared with 0.074 nm)

8 $2Li(s) + O_2(g) \rightarrow 2Li_2O(s)$

9 No. The oxygen molecule needs to be converted into atoms. Endothermic, +248 kJ mol^{-1}

10 $[(2 \times 161) + (2 \times 519) + 248 + 702] =$ +2310 kJ mol^{-1}

11 The lattice enthalpy change for $2Li^+(g) + O^{2-}(g)$ $\rightarrow Li_2O(s) = -596 - (+2310) = -2906$ kJ mol^{-1}

13 Smaller. The ions are larger, and chlorine has only a single negative charge.

14 Sodium chloride has virtually 100% ionic character. Lithium oxide, although mainly ionic, possesses some covalent character. This is due to the greater charge density of the lithium ion compared with the sodium ion.

Chapter 5

page 98

1 −103 kJ mol^{-1}

2 a By definition, the standard enthalpy change of all elements is zero. **b** −975 kJ

3 −110 kJ mol^{-1} **4** −847 kJ mol^{-1}

page 101

1 −242 kJ mol^{-1} and −530 kJ mol^{-1}

For $H_2 + ½ O_2 = 13\,400$ kJ per 1 kg of fuel; for $H_2 + F_2 = 13\,250$ kJ per 1 kg of fuel

2 a −101 kJ   **b** −132.8 kJ

c The enthalpy given out when iodine gas forms a solid is more than the enthalpy change to convert liquid bromine into gaseous bromine.

page 102

1 −135 kJ mol^{-1} **2** The enthalpies of vaporization for both hexane and hex-1-ene

page 104

1 The lattice enthalpy change for $CaCl_2$ is much larger than for CaCl   **2** −603 kJ mol^{-1}

page 105

1 The fluoride ion is smaller than the chloride ion.

2 AgF has some covalent character. 14.9%, AgCl has even more covalent character.

page 107

1 Entropy change is negative. $\Delta G^{\ominus} = -237$ kJ mol^{-1}. 1745 K (1472 °C)

2 −30 kJ mol^{-1}; spontaneous   **3** −37 kJ mol^{-1}

Chapter 6

page 116

1 b 240 cm³

page 120

1 a zero order with respect to A and second order with respect to B.
b 2   **c** rate = $k[B]^2$
d $k = 75$ mol^{-1} dm³ min^{-1} **e** 4

2 a rate = $k[O_3][NO]$ **b** 2.20×10^7 mol^{-1} dm³ s^{-1}; 7.13×10^{-4} mol dm^{-3} s^{-1}.

page 122

1 a Step 1, as it is the slow step.   **b** Step 1, as it is slow and does not involve free radicals.   **c** rate = $k[NO_2][F_2]$.   **d** The weak F–F bond strength

2 The first step of the reaction will be the rate-determining step and will involve the protonation of propanone.

page 124

1 The increase in temperature is taken from 298 K to 308 K. What would happen if the 10 degree temperature rise was over a different range, e.g. from 498 K to 508 K?

Chapter 7

page 135

3 a It is endothermic.　**b** It will turn more blue.
c It will turn more pink.

page 140

1 2×10^{-2} mol dm^{-3}　　**2 a** 50;
b 0.16 mol dm^{-3}

Chapter 8

page 144

2 a Lowry–Brønsted　　**b** H_2S and H_3O^+
c Arrhenius, neither; Lowry–Brønsted, both; Lewis, base

page 146

1 $Al(OH)_3 + 3HCl \rightarrow AlCl_3 + 3H_2O$
$MgO + 2HCl \rightarrow MgCl_2 + H_2O$
$Mg(OH)_2 + 2HCl \rightarrow MgCl_2 + 2H_2O$
$NaHCO_3 + HCl \rightarrow NaCl + CO_2 + H_2O$
$CaCO_3 + 2HCl \rightarrow CaCl_2 + CO_2 + H_2O$

2 The amine is a base, so reacts with hydrochloric acid to form an ionic salt, which makes it soluble in water, i.e. $C_6H_5NH_2 + HCl \rightarrow C_6H_5NH_3^+Cl^-$.

page 148

1 1.0×10^{-4} mol dm^{-3}　　**2** 4.0

page 151

1 a 3.0　　**b** 3.0　　**c** 12.3
2 a 1.0×10^{-2} mol dm^{-3}　　**b** 2.0×10^{22} mol dm^{-3}
c 1.7　　**d** 12.3

page 153

1 a $C_2H_5COOH(aq) \rightleftharpoons C_2H_5COO^-(aq) + H^+(aq)$
b $10^{-4.78}$ or 1.35×10^{-5} mol^2 dm^{-6}
c $K_a = [C_2H_5COO^-][H^+] / [C_2H_5COOH]$
d 3.1, the equilibrium concentration of the acid is the same as the initial concentration

2 11.1, ammonia (as it is the weakest base)

page 155

1 KNO_3, neutral, Na_2CO_3, basic; $(NH_4)_2SO_4$, acidic; $Fe_2(SO_4)_3$, acidic

2 KCN is the salt of a strong base with a weak acid. In solution HCN is produced, which is less soluble in hot water and so may be evolved as a gas.

page 158

1 4.9 (it is the half-equivalence point so pH = pK_a)
2 8.3

page 162

1 The end point refers to the point at which the indicator changes colour; the equivalence point refers to the addition of stoichiometric amounts of reactants.

Chapter 9

page 166

1 (HNO_3) +5, (N_2H_4) −2, (NO_2^+) +5, (N_2) 0, (NO_2^-) +3, (NO_2) +4, (N_2O_4) +4, (N_2O) +1, (NO) +2

2 (K_2CrO_4) +6, (CrO_3) +6, ($[Cr(H_2O)_6]^{3+}$) +3, ($K_2Cr_2O_7$) +6, ($[Cr(H_2O)_4Cl_2]^+$) +3

3 a decreases from 0 to −1　　**b** no change (stays at −1)　　**c** increase from −1 to 0

page 169

1 a $2I^- \rightarrow I_2 + 2e^-$ and $Br_2 + 2e^- \rightarrow 2Br^-$. Overall: $Br_2 + 2I^- \rightarrow I_2 + 2Br^-$.
b $Zn \rightarrow Zn^{2+} + 2e^-$ and $2H^+ + 2e^- \rightarrow H_2$. Overall: $Zn + 2H^+ \rightarrow Zn^{2+} + H_2$.
c $SO_3^{2-} + H_2O \rightarrow SO_4^{2-} + 2H^+ + 2e^-$ and $MnO_4^- + 8H^+ + 5e^- \rightarrow Mn^{2+} + 4H_2O$.
Overall: $5SO_3^{2-} + 2MnO_4^- + 6H^+ \rightarrow 5SO_4^2 + 2Mn^{2+} + 3H_2O$

2 $2I^- \rightarrow I_2 + 2e^-$ and $H_2O_2 + 2H^+ + 2e^- \rightarrow 2H_2O$.
Overall: $2I^- + H_2O_2 + 2H^+ \rightarrow I_2 + 2H_2O$
$H_2O_2 \rightarrow O_2 + 2H^+ + 2e^-$ and $MnO_4^- + 8H^+ + 5e^- \rightarrow Mn^{2+} + 4H_2O$
Overall: $5H_2O_2 + 2MnO_4^- + 6H^+ \rightarrow 5O_2 + 2Mn^{2+} + 8H_2O$

page 170

1 Mn > Cr > Ni > Sn

2 $Zn + Fe^{2+} \rightarrow Zn^{2+} + Fe$. Zn is above iron in the reactivity series and so displaces Fe^{2+} to Fe

3 a $Fe + Sn^{2+} \rightarrow Fe^{2+} + Sn$. Fe is above Sn in the reactivity series and so reduces Sn^{2+} to Sn.
b Zinc is poisonous.

page 173

1 Zn is negative, Fe is positive, electrons flow from zinc to iron　　$Zn + Fe^{2+} \rightarrow Zn^{2+} + Fe$. e.m.f. = 0.32 V

2 $Mg + Ni^{2+} \rightarrow Mg^{2+} + Ni$　　e.m.f. = 1.45 V

page 175

1 In the solid state the ions are not free to move. (+) brown gas evolved, (−) silver metal deposited.
$2Br^- \rightarrow Br_2 + 2e^-$ reduction, (−) $Pb^{2+} + 2e^- \rightarrow Pb$ oxidation, molar ratio: 1:1

2 Oxidation always occurs at the anode.

page 181

1 0.90 V. Mn + Co^{2+} → Mn^{2+} + Co

2 Cr$_2$O$_7$$^{2-}$ + 6Cl$^-$ + 14H$^+$ → 2Cr^{3+} + 3Cl$_2$ + 7H$_2$O $E^{\ominus}$total = − 0.03 V, so it is not spontaneous. However, $E^{\ominus}_{\text{total}}$ is only just negative, and if the conditions change it may become positive and the reaction become spontaneous.

3 As $E^{\ominus}$ for fluorine is more positive than $E^{\ominus}$ for water, fluorine will oxidize water, and bubbles of oxygen will be observed. F$_2$ + 2H$_2$O → O$_2$ + 2H$^+$ + 2F$^-$

page 185

1 0.327 g   **2** 1.93 × 10^4 s (5.36 h), its mass would have decreased by 6.4 g

Chapter 10

page 195

1 2,3-dimethylbutane   **2** Free radical substitution, bromoethane   **3**

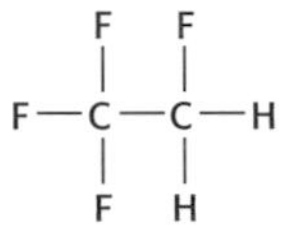

page 198

1 Ethanol only has a very short non-polar chain; cholesterol has a large non-polar group.   **3** Ester

page 204

1

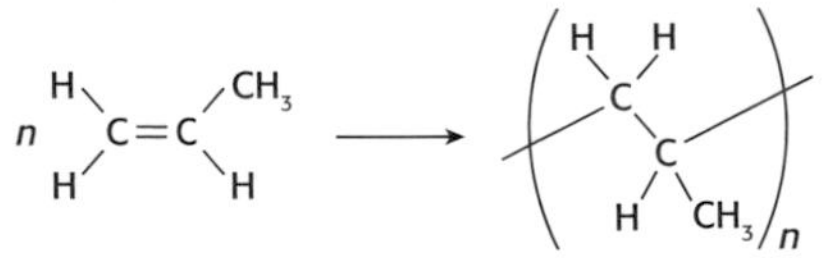

page 212

1 a

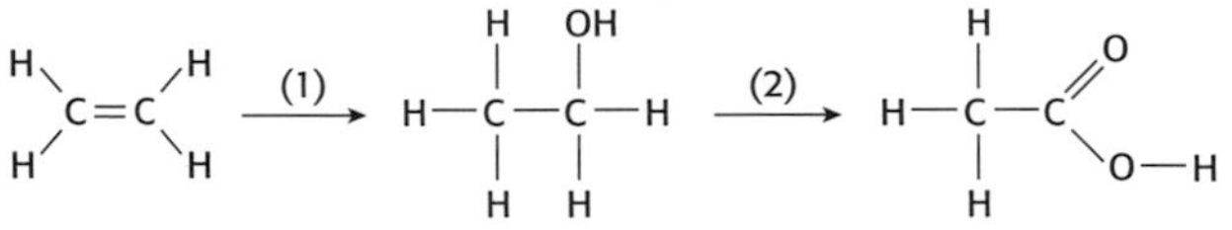

i H$_2$O and heat using concentrated H$_2$SO$_4$ as catalyst   ii H$^+$/ K$_2$Cr$_2$O$_7$ and heat

b

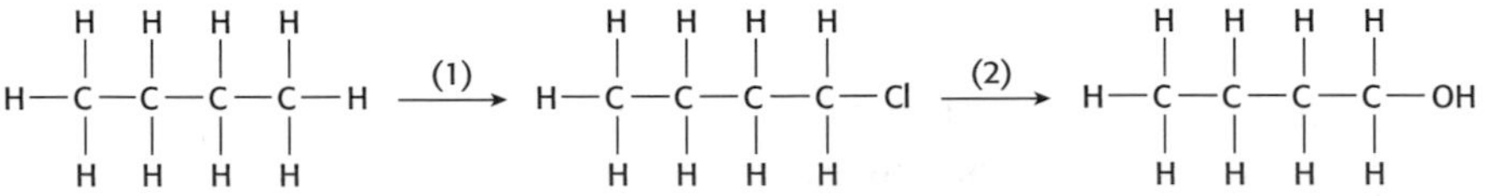

i Cl$_2$ in ultraviolet light   ii warm dilute aqueous NaOH solution

c

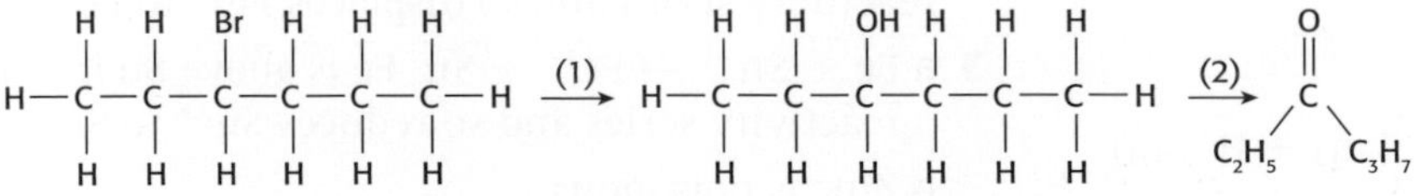

i warm dilute NaOH(aq)   ii H$^+$/ K$_2$Cr$_2$O$_7$ and heat

d

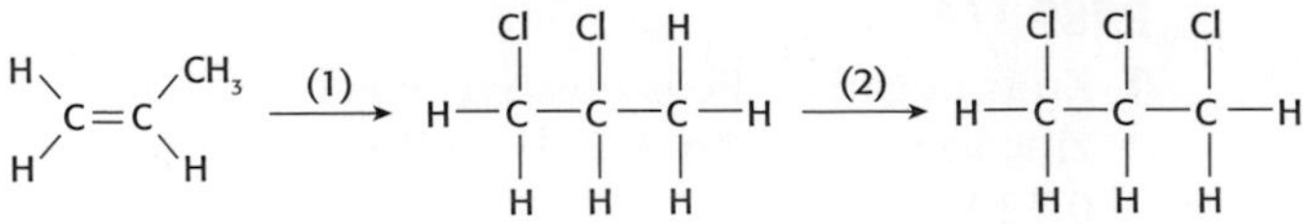

i Cl$_2$, room temperature   ii Cl$_2$ in ultraviolet light

e

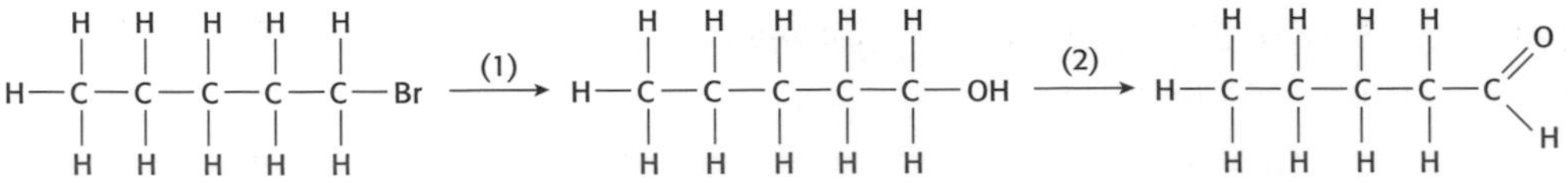

i warm dilute NaOH(aq)   ii H$^+$/ K$_2$Cr$_2$O$_7$ and heat   **2** 8.2%

page 214

1

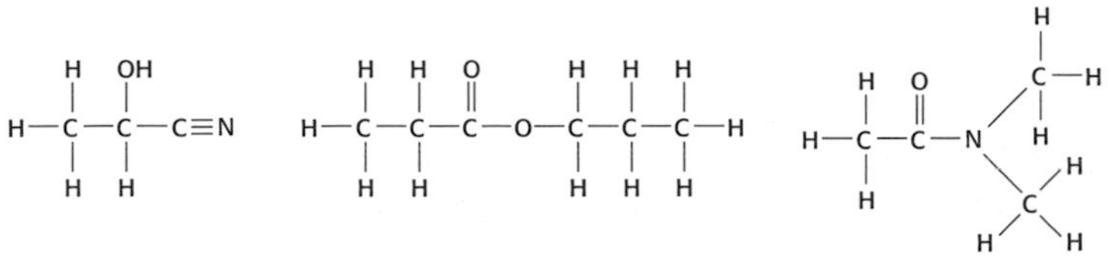

H—C(H)(H)—C(OH)(H)—C≡N   H—C(H)(H)—C(H)(H)—C(=O)—O—C(H)(H)—C(H)(H)—C(H)(H)—H   H—C(H)(H)—C(=O)—N with two CH3 groups (N(CH3)2 type)

2 methyl butanoate, 2-methylbutanenitrile, 2,3-diaminobutane

page 221

1 caffeine, diazepam, paracetamol

page 223

1 a H₂N—C(CH₃)(H)—C(=O)—N(H)—C(CH₂OH)(H)—C(=O)—N(H)—C(CH₂—C₆H₄—OH)(H)—C(=O)—OH

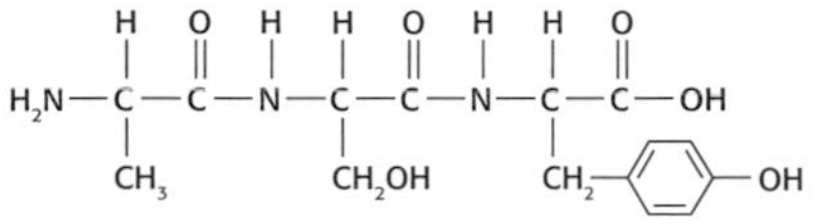

b 5 more (six in total)

2 (O—CH₂—CH₂—CH₂—O—C(=O)—C(=O))ₙ

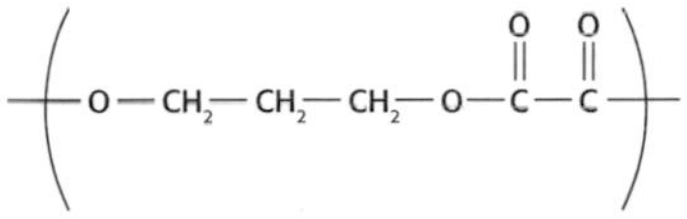

page 230

1 a

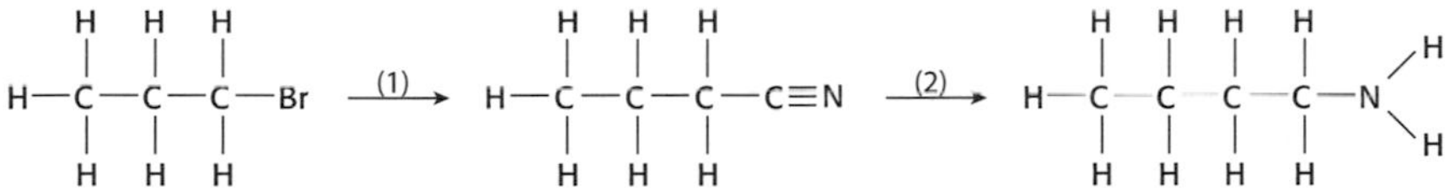

H—C(H)(H)—C(H)(H)—C(H)(H)—Br —(1)→ H—C(H)(H)—C(H)(H)—C(H)(H)—C≡N —(2)→ H—C(H)(H)—C(H)(H)—C(H)(H)—C(H)(H)—N(H)(H)

i HCN   ii H₂ / Ni

b

H—C(H)(H)—C(H)(H)—OH —(1)→ H—C(H)(H)—C(=O)(OH) —C₂H₅OH (2)→ H—C(H)(H)—C(=O)—O—C₂H₅

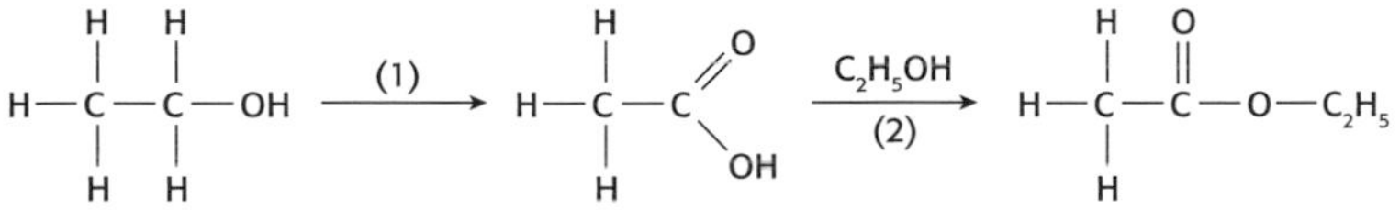

i H⁺/ K₂Cr₂O₇ and heat   ii C₂H₅OH in the presence of H₂SO₄ (conc.) and warm

c

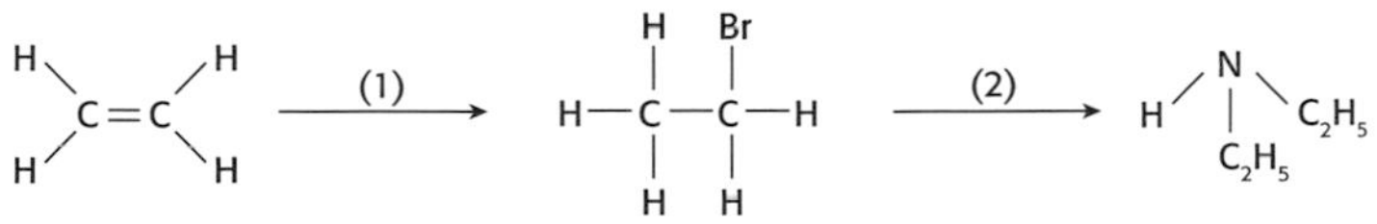

CH₂=CH₂ —(1)→ H—C(H)(H)—C(H)(Br)—H —(2)→ H—N(C₂H₅)(C₂H₅)

i addition of HBr   ii react with NH₃

d

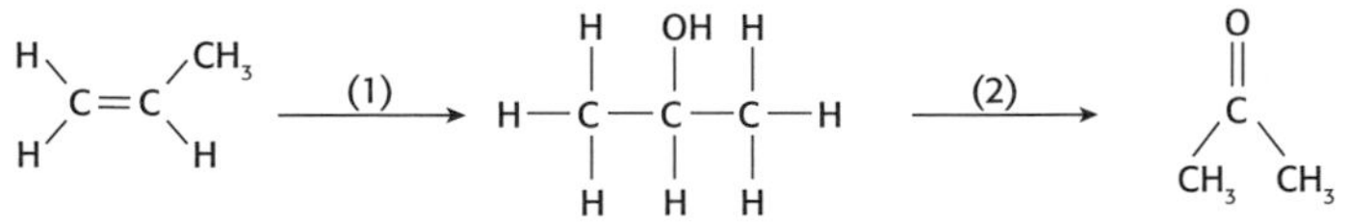

CH₂=CH—CH₃ —(1)→ H—C(H)(H)—C(OH)(H)—C(H)(H)—H —(2)→ C(=O)(CH₃)(CH₃)

i H₂O and heat using concentrated H₂SO₄ as catalyst   ii H⁺/ K₂Cr₂O₇ and heat

e

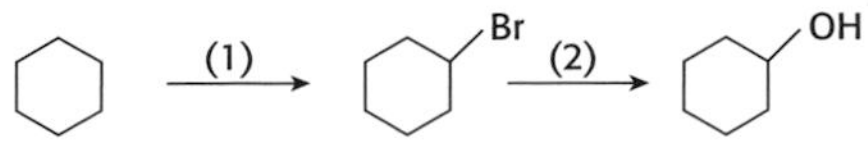

cyclohexane —(1)→ bromocyclohexane —(2)→ cyclohexanol

i Br₂ in ultraviolet light   ii warm dilute NaOH(aq)

2 Free radical substitution: Cl₂ and ultraviolet light
Nucleophilic substitution: warm dilute NaOH(aq) then H⁺/ K₂Cr₂O₇ and heat

Chapter 14

page 277

1 a ethanol   **b** propanal   **c** methyl methanoate   **d** propan-2-ol

2 13.3 ppm, so the water is within permitted limit.

page 283

1 a $C_2H_5COOCH_3$ triplet(3), quartet(2) singlet(3) $HCOOC_3H_7$ singlet(1), triplet(2), complex(2), triplet(3)

b CH$_3$COOC$_2$H$_5$ singlet(3), quartet(2), triplet(3)
C$_3$H$_7$COOH triplet(3), complex(2), triplet(2),
singlet(1)

c CH$_3$COCH$_2$CH$_2$CH$_3$ singlet(3), triplet(2), complex(2), triplet(3) CH$_3$CH$_2$COCH$_2$CH$_3$ triplet(3),
quartet(2) (ratio is 3:2, actual number is 6:4)

Chapter 15

page 285

a 6

page 291

1

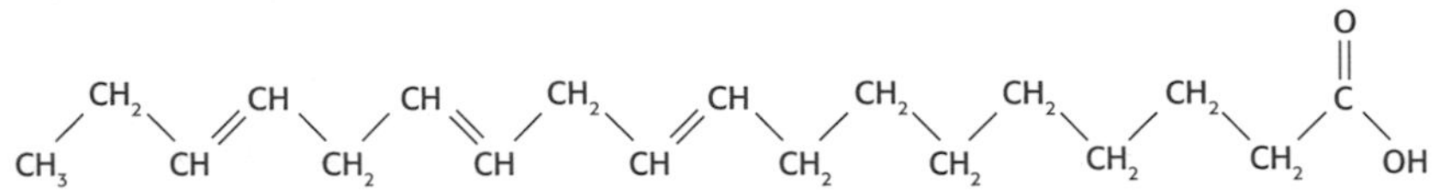

2 3

page 294

1 Present in both: ketone, alkene. Present in testosterone only: alcohol

2 Anabolic steroids detected using chromatography (GLC or HPLC) linked to a mass spectrometer.

Chapter 17

page 336

3 Ganclovir and lysergic acid diethylamide, LSD, are chiral molecules

Chapter 18

page 333

1 363 kJ mol^{-1}, which is between the values for O–O and O=O

Chapter 20

page 379

1 2-chloropentane

2 The electrophile in both cases is Br$^+$, but in water OH$^-$ rather than Br$^-$ will react with the carbocation intermediate.

5 H$_3$PO$_4$, unlike H$_2$SO$_4$, is not an oxidizing agent so is less likely to form other products.

6

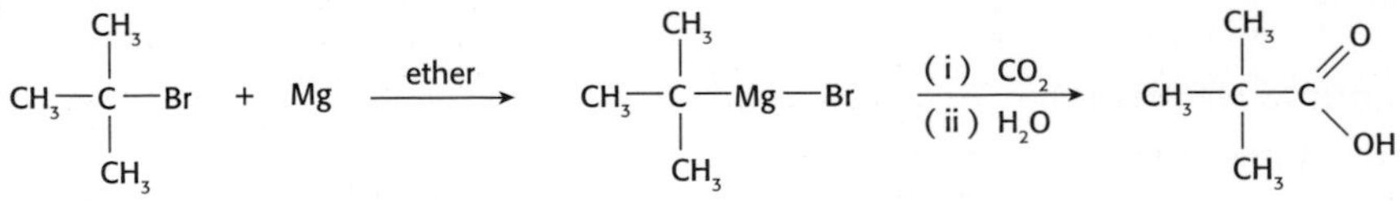

page 386

1 a i CH$_3$Br and AlBr$_3$ to form methylbenzene then concentrated HNO$_3$/H$_2$SO$_4$

ii Concentrated HNO$_3$/H$_2$SO$_4$ to form nitrobenzene then CH$_3$Br and AlBr$_3$

iii C$_2$H$_5$Cl and AlCl$_3$ to form ethylbenzene then C$_2$H$_5$COCl and AlCl$_3$

b i Cl$_2$ and ultraviolet light then warm NaOH(aq)

2 Al^{3+}(aq) + PO$_4^{3-}$(aq) → AlPO$_4$(s) [PO$_4^{3-}$(aq)] = 3.74 × 10^{-11} mol dm^{-3}

Chapter 19

page 360

1 A: alkene, B: (secondary) amine

page 371

1 a Anthocyanin (flavonoid); **b** coloured; **c** soluble **2** Oranges

3 1-iodo-2-chlorobutane

4 Nucleophilic substitution with CN$^-$ followed by acid hydrolysis or conversion to the Grignard reagent C$_2$H$_5$–Mg–Br followed by reaction with CO$_2$ then water.

ii Br$_2$ and ultraviolet light then Br$_2$ and AlBr$_3$

iii CH$_3$COCl and AlCl$_3$ then react product with 2,4-DNPH

2 Electrophile: use a halogen carrier, e.g. R–Cl and AlCl$_3$ with benzene

Nucleophile: Form a Grignard reagent, e.g. R–Br with Mg in ether then react with CO$_2$ and water

Index

Index entries are arranged in letter-by-letter sequence. Page numbers in bold refers to tables; page numbers in italics refers to figures.